Enhancing Sustainable Performance With Low-Carbon Supply Chains

Chafik Bensfia
Mohammed Premier University, Morocco

Badr Bentalha
Sidi Mohammed Ben Abdellah University, Morocco

Lhoussaine Alla
Sidi Mohamed Ben Abdellah University, Morocco

IGI Global
Scientific Publishing
Publishing Tomorrow's Research Today

Vice President of Editorial	Melissa Wagner
Director of Acquisitions	Mikaela Felty
Director of Book Development	Jocelynn Hessler
Production Manager	Mike Brehm
Cover Design	Jose Rosado

Published in the United States of America by
IGI Global Scientific Publishing
701 East Chocolate Avenue
Hershey, PA, 17033, USA
Tel: 717-533-8845
Fax: 717-533-7115
Website: https://www.igi-global.com E-mail: cust@igi-global.com

Library of Congress Cataloging-in-Publication Data

Names: Bensfia, Chafik, 1983- editor | Bentalha, Badr, 1985- editor | Alla, Lhoussaine, 1969- editor
Title: Enhancing sustainable performance with low-carbon supply chains / edited by: Chafik Bensfia, Badr Bentalha, Lhoussaine Alla.
Description: Hershey, PA : IGI Global Scientific Publishing, [2026] | Includes bibliographical references and index. | Summary: "The text addresses the integration of intelligence and sustainability in supply chains. Balanced between application and theory, the text covers the latest technological, organizational and regulatory developments in the supply chain"-- Provided by publisher.
Identifiers: LCCN 2025013361 (print) | LCCN 2025013362 (ebook) | ISBN 9798337330129 hardcover | ISBN 9798337330136 paperback | ISBN 9798337330143 ebook
Subjects: LCSH: Supply chain management | Industrial management--Environmental aspects
Classification: LCC HD38.6 .E54 2025 (print) | LCC HD38.6 (ebook) | DDC 658.7--dc23/eng/20250929
LC record available at https://lccn.loc.gov/2025013361
LC ebook record available at https://lccn.loc.gov/2025013362
British Cataloguing in Publication Data
A Cataloguing in Publication record for this book is available from the British Library.

Table of Contents

Detailed Table of Contents

The sustainability of supply chains has become crucial in the face of environmental challenges and heightened global competition. In this context, artificial intelligence is emerging as a potential catalyst for converging economic performance requirements and current environmental concerns. Transformation towards sustainable supply chains with reduced carbon footprints and territorial competitiveness calls for a thorough re-reading of theories on the role of AI. This chapter proposes an integrative framework to analyze the role of AI as a catalyst for this sustainable and resilient transition with a view to enhanced territorial competitiveness. This model is based on a critical review of the literature. AI, via its vast forecasting, optimization and automation capabilities, facilitates carbon reduction and enhances competitiveness. The proposed framework identifies causal mechanisms and moderating factors. AI is a powerful but conditional catalyst, requiring strategic alignment, organizational capabilities and ethical governance. Managerial and political implications are discussed.

This study addresses the climate emergency by assessing the environmental performance of 270 Moroccan transport organizations (2019–2023) using four machine learning algorithms: Random Forest, XGBoost, SVM, and PLS. Among them, XGBoost achieved the best results with an R^2 of 0.88 and F1-score of 0.83 in predicting fuel consumption and classifying environmental performance. The analysis identified route optimization and adoption of low-emission vehicles as key factors influencing environmental outcomes. The study also highlights differences in these factors' effectiveness across organization types (public vs. private) and sizes (SMEs vs. large firms). Ultimately, the research proposes a robust, data-driven framework to enhance environmental performance and guide public policies and managerial strategies towards greener logistics and decarbonized transport in Morocco.

The abstract discusses sustainable procurement, which aligns purchasing with sustainable development principles to support goals like the UN SDGs. It examines strategies to integrate sustainable procurement with sustainability frameworks, emphasizing economic, social, and environmental impacts. Benefits include advancing SDGs related to poverty, health, education, and responsible consumption, while challenges involve policy support, supply chain collaboration, and performance measurement. Case studies highlight successful strategies, and the paper calls for ongoing dialogue among stakeholders and further research to maximize sustainable procurement's impact.

In recent years, the transition towards circular supply chains has gained significant momentum as organizations strive to enhance sustainability and reduce waste. The adoption of digital tools and technologies plays a pivotal role in enabling effective traceability within these circular systems. This article explores the innovative tools and technologies that facilitate enhanced traceability in circular supply chains, highlighting their importance in ensuring transparency, accountability, and sustainability. The study delves into various digital solutions, such as blockchain, IoT, and big data analytics, that provide real-time insights into product lifecycles, thereby allowing organizations to monitor and manage resources efficiently. By examining case studies and best practices, this article aims to provide a comprehensive understanding of how digital technologies can transform circular supply chains and contribute to a more sustainable future. Ultimately, this research serves as a foundation for both academia and industry.

The contribution of port operational activities to climate change has made it a research focus of increasing concern. To identify the news and characteristics of research on carbon emissions in the maritime transport sector, this study used the "Scopus" and "Web of Science" databases for the period 2020-2024. In addition, we identify the contributions of the most cited review articles. The results of the analysis highlight that port carbon emission reduction strategies involve operational optimization, equipment transformation and energy management and that the inherent gaps between ports present challenges in establishing a standard methodology at the global level. Future research avenues have been suggested, such as comprehensive assessment of emission sources, harmonization of estimation methods, equipment-integrated planning, port automation and electrification, and energy management. This study guides the progress of port carbon emission assessment and reduction, thus enabling port operators to manage production efficiently.

Because of climate change, companies are now moving toward carbon-free supply chains, and circular models appear to be a good answer. Such supply chains focus on using materials again, recycling them, and adding renewable energy, as this leads to less carbon output and efficient use of resources. They are not the same as standard supply chains since they strive to reduce waste and reuse things as much as possible. It reviews how circular economic models can help reach sustainability goals and also identifies the problems in using them and any new technologies involved. Although it is hard to fix some problems in infrastructure, policies, and stakeholder cooperation, firms can use AI, blockchain, and IoT to improve their circular operations. Product longevity, the process of handling products returned after use, and sustainable design raise the overall sustainability of the company. Thanks to effective government policies and more demand for eco-friendly items, circular supply chains help the economy become carbon neutral and stronger.

This investigation uses a sample size of 250 firms from a variety of sectors to investigate the association among waste recovery, innovative business models, and reverse logistics. Using a mixed-procedures approach, surveys were assimilated to gather quantitative data, and qualitative interviews were manifested to provide additional in-depth understanding. With innovative business models acting as a partial mediator in this connection, major outcomes manifest that efficient reverse logistics greatly augments waste recovery rates. Outcomes manifest that corporates that incorporate these components have augmented operational influence and sustainability outcomes. The investigation comes to the conclusion that optimizing trash recovery operations manifests a comprehensive strategy that strikes a balance among logistics and innovation. Future investigation will investigate certain cutting-benefit business models that augment trash recovery, how new knowhow influence reverse logistics, and how regulatory frameworks influence sustainable practices.

Given the context of managerial transformation and the growing importance of sustainable development imperatives, corporate social responsibility has become a key strategic lever in the transition to global, sustainable performance. The aim of the study is to assess the impact of CSR management on the evolution of corporate performance towards the integral dimension between economic, social and environmental aspects. To this end, a quantitative empirical study was carried out on a sample of 80 Moroccan companies involved in CSR actions with a CGEM label. Data collection was based on a structured questionnaire designed to identify the differential effect of CSR management practices on overall performance. The results show that certain CSR management practices have very significant effects in the transition from performance to globality and sustainability. These include responsible governance, consumer protection, labor relations and conditions, and community involvement, all of which strengthen management systems and organizational resilience.

Mass retail supply chains are increasingly expected to balance high performance with sustainable practices. Retailers must meet economic and operational goals while addressing growing environmental and social responsibilities. This dual pressure challenges traditional sourcing and procurement strategies, which remain cost-driven. Despite interest in sustainable supply chain management, limited research explores how sustainable sourcing and procurement drive performance in the mass retail sector. This study fills that gap using an exploratory qualitative approach based on semi-structured interviews with retail supply chain stakeholders. Thematic analysis reveals emerging practices, strategic shifts, and tensions between cost constraints and long-term sustainability. Findings show that when strategically aligned, sustainable procurement can enhance performance. The study also highlights the need for better internal coordination and clearer sustainability frameworks. It offers practical insights and recommendations to support more resilient and responsible mass retail supply chains.

The main goal of this investigation is to investigation the association among supply chain strategies, supply chain integration, as well as logistics proxies. It solves the issue of how corporates can optimize supply chain operations to augment efficiency and responsiveness in competitive market places. The investigation uses a quantitative investigations approach that uses surveys to collect data from supply chain specialists in various industries. Statistical analysis, assimilating regression and factor analysis, is carried out to investigation the association among proxies.

Preface

The growing urgency of climate change, resource depletion, and social equity concerns has compelled supply chains across the globe to evolve beyond traditional models of efficiency and cost minimization. In this transformative era, sustainable supply chains—particularly those that prioritize low-carbon operations—are not just aspirational but necessary. *Enhancing Sustainable Performance With Low-Carbon Supply Chains* emerges at this intersection of innovation, environmental responsibility, and strategic supply chain management.

This book presents a comprehensive and interdisciplinary collection of research efforts that explore how emerging technologies, policy interventions, business models, and circular strategies can collectively transform supply chains into engines of sustainable development. With a focus on diverse contexts, particularly from Morocco, India, Ghana, Ethiopia, and other developing regions, the chapters provide empirical studies, conceptual frameworks, and case analyses that illuminate practical pathways for decarbonizing logistics, procurement, and distribution networks.

A core theme of this volume is the catalytic role of digital technologies—such as artificial intelligence, blockchain, and big data—in enabling transparency, traceability, and operational efficiency within supply chains. Equally critical is the exploration of innovative governance models, public sector accountability mechanisms, and community-based economic participation (such as women's cooperatives), which are pivotal in localizing sustainability solutions.

From AI-enabled green logistics and decarbonized port operations to circular economic models and CSR-driven organizational performance, this book bridges theory and practice to support researchers, policymakers, supply chain professionals, and sustainability advocates. Whether through predictive analytics, ethical procurement, or grassroots innovation, the chapters together advocate for a resilient and environmentally responsible supply chain architecture.

CHAPTER OVERVIEW

1. The integration of artificial intelligence as a strategic driver of carbon reduction and territorial competitiveness in sustainable supply chains.
2. Evaluation of machine learning techniques to enhance green logistics and decarbonization among transport organizations in Morocco.
3. Strategic alignment of sustainable procurement with global sustainability goals and digital innovations for accountability in the public sector.
4. Exploration of smart technologies—including AI, blockchain, and robotics—for transparent and circular logistics systems.
5. A systematic review of strategies and challenges for achieving carbon-neutral port management.
6. Examination of how circular economy models contribute to the development of carbon-free supply chains.
7. Analysis of reverse logistics and its mediation by innovative business models in enhancing waste recovery in Ethiopia.
8. Investigation into the role of corporate social responsibility in fostering global and sustainable corporate performance.
9. Qualitative study of sustainable sourcing and procurement strategies within mass retail supply chains.
10. Empirical exploration of the mediating role of supply chain integration in logistic sustainability in Ethiopia's coffee industry.
11. Case study on strengthening local supply chains through platform-based models focusing on women's cooperatives in disaster-affected areas.
12. Empirical study from Morocco on how green supply chain management practices influence environmental performance.
13. Bibliometric analysis of the relationship between blockchain technology and supply chain transparency.
14. Exploratory study of the relationship between sustainability practices and supply chain resilience in the pharmaceutical sector of Morocco.

This book seeks to inform and inspire efforts to design supply chains that are not only efficient and adaptive but also deeply attuned to the imperatives of sustainability, equity, and global resilience. Through the diverse contributions compiled herein, we hope to support the transition toward a more sustainable, carbon-conscious global supply network.

Chafik Bensfia
Mohammed Premier University, Morocco

Badr Bentalha
Sidi Mohamed Ben Abdellah University, Morocco

Lhoussaine Alla
Sidi Mohamed Ben Abdellah University, Morocco

Chapter 1
AI as a Catalyst for Sustainable Supply Chains:
From Carbon Reduction to Territorial Competitiveness

Badre Meziane
https://orcid.org/0000-0003-3097-4909
National School of Business and Management, Sidi Mohamed Ben Abdellah University, Morocco

Lhoussaine Alla
https://orcid.org/0000-0002-7238-1792
National School of Applied Sciences, Sidi Mohamed Ben Abdellah University, Morocco

Badr Bentalha
https://orcid.org/0000-0003-1339-542X
National School of Business and Management, Sidi Mohamed Ben Abdellah University, Morocco

ABSTRACT

The sustainability of supply chains has become crucial in the face of environmental challenges and heightened global competition. In this context, artificial intelligence is emerging as a potential catalyst for converging economic performance requirements and current environmental concerns. Transformation towards sustainable supply chains with reduced carbon footprints and territorial competitiveness calls for a thorough re-reading of theories on the role of AI. This chapter proposes an integrative framework to analyze the role of AI as a catalyst for this sustainable

DOI: 10.4018/979-8-3373-3012-9.ch001

and resilient transition with a view to enhanced territorial competitiveness. This model is based on a critical review of the literature. AI, via its vast forecasting, optimization and automation capabilities, facilitates carbon reduction and enhances competitiveness. The proposed framework identifies causal mechanisms and moderating factors. AI is a powerful but conditional catalyst, requiring strategic alignment, organizational capabilities and ethical governance. Managerial and political implications are discussed.

1. INTRODUCTION

Today's supply chains are faced with a double dilemma: on the one hand, increased and fierce competitiveness demands greater responsiveness and flexibility, and on the other, environmental challenges exacerbated by the growing importance of ecological movements, which are increasingly shaping consumer habits. Indeed, from supplier to customer, via a distribution chain of varying length, supply chains are responsible for a very large proportion of greenhouse gas emissions (notably CO_2), but they are also vectors for value creation and are at the heart of contemporary international trade. Whether in terms of image for companies, or competitiveness for regions, managing the sustainability of supply chains has become a key issue.

It is in this context that the use of the various technologies offered by artificial intelligence (AI) offers interesting prospects for catalyzing efforts to transform supply chains towards sustainability and ecological resilience, combined with economic and financial imperatives: the use of AI to make these chains “greener” is an increasingly popular area of research (Zejjari and Benhayoun, 2024). Thanks to its capacity to acquire, process, analyze and store data, AI is an important vector for transforming logistics practices and processes, as well as the related business models. AI is increasingly being used in virtually all sectors to reap its benefits in terms of cost and time reduction, as well as resource and process optimization. The transport and logistics sector are not lagging behind, and is one of the sectors where the use of AI technologies is ubiquitous, as Figure 1 shows (Chui et al., 2018) (although to be taken with caution, as those are 2018 statistics covering a time horizon of 3 years ahead).

In the face of global warming, reducing the carbon footprint of greenhouse supply chains is becoming a necessity for companies, who are encouraged - and sometimes forced - to make efforts in this direction. Besides, territorial competitiveness now includes an environmental component that goes beyond the economic and social dimensions traditionally attributed to it. This is precisely where AI comes into play, by optimizing the economic, social and environmental performance of supply chains, leading to sustainable regional competitiveness. The coexistence of these 2 seemingly contradictory objectives is, however, necessary to take into account the changes in the environment, where consumers now have a great deal of negotiating power, and where territories must imperatively think about the ecological component when drawing up their local development strategies.

Figure 1. Future AI spending over next 3 years (%)

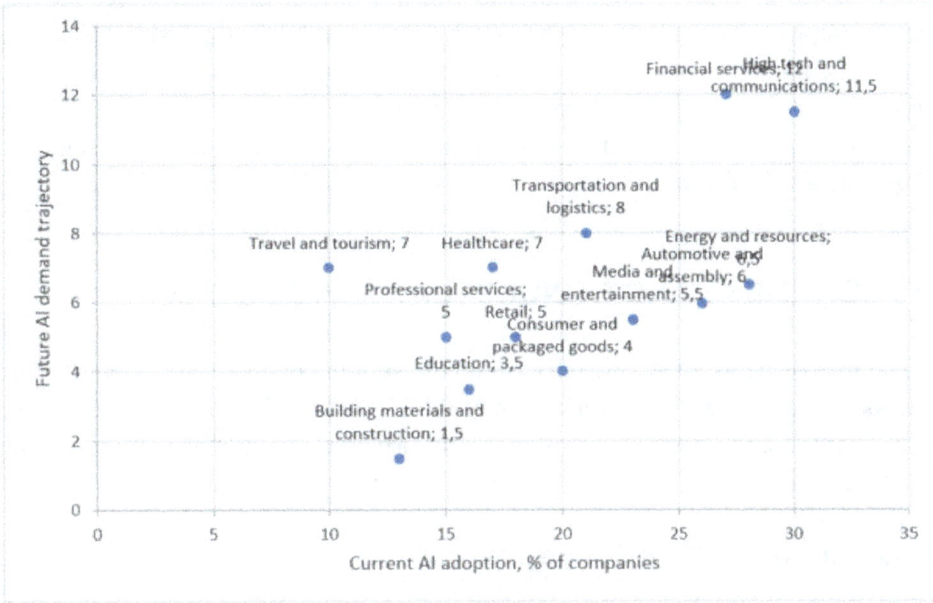

Source: Chui et al., 2018

However, the emerging nature of AI technologies in supply chains can lead to disappointment due to operators' over-ambitious expectations, as well as a lack of understanding of the systemic implications of these technologies (Sodhi et al, 2022). A rigorous theoretical framework is therefore needed to conceptualize the role of AI in supply chain optimization, specifying the mechanisms of influence, the implications as well as the conditions for effectiveness.

Recent research has, for the most part, focused on the economic and financial implications of using AI to promote the sustainable transition of supply chains, to the detriment of other environmental, societal and ethnic considerations (Bensfia and Bouklata, 2022). The latter are often under-explored, which constitutes a gap in our understanding of the mechanisms by which AI can fully play a catalytic role in supply chains in the context of building territorial competitiveness, and in particular its impact on improving decision-making, optimizing resources and processes, and creating value.

This chapter aims to set out a conceptual framework reframing the use of AI as a catalyst for green supply chains, and their role in promoting territorial competitiveness. It is based on a critical analysis of existing literature to account for the interactions between AI and the sustainability of territories through cleaner supply chains.

Following this introduction, a section will present the theoretical and conceptual framework and detail the theories underlying the concepts studied. This is followed by an analysis of AI technologies that can transform supply chains and reduce their carbon footprint. Next, the relationship between environmental sustainability and territorial competitiveness will be established, identifying the role of AI in facilitating this convergence. The final section will propose an integrative theoretical framework conceptualizing the role of AI as a catalyst for a sustainable ecological and logistical transition.

2.THEORETICAL AND CONCEPTUAL FRAMEWORK

The role of AI as a catalyst for sustainable supply chains can only be fully understood through a careful examination of the various established theories through the 3 essential components of this research: artificial intelligence, supply chains and territorial competitiveness.

2.1 Definition of Key Concepts:

Artificial intelligence (AI) is defined as a set of technologies enabling information systems to perform tasks without human intervention, such as learning, reasoning or decision-making. In the specific context of supply chains, AI can represent technologies capable of making predictions, machine learning, natural language processing or optimization decisions. Table 1 shows some of the definitions encountered in our theoretical corpus:

Table 1. IA definitions

Authors	Year	Definition
Teixeira et al.	2025	A set of computational technologies that enable machines to perform tasks traditionally requiring human intelligence, including machine learning, natural language processing and computer vision, to optimize supply chain management.
Shamsuddoha el al.	2025	A transformative technology that revolutionizes supply chains through intelligent automation and real-time information flow, capable of learning, reasoning and adapting to environmental changes.
Riad et al.	2024	A set of advanced computational technologies that simulate human intelligence to enhance supply chain resilience through predictive analysis, optimization and automation of decision-making processes
Belhadi et al.	2024	An innovation engine that enhances supply chain resilience and performance by adapting to a dynamic environment, characterized by its ability to learn and optimize processes autonomously.
Hendriksen, C	2023	A disruptive technology that transforms supply chain management by automating complex decision-making processes and generating predictive insights from vast data sets.
Sodhi et al.	2022	An emerging supply chain technology that uses advanced algorithms to analyze complex data, automate processes and generate forecasts, although its initial adoption may face implementation and integration challenges.

Source: The authors

Sustainable supply chains represent a major conceptual shift away from traditional supply chains, as environmental, social and economic concerns are integrated into the design, operation and evaluation of logistics networks (Zejjari and Benhayoun, 2024). Known as the three-dimensional approach or the Triple Bottom Line (economic, environmental and social), the underlying intention is to create long-term value for all stakeholders while avoiding deleterious impacts on the ecosystem and human well-being (Bentalha et al., 2019).

Carbon reduction refers to the systemic reduction of greenhouse gas emissions produced by supply chain activities, including the extraction, production, transportation, storage, distribution and related natural balance and disposal of materials. This paradox has become most urgent in the agricultural sector, where supply chains are both threatened by and act as a driving force for climate change (Elufioye et al., 2024).

Traditionally, only a few data sources are used in classic data processing approaches, such as global demand trends or price levels. In contrast, the new "Smart" approaches involve a combination of tools and methods using both the Internet of Things (IoT) to acquire and organize the data needed for decision-making, and a human approach based on the Internet of People (Boute and Udenio, 2022). But as the amount of data increases exponentially (due in part to the high computing capacities of connected objects), the processing of this data becomes complicated, and rationality is considerably reduced. This is where Deep Learning comes in, enabling the machine to "mimic" rational human behavior through a neural network, and make decisions that are reputed to be well thought-out and efficient (Agrawal et al, 2022).

Empirical studies have shown that the transformative potential of AI technologies in supply chain optimization is evident in virtually all operational areas relating to flow management operations, as they enable companies to significantly improve productivity, customer satisfaction by offering products that exactly meet their expectations, as well as process and resource optimization. Figure 2 shows the impact of AI integration on improving certain aspects of supply chain management (Yandrapalli, 2023).

Figure 2. AI's impact on supply chain management

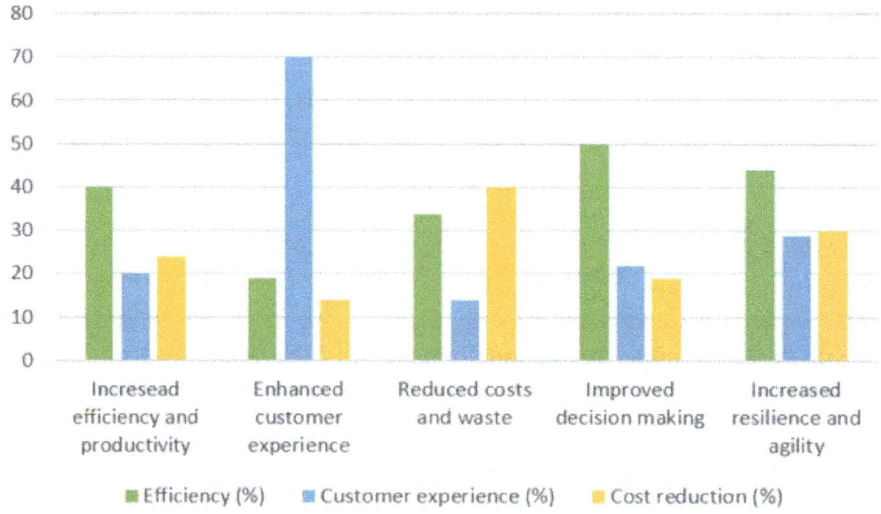

Source: Yandrapalli, 2023

The impact of AI technologies is evident in the transformation of supply chains by improving operational efficiency, customer experience and the perception of value created, while helping to reduce costs and lead times. AI has been shown to increase process fluidity by up to 40% by mobilizing intelligent prediction and automation (Redżeb, 2024). On the other hand, improving the customer experience seems to benefit most from AI integration, with a notable increase of around 70%, with services such as real-time parcel tracking and personalized recommendations based on customer preferences.

Finally, the competitiveness of a region is a multi-dimensional concept that goes beyond the simple economic performance of a region. It refers to a territory's ability to attract and retain sustainable economic activities, innovate, develop human capital and maintain or improve the quality of life of its inhabitants, while enhancing its natural and cultural capital. In the supply chain field, territorial competitiveness is linked to its ability to integrate into global logistics networks and maximize positive local returns while minimizing negative externalities.

2.2. Theories Underlying The Integration of AI in Supply Chains:

Different theoretical frameworks can be applied to the analysis of AI integration in sustainable supply chains. According to the resources and competence's view, AI can serve as a strategic resource if it is rare, special, difficult to imitate and non-substitutable for organizations that thereby gain a long-term competitive advantage. In this framework, AI is not just a technological gimmick, but an organizational competency that can fundamentally transform supply chain operations (Alla et al., 2022).

Similarly, transaction cost theory provides a relevant framework, since AI appears to have a significant impact on the costs of information seeking, negotiation and control of inter-organizational relationships that occur in supply chains. According to Weisz et al, AI can even promote collaboration in the supply chain by increasing the quantity and quality of information shared and strengthening trust between peers (Onukwulu et al., 2023).

The other relevant theoretical framework would be the theory of complex adaptive systems, for which it would be useful to think of supply chains as dynamic non-linear networks in which AI can facilitate the detection of emergent patterns, adaptation to disturbances and self-organization, and much more. More specifically, this perspective helps to understand how AI can contribute to the resilience of supply chains in the face of external disruptions. for instance, the established possibility of supply chains being loosely coupled and constituting their own unit has been demonstrated for supply chain disruption cause of the SARS epidemic (Dey et al., 2024).

Finally, stakeholder theory provides a framework for addressing the question of the value created by AI through supply chain efficiency for a greater number of actors than just shareholders, realizing a primarily financial number. This perspective is crucial for grasping how AI's ability to optimize supply chains is used as a factor of territorial superiority, as it involves a multiplicity of actors and opinions.

2.3. Existing Conceptual Models Of Supply Chain Sustainability:

Academic literature has developed several conceptual models for understanding the sustainability and resilience of supply chains in today's context. The "Triple Bottom Line" model (economic, environmental and social) (Elkington, 1998) is a fundamental framework that has been progressively enriched to include additional dimensions such as resilience and governance (Attah et al., 2024).

The Resources Based Vision (RBV) can also explain how effective resource management can create a sustainable competitive advantage for organizations. In the context of sustainable supply chains, this concept converges with intelligent management of flows of all kinds (goods transport, waste, delivery, etc.) and shows that sustainable sourcing practices can effectively improve economic performance while reducing the environmental impact of logistics operations (Barney, 1991).

According to stakeholder theory (Freeman, 2010) companies need to take into account the interests of their various stakeholders, such as suppliers, employees, and especially the community and customers, who are increasingly demanding in terms of compliance with environmental standards. In this way, companies foster innovation and collaboration, which will have a positive effect on economic, social and societal performance through the promotion of sustainable practices.

Another recent theoretical model conceptualizes AI implementation and supply chain optimization, emphasizing resilience as a key dimension of sustainability (Riad et al., 2024). This model identifies various essential levers of action to strengthen the ability of supply chains to absorb shocks, adapt to change and transform in the face of emerging challenges.

AI-enabled supply chain optimization can also be approached by identifying pathways to efficiency and resilience (Abaku et al., 2024). This model conceptualizes AI as a facilitator of more informed, faster and more adaptable decision-making processes in logistics flow management.

The Sustainable Supply Chain Maturity Model also offers a useful conceptual framework for understanding how AI can accompany organizations' progress towards more sustainable and resilient practices. This model generally identifies several stages, from reactive compliance to proactive innovation, and can be enriched to incorporate the inherent technological dimension as a factor in accelerating this transformation (Guo and Mantravadi, 2025).

3. METHODOLOGY

This chapter is based on a three-stage theoretical triangulation approach. The first step consists in identifying 3 corpuses relating to the key concepts of our research: theories relating to supply chains, theories of artificial intelligence in its organizational conception, and theories relating to competitiveness and territorial sustainability. Next, the examination phase consists of an in-depth analysis of the identified corpora in order to extract general postulates as well as possible causal relationships between key concepts. Finally, the articulation phase involves proposing a unifying theoretical framework that integrates the different conceptions and transcends them into a single theoretical framework (Naeem et al., 2023).

The development of this theoretical framework is based on a literature review structured according to the PRISMA method and comprising 87 scientific articles covering the period 2015-2025, with a particular focus on articles published from 2021 onwards to reflect the rapid evolution that AI technologies have undergone recently. We proceeded to establish relationships between the various theories and empirical results in our corpus, while taking care to respect a certain internal coherence through an effort of conceptual articulation to avoid redundancies.

We have also set an internal consistency criterion to eliminate logical contradictions between the different elements of the corpus. Nevertheless, we recognize that our research has the limitation of lacking empirical validation, which requires further work, both quantitative and qualitative, to test the validity of the model's hypotheses. Although we have tried to be as exhaustive as possible, this research remains partial insofar as we have not been able to integrate all the research inherent to our subject, so abundant is the literature on this subject and so rapid are the technological evolutions that give rise to new analyses and theories all the time.

We relied primarily on the SCOPUS database for its coverage of publications in management journals and its multidisciplinary indexing capability. SCOPUS covers the fields of management science, computer science and territorial economics, providing us with a multi-disciplinary theoretical corpus that we could use for our research. We combined keywords related to our research and connected them with Boolean operators: “Artificial Intelligence” OR “AI” OR “Machine Learning”) AND (“Supply Chain” OR “Logistics” OR “Value Chain”) AND (“Territorial Competitiveness” OR “Regional Development” OR “Sustainability”). We then carried out a series of iterations to fine-tune the results obtained, et also queried the Web of Science and Google Scholar databases to identify a number of works that were relevant to our research and had a high number of citations. Bibliographic data were analyzed and processed using the Bibliometrix package of the R language, and bibliographic management was visualized using Zotero software.

4. RESULTS:

Having reviewed the literature and identified the articles relevant to this research, the topics addressed by the researchers cover virtually all recent issues in intelligent supply chain management, with a particular focus on 3 components ((Bahuguna et al., 2023): the types of technologies that can be applied at supply chain management level, the issue of carbon emissions in the context of modern supply chains and their mitigation through the use of AI (, and finally the impact of such an approach on the territory as a whole to improve its competitiveness and attractiveness.

4.1. AI as a Lever For Transforming Supply Chains:

Today, AI is considered one of the most promising technological levers for radically transforming supply chains.

4.1.1. Typology of AI Technologies Applicable To Supply Chains:

The ecosystem of AI technologies capable of transforming supply chains is fairly broad, and has been constantly evolving in recent years. Machine Learning is undeniably a fundamental category, encompassing algorithms capable of learning autonomously from supplied data without being explicitly programmed. In supply chains, these technologies are particularly useful for predicting demand, anomalies in logistics flows and dynamic inventory optimization. Deep Learning, a subset of machine learning based on multi-layer artificial neural networks, can be used to tackle much more complex problems, particularly in the presence of unstructured or very large quantities of data (Bahuguna et al., 2023).

Natural Language Processing (NLP) represents another major subset of AI, enabling computer systems to automatically interpret, generate and manipulate human language in various forms. In the context of supply chains, these technologies facilitate document analysis, communication with partners, or the extraction of relevant information from structured and unstructured textual sources. In this sense, large language models (LLMs) offer particularly promising prospects for collaborative supply chain management (Phuengrod et al., 2025).

Computer vision, which enables machines to interpret and understand visual information, is also finding numerous applications in supply chains, notably for quality control, object recognition in warehouses, or the monitoring of logistics infrastructures. This technology, combined with advanced robotics, paves the way for intelligent automation of physical operations in supply chains.

Decision support technologies such as AI-powered expert systems form another important subset. Based on explicit rules and/or learning models, these systems automate complex decision-making or at least assist human decision-makers in uncertain, multi-objective or random contexts. These technologies can considerably improve prediction, inventory and flow management, and overall logistics in supply chains. (Aileni, 2025).

Finally, AI-based advanced optimization technologies, especially reinforcement learning or genetic algorithms, can solve complex optimization problems that often characterize modern supply chains, enhancing supply chain resilience and sustainability (Shrivastav, 2022).

4.1.2. AI's Transformative Capabilities: Forecasting, Optimization, Automation:

AI overcomes the limitations of conventional statistical forecasting methods by integrating a large number of variables and heterogeneous data. AI-based forecasting capabilities are now revolutionizing inventory optimization by identifying non-linear correlations as well as increasingly complex patterns that conventional approaches sometimes fail to detect (Nweje and Taiwo, 2025). This improved predictive capacity enables us to reduce safety stocks, for example, while maintaining or even improving the level of service and added value created for the customer (Vandana et al.; 2024).

Thanks to a very high processing and analysis capacity, AI-based optimization algorithms can handle multiple objectives, constraints and decision variables at the same time, enabling a holistic approach to supply chain optimization. AI can be used here to improve supply chain optimization while respecting performance constraints, such as environmental sustainability. This capability is particularly useful in today's context, where supply chains need to reconcile potentially conflicting objectives such as cost reduction on the one hand, and carbon footprint minimization and resilience maximization on the other (Eyo-Udo, 2024).

Beyond traditional automation, which is limited to predefined, repetitive tasks and constantly requires some human intervention, AI-based automation can adapt to changing situations and learn from its experiences thanks to its Deep Learning capabilities (Shamsuddoha et al., 2025). This developed form of automation can be applied to decision-making processes as well as physical operations.

4.1.3. Analysis of Theoretical And Potential Use Cases:

With the aim of converging towards integrated planning, AI makes it possible to go beyond traditional functional models to simultaneously optimize all the links that make up the supply chain, from forecasting demand quantities to planning production and distribution (Hossain et al., 2025). AI can significantly improve planning for sustainable supply chains, by integrating environmental and social considerations often absent from planning decisions.

AI technologies can now efficiently analyze large sets of internal and external data on a continuous basis to identify early on weak signals heralding potential disruptions as part of proactive management (Jahin et al., 2024). Indeed, these technologies can enhance the ability to anticipate and react to disruptive events in advance of their occurrence, thereby reducing infrastructure maintenance and re-commissioning costs (Ejjami and Boussalham, 2024). AI can be used for predictive maintenance and risk mitigation in resilient supply chains, in contrast to curative maintenance (Kalusivalingam et al., 2022). These approaches make it possible, to extend asset life, optimize maintenance interventions and reduce downtime as part of a SMED (Single-Minute Exchange of Die) approach, thus contributing to improved efficiency and sustainability of logistics operations.

Transparency and traceability in the supply chain represent a promising potential use case. Combined with other innovations such as Blockchain or IoT, AI technologies make it possible to verify the origin, production conditions and journey of products throughout the supply chain (Sah and Shaikh, 2024).

4.1.4. Current Limitations And Implementation Challenges:

Despite its transformative potential, the integration of AI into supply chains faces several challenges and constraints of different kinds. Table 2 presents the major challenges faced by organizations when introducing AI into supply chain optimization (Hangl et al., 2022):

Table 2. Categories of obstacles to AI integration in supply chains

Category	Definition
Technical challenges	Limitations related to data access, quality or processing, as well as system interoperability.
Organizational and human challenges	Human, cultural and management factors holding back AI adoption.
Economic and financial challenges	High costs and uncertainty about the return on investment of AI technologies.
Ethical and regulatory challenges	Problems of data protection, transparency and algorithmic accountability.
Challenges of complexity and uncertainty	Volatile environments make it difficult to apply AI systems effectively.

Source: The authors

Technical challenges are naturally the first hurdles to be overcome by those involved. Data quality and availability often represent a major challenge, as many organizations lack the necessary infrastructure to collect, store and process the data required by advanced AI systems. Inter-compatibility between different systems and platforms based on heterogeneous technical environments also poses a problem, particularly in supply chains involving a multitude of players with different levels of technological maturity.

Cultural enablers (Cadden et al., 2022) act as a mediator in the integration of AI into the supply chain, illustrating the prominent role of human and organizational factors: Lack of adequate skills coupled with a systematic reluctance to change are classic challenges to the adoption of these technologies.

The investments in infrastructure and skills required to deploy AI solutions can be substantial, with a return on investment that is sometimes difficult to achieve, risking disappointment due to over-optimistic expectations or insufficient understanding of their economic implications (Sodhi et al., 2022).

The use of AI raises ethical issues, particularly relating to the protection of user data, which is the raw material of automated learning processes, but also to the determination of responsibilities in the event of error or bias, and fairness in the distribution of benefits and risks (Vann Yaroson et al., 2025).

The effectiveness of AI technologies depends largely on the stability of the environment in which they evolve. Volatility is undoubtedly the most appropriate character to describe supply chains, given the complexity and large number of operators, flows and variables to be taken into consideration... efficient which can limit the performance of AI in this sense, especially if the analysis is based on historical data.

4.2. Reducing the Carbon Footprint Of Supply Chains Through AI:

AI can effectively contribute to reducing the carbon footprint of supply chains, through logistics flow optimization, predictive modeling, as well as carbon footprint measurement and monitoring approaches.

4.2.1. Mechanisms for Carbon Reduction Via Logistics Flow Optimization:

Since the 1970s, supply chains have become responsible for a significant proportion of global greenhouse gas emissions, particularly CO_2, through freight transport (sea, air and land freight), increasingly energy-intensive logistics facilities, and production and distribution processes. AI offers several mechanisms for optimizing these logistics flows and reducing their carbon footprint (Bensfia, 2025).

AI algorithms can analyze a multitude of variables (traffic fluidity, weather conditions, vehicle capacity, etc.) in real time to determine the most efficient routes in terms of carbon emissions, and thus optimize them to minimize energy consumption, delivery times and the production of greenhouse gases (Jones, 2025). In the same vein, AI can optimize the grouping of goods to maximize vehicle fill rates and thus reduce the total number of journeys required to move goods. This approach often requires complex coordination between different logistics resources, and is particularly well suited to AI's multi-constraint optimization capabilities. AliExpress is particularly successful in this respect, with an AI-based logistics operations management system that ensures optimal container scheduling to reduce costs, lead times and gas emissions (Xue and Te Chuan, 2024).

AI can also analyze geographic, economic and environmental data to optimize the location of warehouses and distribution centers to minimize total distances traveled and associated emissions. This optimization can explicitly include environmental constraints, such as proximity to low-emission transport modes or the availability of renewable energy sources.

Dynamic management of logistics capacity can be a challenge for AI technologies, in that they can adjust the allocation of logistics resources in real time according to fluctuations in demand and operational constraints, so as to avoid overcapacity and energy inefficiencies. This smart adaptive approach matches deployed capacities with actual needs, reducing resource wastage and unnecessary emissions (Bentalha, 2022).

4.2.2. Predictive Modeling And Emissions Reduction:

AI tools are revolutionizing demand forecasting by integrating a multitude of variables and identifying complex patterns that traditional approaches cannot detect, thus improving predictive accuracy (Nweje and Taiwo, 2025). More accurate forecasts make it possible to reduce costly safety stocks, avoid out-of-stocks and overstocks with all their associated hidden and apparent costs, and optimize production and distribution, with direct benefits in terms of carbon emissions.

One of AI's greatest contributions to supply chain optimization is undoubtedly its ability to model and forecast the carbon emissions associated with different logistics decisions, enabling this variable to be integrated into planning and decision-making processes in real time. This capability is particularly useful for assessing ex ante the environmental impact of different supply chain strategies or configurations, even before they are implemented.

Proactive disruption prevention is a significant application, as AI algorithms can continuously analyze data from a variety of sources to identify early warning signals of potential disruptions that could potentially reduce supply chain efficiency and increase its carbon footprint. This early detection should implement measures to mitigate their impact before disruptions fully materialize. In this sense, predictive maintenance practices for logistics equipment can optimize maintenance interventions, predict and reduce unplanned breakdowns, and extend asset life (Ejjami and Boussalham, 2024). These benefits naturally translate into a significant reduction in the carbon footprint associated with production processes and equipment replacement, as well as improved energy efficiency of facilities provided, they are maintained in optimum condition.

4.2.3. AI Approaches To Measuring And Monitoring Carbon Footprints:

Through natural language processing and machine learning, AI can extract and integrate environmental data from often diverse and heterogeneous sources (industry reports, government databases, information systems etc.) to build a comprehensive and up-to-date picture of the supply chain's carbon footprint. In the same vein, AI systems can model and predict with great accuracy the carbon emissions of a product's entire lifecycle, from raw material extraction to end-of-life. This holistic approach makes it possible to identify the production stages responsible for the greatest gas emissions, and thus limit the transfer of pollution between different stages in the product life cycle. This notion of carbon traceability makes it possible to track emissions at every stage of the supply chain, offering transparency on the carbon footprint of the production process, and respecting the company's societal commitments to its consumers, investors and regulators.

Dynamic emissions scenario analysis is a fourth important approach, as AI systems can simulate and analyze the impact of different decisions on the carbon footprint of the supply chain, enabling proactive and adaptive emissions management. This capability is particularly valuable in a context of uncertainty and rapid change, where carbon reduction strategies need to be constantly adjusted in line with changing external and internal conditions.

4.3. From Environmental Sustainability To Territorial Competitiveness:

The transition from a focus on carbon footprint reduction to a broader perspective of territorial competitiveness is driven by major changes in the international economic, social and environmental environment, and represents a major evolution in the conceptualization of sustainable and resilient supply chains.

4.3.1. Concept of Territorial Competitiveness In The Context Of Supply Chains:

Territorial competitiveness is the ability of a territory to attract and retain economic, financial and tourism activities, generate innovation, develop human capital and maintain or improve the quality of life of its inhabitants, within the constraints of preserving its natural and cultural capital. In the specific context of supply chains, this concept takes on a particular dimension (Bensfia and Bouklata, 2023).

Thanks to the globalization of trade, modern supply chains have become global in scope, enabling more extensive exchanges than ever before, but also rooted in specific territories through their physical infrastructures, human resources and interactions with local ecosystems. This duality generates both challenges and opportunities for territorial competitiveness. On the one hand, sustained integration into global logistics networks necessarily exposes territories to increased competition from other territories and regions. On the other, it offers new opportunities for access to new markets, financial resources and knowledge that can make local development more sustainable and resilient.

In this context, AI can play an innovative role in enabling territories to benefit from their integration into global supply chains while minimizing the underlying risks and costs (Shamsuddoha et al., 2025). AI can indeed help territories develop sustainable and distinctive competitive advantages based on the efficiency, innovation and sustainability of their supply chains.

Territorial competitiveness in the specific context of supply chains can be conceptualized according to several key concepts: the ability to attract and retain logistics investment, the ability to foster innovation in logistics processes, the quality and sustainability of logistics infrastructures, the efficiency of material and information flows in terms of cost and lead times, and resilience in the face of environmental disruption.

4.3.2. AI's Contribution To The Resilience Of Local Ecosystems:

The resilience of local territorial ecosystems, both economically and environmentally, has unquestionably become an essential component of territorial competitiveness in an environment characterized by growing uncertainty and volatility. For example, systemic risk management is now at the heart of AI technologies that can analyze large heterogeneous datasets to identify vulnerabilities in local ecosystems related to supply chain activities and prepare specific action plans (Mittal et al., 2023). AI-based assessment systems for supply chain vulnerabilities and resilience to external shocks (natural disasters, pandemics, geopolitical instability...) and demonstrating that these approaches can strengthen territories' capacity to anticipate and adapt.

On the other hand, the intelligent diversification of economic activities represents an imperative for territories with a view to reducing systemic risks. AI can analyze the complex interdependencies and complementarities between different sectors and activities within a territory, and then implement diversification strategies that maximize resilience while exploiting potential synergies. This approach reduces the risk of dependence on a single sector in the event of a severe external shock (Mehta and Sharma, 2023).

From an ecological perspective, the identification and exploitation of circular economy opportunities on a territorial scale is greatly facilitated by AI, by meticulously analyzing the flows of materials, energy and water between different economic players. This circularity contributes to both economic resilience (reducing import dependency via local production policies) and environmental resilience (reducing pressure on local ecosystems and preserving local territorial resources). Indeed, adaptive management of natural resources in supply chains, through real-time monitoring, promotes predictive modeling and dynamic optimization of these resources. This approach is particularly well-suited to renewable but stressed resources such as water, forests and fisheries, whose sustainable management determines the resilience and even the long-term survival of territories.

4.3.3. AI-Facilitated Collaborative Dynamics At Territorial Level:

Territorial competitiveness depends on a number of variables, not least the collaborative dynamics between different players at territorial level. The implementation of collaborative platforms that facilitate communication and collaboration can benefit from the contribution of AI in the sense that it optimizes supply chain processes. The transformative potential of this approach is particularly useful in terms of information sharing and trust between the various players (Weisz et al., 2025).

The dynamic orchestration of business ecosystems represents a second important lever. AI systems can also help to locate and exploit, or even create, complementarities between different local players, in such a way as to facilitate the emergence and evolution of integrated territorialized business ecosystems. This intelligent approach improves interactions between the various players and maximizes the creation of collective territorial value.

Another area where AI can benefit territorial competitiveness is in the optimized pooling of resources and infrastructures. Indeed, AI can promote the sharing and mutual use of idiosyncratic logistics infrastructures, tangible and intangible resources, as well as collaborative capabilities between different territorial players, improving overall territorial efficiency in the process, while reducing costs and the carbon footprint for each stakeholder in supply chains. This intelligent pooling of resources is particularly relevant to territories with limited resources.

AI technologies can also contribute to the identification of best practices and collaborative learning between different territorial actors involved in supply chains, to foster knowledge sharing and thus create routines that will be used later to solve specific problems. This collective learning approach significantly enhances a region's capacity to innovate and adapt within its environment, with all its constraints, thus contributing to its improvement in sustainable and resilient territorial competitiveness.

5. DISCUSSION:

5.1. Integrative Framework: AI as a Catalyst For Sustainable Transition:

Having explored the various dimensions of AI's role in sustainable supply chains, an integrative and conceptual framework is proposed to proposed to establish the role of AI as a catalyst in the transition towards more sustainable and territorially competitive supply chains, examining causal mechanisms, conditions of effectiveness, moderating factors moderating factors and the systemic implications of this relationship (Farooq and Yen, 2024).

5.1.1. Proposal for a Theoretical Model Integrating AI, Sustainability And Competitiveness:

The integrative theoretical framework proposed after analyzing the various contributions of this literature review conceptualizes AI as a catalyst that benefits a certain transition of supply chains towards greater sustainability and territorial competitiveness (Teixeira et al., 2025). This model is articulated around three main components that we felt were most relevant: AI's transformative capabilities in greening supply chain transition, the dimensions of sustainability of these supply chains, and the factors of sustainable and sustained territorial competitiveness, as shown in Figure 3.

The transformative capabilities of AI include early and advanced forecasting based on behavioral models of the various players and predictive and anticipatory management of the necessary resources, multi-objective optimization and intelligent automation through AI agents capable of synthesizing large quantities of data. These capabilities constitute the technological engine of the model, and would help overcome some of the limitations of traditional supply chain management approaches.

Supply chain sustainability encompasses environmental performance (reducing the carbon footprint of transits), economic performance (profitability and value creation) and social performance (working conditions within supply chains, development of key skills, direct impact on territories). From a systemic perspective, these dimensions are interdependent and need to be optimized jointly rather than in isolation.

On the other hand, territorial competitiveness is above all focused on the resilience of local ecosystems, the creation of sustainable and shared value, and the collaborative dynamics between the players that make up the territorial economic fabric. These factors represent the territorial manifestations of a successful transition to more sustainable supply chains.

Figure 3. AI as a Catalyst for Sustainable Supply Chain

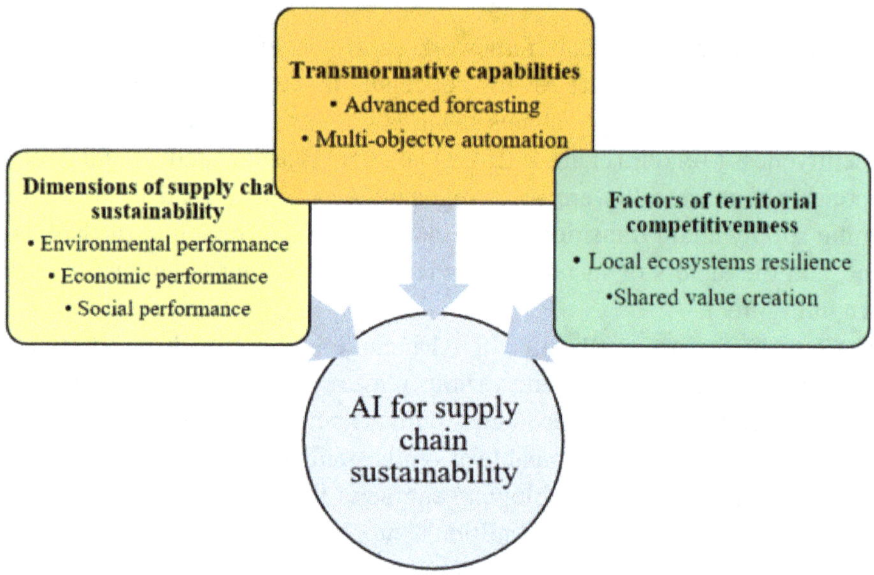

Source: The authors

The proposed theoretical model introduces the idea of an AI that can effectively act as a catalyst for the sustainable transition of supply chains through several mechanisms: by operationalizing sustainability objectives previously considered too complex or costly due to the obsolete technologies that were the norm some years ago (and still remain), by facilitating the identification and exploitation of synergies between different performance dimensions, and by accelerating the learning and adaptation phases necessary for continuous improvement.

5.1.2. Causal Mechanisms Between AI Adoption And Sustainable Performance:

In addition to articulations on essential and explanatory components, this proposed integrative framework identifies causal mechanisms that explain how AI adoption can lead to a notable improvement in sustainable supply chain performance and, ultimately, territorial competitiveness, as shown in Figure 4:

Figure 4. Causal mechanisms of AI and supply chains sustainability

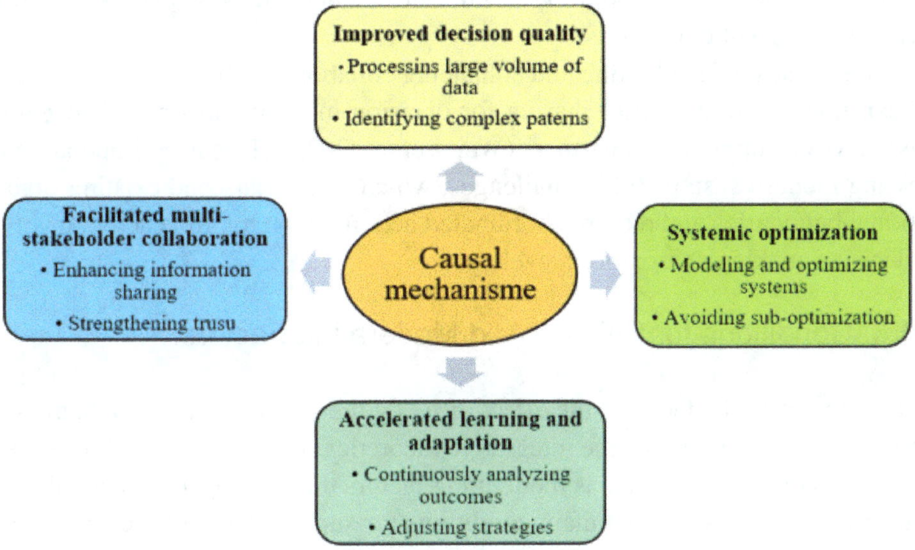

Source: The authors

In the context of improving decision-making quality, AI makes it possible to process very large quantities of heterogeneous and often unstructured data, and to identify complex patterns that traditional analytical approaches or human rationality cannot (or only with difficulty) detect. This analytical and processing capability leads to more considered and accurate decisions regarding the design and management of supply chains, with direct benefits in terms of improved economic and environmental operational performance (Abaku et al., 2024). AI opens up new perspectives towards efficiency and resilience in supply chain optimization, precisely through this improvement in decision-making quality.

Secondly, in today's supply chain context, systemic rather than local optimization is required. AI algorithms can in fact model, optimize and sometimes remake complex systems as a whole, thus avoiding a fragmented approach that would lead to sub-optimization and prevent the identification and exploitation of possible synergies between different links in the supply chain and between different performance dimensions, leading to simultaneous improvements in economic, environmental and social terms.

Concerning the acceleration as well as the improvement of learning and adaptation loops, AI systems can continuously analyze the results of past decisions, locate their impact in terms of efficiency and performance, identify opportunities for improvement and adjust strategies accordingly, thus creating a virtuous cycle of continuous learning and adaptation. This ability is particularly valuable in a context of transition towards sustainability and resilience, which often requires experimentation and the evolution of current practices towards new perspectives dictated by current constraints.

In the context of facilitating and enhancing collaboration between stakeholders, AI can improve information sharing for decision-making and strengthen mutual trust between supply chain partners (Weisz et al., 2025). This is an imperative for tackling modern sustainability challenges, which often transcend existing organizational boundaries and require coordinated action by the multiple actors involved in decision-making at territorial level.

5.1.3. Conditions for Efficacy And Moderating Factors:

Starting from the fact that the effect of AI on the efficiency of sustainable supply chains is not automatic, the integrative theoretical framework identifies several efficiency conditions as well as moderating factors that greatly influence this relationship. Figure 5 shows the main conditions for supply chain efficiency, with an emphasis on the associated mediating factors.

Figure 5. Conditions of effectiveness and moderating factors

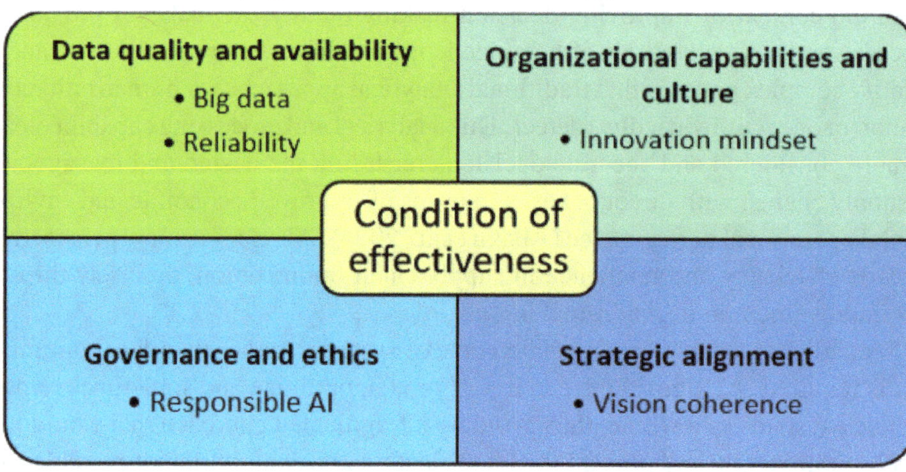

Source: The authors

Data quality and availability are undeniably critical to this process, as any rational decision must be based on reliable data. AI generally requires large volumes of high-quality data to operate efficiently and effectively (Kollia et al., 2021). In the context of sustainable supply chains, this implies traditional operational data, but also and above all heterogeneous environmental and social data, which can be more difficult to collect and standardize because they come from several sources and are not formally structured. The low quality or even absence of such data can significantly limit the effectiveness of AI-based approaches to optimizing sustainable supply chains, as shown in figure 6.

Figure 6. Systemic Implications and Organizational Transformations

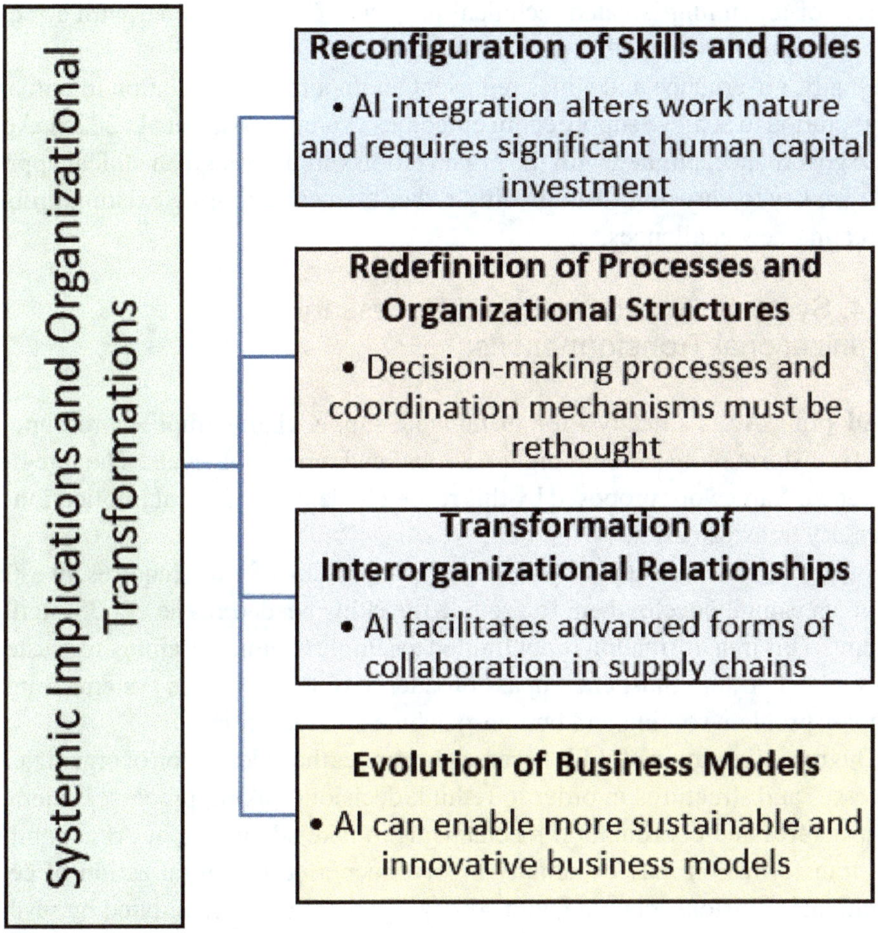

Source: The authors

Cultural enablers as well as organizational capabilities play an influential and mediating role in the integration of AI into the supply chain (Cadden et al., 2022). An organizational culture that advocates and encourages the cultivation of innovation, continuous learning and decision-making based on reliable, quality data is more likely to take full advantage of AI's potential for sustainability.

These factors must converge and serve organizational orientations and objectives in terms of efficiency and sustainable transition, giving greater importance to alignment as an organizational and decision-making imperative. The effectiveness of AI therefore depends closely on its alignment with the broader strategic objectives of the organization and the territory. Without this alignment, AI-based initiatives run the risk of remaining isolated technical projects of limited scope, with a reduced impact on overall sustainable performance.

Finally, governance and ethics represent an important articulation in human-IA collaboration to achieve supply chain outcomes (Vann Yaroson et al., 2025). Appropriate governance, aligned with the organization's strategy, and an ethical approach to AI are key to ensuring sustainability rather than exacerbating existing problems or creating new challenges.

5.1.4. Systemic Implications And Necessary Organizational Transformations:

Adopting AI as a catalyst for sustainable supply chains implies profound systemic transformations, at both organizational and territorial levels. The integrative theoretical framework proposed by this research identifies several implications and necessary transformations.

Integrating AI into supply chains changes the nature of work requires investment in human capital development to provide the skills needed in the era of intelligent systems. This transformation is not limited to simple technical training to master the new tools of AI, but must encompass broader new skills such as systems thinking, complex problem solving and human-machine collaboration.

This reconfiguration of skills necessarily implies the redefinition of organizational processes and structures in order to rethink decision-making processes, hierarchical structures and coordination mechanisms to take advantage of AI capabilities. This transformation may eventually involve increased decentralization of certain operational decisions, coupled with strategic centralization facilitated by visibility throughout the supply chain. It involves a genuine overhaul of business models to make them more sustainable, such as the functionality economy, various collaborative platforms or circular economy ecosystems. These various innovative models have the potential to create new sources of value while reducing environmental footprints and strengthening territorial resilience.

The transformation of inter-organizational relationships represents an important systemic implication in the sense that AI can foster more developed forms of collaboration between different stakeholders in the supply chain, moving beyond traditional transactional relationships towards genuine strategic partnerships and integrated innovation ecosystems. This evolution certainly requires new contractual frameworks, governance mechanisms and the establishment of a certain culture of collaboration.

5.2. Managerial and Political Implications:

The proposed integrative theoretical framework has significant implications for both business leaders and territorial public policy makers. Based on the imperatives and constraints already outlined, concrete recommendations are made for maximizing the benefits derived from AI's role as a catalyst for sustainable, territorially competitive supply chains.

5.2.1. Recommendations for Corporate Decision-Makers:

Faced with the complex challenges of transitioning to more sustainable and resilient supply chains, corporate decision-makers can benefit from some recommendations made to guide them through this transformation.

Companies would benefit from integrating AI technologies into an integrated strategic vision rather than seeing them as mere technological solutions to specific problems (such as transforming their supply chain with the aim of reducing their carbon footprint) (Fosso Wamba et al, 2022). The most significant benefits will be seen when these technologies are deployed as part of a systemic transformation rather than as isolated initiatives.

The effectiveness of AI as a catalyst for supply chain sustainability and resilience is highly dependent on the development of complementary organizational capabilities such as good data governance, decision-making agility under cost and time constraints, cross-functional collaboration between different stakeholders and continuous organizational and individual learning, reflecting the importance of cultural enablers in the integration of AI into the supply chain, due to the fact that investments in these "Soft" dimensions are as capital as the "Hard" technological investments themselves (Cadden et al., 2022) (Mary, 2025).

In the same vein, the adoption of a collaborative, inter-actor and ecosystem-based approach is an important recommendation for private and public decision-makers. Indeed, today's sustainability challenges now transcend organizational boundaries and require increased coordination between multiple actors involved in the value chain (Weisz et al., 2025). Companies should explore AI mechanisms to facilitate this collaboration, notably through dedicated sharing and collaboration platforms, common standards accepted by different communities and collaborative governance mechanisms. AI can precisely improve information sharing and enhance trust in supply chain collaboration, offering concrete and applicable avenues for this eco-systemic approach.

Another recommendation that can be made to company managers is the explicit integration of sustainability and resilience objectives into AI initiatives. Quite often, AI implementation projects in supply chains are evaluated solely on the economic criteria of operational efficiency or cost reduction. Instead, companies should broaden their evaluation criteria to explicitly include environmental and social performance indicators, as well as measures of the territorial impact of supply chain sustainability. This multi-criteria approach helps to align technological initiatives with sustainability and territorial competitiveness objectives.

5.2.2. Guidelines for Territorial Public Policies:

In addition to private sector business leaders, local public policy decision-makers also have a crucial role to play in facilitating and guiding the adoption of AI as a catalyst for sustainable and resilient supply chains. Several orientations can be proposed in this perspective. The development of infrastructure as well as adequate skills is a first fundamental direction towards a transition towards improved territorial competitiveness. Indeed, territories should invest in digital infrastructures capable of hosting AI technologies, massive data platforms and the development of the skills needed to adopt AI in supply chains (Bouklata and Bensfia, 2020). These investments create the basic favorable conditions so that local economic players can take advantage of the opportunities offered by these technologies. Particular attention should nevertheless be paid to the inclusivity of these initiatives, to prevent the digital transition from exacerbating already existing inequalities (Helo and Hao, 2022).

Another important orientation is the facilitation of collaborative dynamics, in the sense that public policies can play a key role in the emergence and strengthening of collaborations between different territorial players involved in supply chains. This may involve, for example, creating collaborative innovation spaces, supporting value-creating pilot projects, or setting up collaborative territorial platforms for sharing and disseminating data essential to decision-making (Andres et al., 2024).

Furthermore, the alignment of incentives and regulations should create a regulatory and incentive environment that encourages the adoption of AI for sustainability objectives in supply chains, clearly setting out the obligations and rights of each stakeholder, in order to create a climate of trust and transparency. This can include mechanisms such as progressive carbon pricing (dependent on the volume of gas emissions), tax incentives for investment in sustainable technologies, or regulatory requirements for transparency on the environmental footprint of products. These instruments need to be designed and implemented in such a way as to stimulate innovation, while ensuring that the benefits of the transition are fairly distributed.

Territories should also contribute to the creation of experimentation spaces where new approaches and simulations combining AI and sustainability in supply chains can be tested under real but controlled and less costly conditions. These "Living Labs" or innovation zones considerably accelerate collective learning, limit the risks associated with innovation, and demonstrate the feasibility of new approaches before they are deployed on a larger scale.

5.2.3. Ethical and Responsible Considerations In The Use of AI in Supply Chains:

Beyond traditional concerns in terms of economic and financial performance, the use of AI in supply chains raises questions about ethical and responsibility practices that need to be addressed incessantly to ensure that these technologies contribute effectively to a sustainable and, above all, equitable transition.

Decision-making processes powered by AI technologies need to be transparent and explainable, in order to guarantee a certain explicability and evaluation on the part of supply chain stakeholders. This transparency is all the more crucial when it has direct environmental, social or territorial implications. On the other hand, equity and inclusiveness explicitly imply that the use of AI in supply chains should not exacerbate already existing inequalities or even create new ones, whether horizontally between different categories of actors, between competing companies (especially small and medium-sized enterprises, which often make up the largest part of the socio-economic fabric of territories) or outright between territories. A form of participative governance can provide a solution to this problem, insofar as decisions concerning the adoption and configuration of AI systems in supply chains would do better to involve the full range of stakeholders, including actors, local territorial communities and representatives of civil society organizations. This participatory approach would not only significantly improve the quality and legitimacy of the decisions taken, and enhance internal cohesion, but also and above all strengthen local ownership of the technologies and their expected benefits (Bensfia and Aidi, 2025).

Paradoxically, ethical considerations should take into account the environmental impact of AI tools themselves in terms of excessive energy consumption for model training operations. These impacts must be explicitly taken into account when assessing the sustainability of introducing AI into supply chains, and at the same time efforts must be made to contain the energy-intensive nature of AI technologies (particularly in terms of cooling and infrastructure maintenance) (Parouty and Arias, 2024).

CONCLUSION

The main theoretical contribution of this chapter lies in the conceptualization of AI as a multi-disciplinary catalyst for accelerating and amplifying the transition to sustainable, territorially anchored supply chains. This conceptualization goes beyond technical approaches that see AI primarily as a tool for optimizing flows and resources, to consider it as a genuine vector for systemic transformation of practices, business models and a reconfiguration of entire territorial ecosystems.

The proposed integrative theoretical framework is built around three main components: the transformative capabilities of AI, the dimensions of supply chain sustainability (under its environmental, economic and social aspects), and the factors of territorial competitiveness, and in particular the resilience of local ecosystems, the creation of shared value. This framework identifies how AI can catalyze sustainable transition through several transmission mechanisms, facilitating the identification and exploitation of synergies between different performance dimensions. These mechanisms by which AI can contribute to reducing the carbon footprint of supply chains are divided between optimization of logistics flows throughout the chain, predictive modeling of emissions through simulations of possible scenarios even before they are implemented, multi-objective optimization integrating several sometimes-contradictory dimensions such as economic and environmental performance, and advanced approaches to measuring and monitoring the carbon footprint (Nitin, 2025).

The conceptualization of the link between environmental sustainability and territorial competitiveness in the context of supply chains has been developed in an approach that goes beyond the traditional vision opposing environmental performance and economic competitiveness, to propose an integrative perspective where AI can facilitate the creation of shared value on a territorial scale.

The possible use of AI technologies as a catalyst for sustainable and territorially competitive supply chains is still emerging and constantly evolving, with new solutions and applications appearing every day. This raises many new questions and opportunities for research and innovation, as well as perspectives for reflection and analysis. For example, evaluative approaches that simultaneously assess the economic, environmental, social and territorial impacts of adopting AI technologies in supply chains are now needed to guide investment and deployment decisions at territorial level. These methods should necessarily integrate multi-scale (from local to global), multi-temporal (from short to long term) and multi-disciplinary perspectives, taking care to include all areas of supply chain management.

The various managerial and political implications of AI as a catalyst for sustainable supply chains are quite vast and multidimensional. Business decision-makers as well as territorial public policy-makers need to ensure a certain complementarity in order to maximize the benefits of adopting these technologies while minimizing their potential risks. Ethical and responsible considerations must be central to the design of AI systems, and a territorial research and innovation plan is needed to continue exploring and exploiting the potential of these approaches.

This chapter aims to have contributed to a better understanding of the conditions for effectiveness as well as the moderating factors influencing the possible relationship between systematic adoption of the tools offered by AI technologies and sustainable, resilient supply chain performance by identifying factors such as data quality, organizational capabilities, strategic alignment and ethical governance, in order to move beyond traditional deterministic views of technology in general to a more nuanced, contextual perspective.

Nevertheless, and despite the efforts made in this chapter, the proposed approach has the limitation of being a theoretical article, and consequently the proposals put forward have not been empirically validated by field studies. Although firmly anchored in the literature analyzed, the causal relationships and mechanisms identified would certainly require empirical validation through in-depth case studies, quantitative analyses or participatory action research.

REFERENCES:

Abaku, E. A., Edunjobi, T. E., & Odimarha, A. C. (2024). Theoretical Approaches to AI in Supply Chain Optimization: Pathways to Efficiency and Resilience. *International Journal of Science and Technology Research Archive*, 6(1), 092-107.

Agrawal, A., Gans, J., & Goldfarb, A. (2022). *Prediction Machines, Updated and Expanded: The Simple Economics of Artificial Intelligence*. Harvard Business Press.

Aileni, V. R. (2025). AI-Driven Supply Chain Optimization: Enhancing Forecasting, Inventory, And Logistics. *Machine Intelligence Research, 19*(1), 347–364.

Alla, L., Bentalha, B., & Bouhtati, N. (2022, May). Assessing supply chain performance in the covid 19 context: a prospective model. In 2022 14th International Colloquium of Logistics and Supply Chain Management (LOGISTIQUA) (pp. 1-6). IEEE. DOI: 10.1109/LOGISTIQUA55056.2022.9938083

Andres, B., Fernandez-Garcia, D., Urze, P., & Araujo, E. (2024). Artificial Intelligence Use in Collaborative Network Processes.

Attah, R. U., Garba, B. M. P., Gil-Ozoudeh, I., & Iwuanyanwu, O. (2024). Enhancing Supply Chain Resilience Through Artificial Intelligence: Analyzing Problem-Solving Approaches in Logistics Management. *International Journal of Management & Entrepreneurship Research, 5*(12), 3248–3265.

Bahuguna, D., Kaur, J., & Singh, B. (2023). Artificial Intelligence's Integration in Supply Chain Management: A Comprehensive Review. *European Economics Letter, 13*(3), 1512–1527.

Barney, J. (1991). Firm Resources and Sustained Competitive Advantage. *Journal of Management, 17*(1), 99–120. DOI: 10.1177/014920639101700108

Belhadi, A., Mani, V., Kamble, S. S., Khan, S. A. R., & Verma, S. (2024). Artificial Intelligence-Driven Innovation for Enhancing Supply Chain Resilience and Performance Under the Effect of Supply Chain Dynamism: An Empirical Investigation. *Annals of Operations Research, 333*(2-3), 627–652. Https://Doi.Org/10.1007/S10479-021-03956-X. DOI: 10.1007/s10479-021-03956-x PMID: 33551534

Bensfia, C. (2025). Collective Intelligence of Port Territory Stakeholders: Case of Moroccan Ports. In Utilizing Technology to Manage Territories (p. 335-370). IGI Global

Bensfia, C., & Aidi, S. (2025). Port Community Systems and Organizational Performance: A Socio-technical Approach to Evaluation. Industrial and Logistics Systems Design and Efficient Operation, 42-52. https://doi.org/DOI: 10.1007/978-3-031-87309-6_4

Bensfia, C., & Bouklata, A. (2022). Validity of the TAM Model in the context of electronic payment ofport dues in Morocco. 1-5

Bensfia, C., & Bouklata, A. (2023). Evaluation of the electronic port right of way payment information system. *Management Systems in Production Engineering*, *31*(3), 254–262. DOI: 10.2478/mspe-2023-0028

Bentalha, B., Hmioui, A., & Alla, L. (2019, October). The digitalization of the supply chain management of service companies: a prospective approach. In *Proceedings of the 4th International Conference on Smart City Applications* (pp. 1-8). DOI: 10.1145/3368756.3369005

Bouklata, A., & Bensfia, C. (2020). Digitalization of port passage procedures: Focus on the transit time of Goods. 1-6.

Boute, R. N., & Udenio, M. (2022). AI in Logistics and Supply Chain Management. In *Global Logistics and Supply Chain Strategies for the 2020s: Vital Skills for the Next Generation* (pp. 49–65). Springer International Publishing.

Cadden, T., Dennehy, D., Mantymaki, M., & Treacy, R. (2022). Understanding The Influential and Mediating Role of Cultural Enablers of AI Integration to Supply Chain. *International Journal of Production Research*, *60*(14), 4592–4620. Https://Doi.Org/10.1080/00207543.2021.1946614. DOI: 10.1080/00207543.2021.1946614

Chui, M., Manyika, J., & Miremadi, M. (2018). What AI Can and Can't Do (Yet) for Your Business. Mckinsey Quarterly, 1(97-108), 1.

Dey, P. K., Chowdhury, S., Abadie, A., Vann Yaroson, E., & Sarkar, S. (2024). Artificial Intelligence-Driven Supply Chain Resilience in Vietnamese Manufacturing Small- And Medium-Sized Enterprises. *International Journal of Production Research*, *62*(15), 5417–5456. Https://Doi.Org/10.1080/00207543.2023.2179859. DOI: 10.1080/00207543.2023.2179859

Ejjami, R., & Boussalham, K. (2024). Resilient Supply Chains in Industry 5.0: Leveraging AI For Predictive Maintenance and Risk Mitigation. *International Journal for Multidisciplinary Research*, *6*(4).

Elkington, J. (1998). Partnerships From Cannibals with Forks: The Triple Bottom Line Of 21st-Century Business. *Environmental Quality Management*, *8*(1), 37–51. DOI: 10.1002/tqem.3310080106

Elufioye, O. A., Ike, C. U., Odeyemi, O., Usman, F. O., & Mhlongo, N. Z. (2024). Ai-Driven Predictive Analytics in Agricultural Supply Chains: A Review: Assessing the Benefits and Challenges of Ai in Forecasting Demand and Optimizing Supply in Agriculture. *Computer Science & IT Research Journal*, *5*(2), 473–497. DOI: 10.51594/csitrj.v5i2.817

Eyo-Udo, N. (2024). Leveraging Artificial Intelligence for Enhanced Supply Chain Optimization. Open Access Research Journal of Multidisciplinary Studies, 7(2), 001-015.

Farooq, M., & Yen, Y. Y. (2024). Artificial Intelligence in Supply Chain Management: A Comprehensive Review and Framework for Resilience and Sustainability. *In Review*. Https://Doi.Org/10.21203/Rs.3.Rs-3878218/V1 DOI: 10.21203/rs.3.rs-3878218/v1

Fosso Wamba, S., Queiroz, M. M., Guthrie, C., & Braganza, A. (2022). Industry Experiences of Artificial Intelligence (AI): Benefits and Challenges in Operations and Supply Chain *Management. Production Planning and Control*, *33*(16), 1493–1497. DOI: 10.1080/09537287.2021.1882695

Freeman, R. E. (2010). *Strategic Management: A Stakeholder Approach*. Cambridge University Press. DOI: 10.1017/CBO9781139192675

Guo, D., & Mantravadi, S. (2025). The Role of Digital Twins in Lean Supply Chain Management: Review and Research Directions. *International Journal of Production Research*, *63*(5), 1851–1872. Https://Doi.Org/10.1080/00207543.2024.2372655. DOI: 10.1080/00207543.2024.2372655

Hangl, J., Behrens, V. J., & Krause, S. (2022). Barriers, Drivers, And Social Considerations for AI Adoption in Supply Chain Management: A Tertiary Study. *Logistics (Basel)*, *6*(3), 63. Https://Doi.Org/10.3390/Logistics6030063. DOI: 10.3390/logistics6030063

Helo, P., & Hao, Y. (2022). Artificial Intelligence in Operations Management and Supply Chain Management: An Exploratory Case Study. *Production Planning and Control*, *33*(16), 1573–1590. Https://Doi.Org/10.1080/09537287.2021.1882690. DOI: 10.1080/09537287.2021.1882690

Hendriksen, C. (2023). Artificial Intelligence for Supply Chain Management: Disruptive Innovation or Innovative Disruption? *The Journal of Supply Chain Management*, *59*(3), 65–76. Https://Doi.Org/10.1111/Jscm.12304. DOI: 10.1111/jscm.12304

Hossain, M. S., Sikdar, M. S. H., Chowdhury, A., Bhuiyan, S. M. Y., & Mobin, S. M. (2025). AI-Driven Aggregate Planning for Sustainable Supply Chains: A Systematic Literature Review of Models, Applications, And Industry Impacts. *American Journal of Advanced Technology and Engineering Solutions*, *1*(01), 382–437. DOI: 10.63125/3jdpkd14

Jahin, M. A., Naife, S. A., Saha, A. K., & Mridha, M. F. (2024). AI In Supply Chain Risk Assessment: A Systematic Literature Review and Bibliometric Analysis. *Clinical Orthopaedics and Related Research.*

Jones, J. (2025). AI-Driven Logistics: Understanding the Transformational Impact on Global Supply Chain Networks. *Business, Economics and Management.* Https://Doi.Org/10.20944/Preprints202502.0095.V1

Kalusivalingam, A. K., Sharma, A., Patel, N., & Singh, V. (2022). Enhancing Supply Chain Resilience Through AI: Leveraging Deep Reinforcement Learning and Predictive Analytics. *International Journal of Artificial Intelligence and Machine Learning*, *3*(9).

Kollia, I., Stevenson, J., & Kollias, S. (2021). AI-Enabled Efficient and Safe Food Supply Chain. *Electronics (Basel)*, *10*(11), 1223. Https://Doi.Org/10.3390/Electronics10111223. DOI: 10.3390/electronics10111223

Mary, B. J. (2025). The Synergistic Role of Cross-Functional Collaboration and AI-Powered Forecasting in Enhancing Supply Chain Agility, Responsiveness, and Strategic Resilience.

Mehta, A., & Sharma, R. (2023). *Innovative Strategies in Supply Chain Management.*

Mittal, U., & Panchal, D. (2023). AI-Based Evaluation System for Supply Chain Vulnerabilities and Resilience Amidst External Shocks: An Empirical Approach. *Reports in Mechanical Engineering*, *4*(1), 276–289. Https://Doi.Org/10.31181/Rme040122112023m. DOI: 10.31181/rme040122112023m

Naeem, M., Ozuem, W., Howell, K., & Ranfagni, S. (2023). A Step-By-Step Process of Thematic Analysis to Develop a Conceptual Model in Qualitative Research. *International Journal of Qualitative Methods*, *22*, 16094069231205789. DOI: 10.1177/16094069231205789

Nweje, U., & Taiwo, M. (2025). Leveraging Artificial Intelligence for Predictive Supply Chain Management, Focus on How AI- Driven Tools Are Revolutionizing Demand Forecasting and Inventory Optimization. *International Journal of Science and Research Archive*, *14*(1), 230–250. Https://Doi.Org/10.30574/Ijsra.2025.14.1.0027. DOI: 10.30574/ijsra.2025.14.1.0027

Onukwulu, E. C., Agho, M. O., & Eyo-Udo, N. L.Ekene Cynthia OnukwuluMercy Odochi AghoNsisong Louis Eyo-Udo. (2023). Developing A Framework For AI-Driven Optimization of Supply Chains in Energy Sector. *Global Journal of Advanced Research and Reviews*, *1*(2), 82–101. DOI: 10.58175/gjarr.2023.1.2.0064

Parouty, J. L., & Arias, S. (2024, December). La Face Rose, Grise Et Sombre De l'Intelligence Artificielle! In *JRES (Journées réseaux de l'enseignement et de la recherche) 2024*.

Phuengrod, S., Wannapiroon, P., & Nilsook, P. (2025). The Intelligent Collaborative Supply Chain Management with Large Language Models. *Journal of Theoretical and Applied Information Technology*, *103*(5).

Redżeb, K. (2024). The Transformative Role of AI In Modern Supply Chains: A Study on Collaboration and Efficiency.

Riad, M., Naimi, M., & Okar, C. (2024). Enhancing Supply Chain Resilience Through Artificial Intelligence: Developing A Comprehensive Conceptual Framework for AI Implementation and Supply Chain Optimization. *Logistics (Basel)*, *8*(4), 111. Https://Doi.Org/10.3390/Logistics8040111. DOI: 10.3390/logistics8040111

Sah, B. P., & Shaikh, S. (2024). AI-Driven Iot and Blockchain Integration in Industry 5.0: A Systematic Review of Supply Chain Transformation. *Innovatech Engineering Journal*, *1*(01), 99–116. DOI: 10.70937/itej.v1i01.12

Shamsuddoha, M., Khan, E. A., Chowdhury, M. M. H., & Nasir, T. (2025). Revolutionizing Supply Chains: Unleashing The Power Of AI-Driven Intelligent Automation and Real-Time Information Flow. *Information (Basel)*, *16*(1), 26. Https://Doi.Org/10.3390/Info16010026. DOI: 10.3390/info16010026

Shrivastav, M. (2022). Barriers Related to AI Implementation in Supply Chain Management. *Journal of Global Information Management*, *30*(8), 1–19. Https://Doi.Org/10.4018/JGIM.296725. DOI: 10.4018/JGIM.296725

Sodhi, M. S., Seyedghorban, Z., Tahernejad, H., & Samson, D. (2022). Why Emerging Supply Chain Technologies Initially Disappoint: Blockchain, Iot, And AI. *Production and Operations Management*, *31*(6), 2517–2537. Https://Doi.Org/10.1111/Poms.13694. DOI: 10.1111/poms.13694

Teixeira, A. R., Ferreira, J. V., & Ramos, A. L. (2025). Intelligent Supply Chain Management: A Systematic Literature Review on Artificial Intelligence Contributions. *Information (Basel)*, *16*(5), 399. Https://Doi.Org/10.3390/Info16050399. DOI: 10.3390/info16050399

Vandana, M., Naveena, M., Ellaturu, N., Kumari, T. L., Bambuwala, S., & Rajalakshmi, M. (2024). Ai-Driven Solutions for Supply Chain Management. *Journal Of Informatics Education and Research*, *4*(2), 1526–4726.

Vann Yaroson, E., Abadie, A., & Roux, M. (2025). Human-Artificial Intelligence Collaboration in Supply Chain Outcomes: The Mediating Role of Responsible Artificial Intelligence. *Annals Of Operations Research*. Https://Doi.Org/10.1007/S10479-025-06534-7

Weisz, E., Herold, D. M., Ostern, N. K., Payne, R., & Kummer, S. (2025). Artificial Intelligence (AI) For Supply Chain Collaboration: Implications on Information Sharing and Trust. *Online Information Review*, *49*(1), 164–181. Https://Doi.Org/10.1108/OIR-02-2024-0083. DOI: 10.1108/OIR-02-2024-0083

Xue, W., & Te Chuan, L. (2024). The Application of Artificial Intelligence in China's Cross Border E-Commerce Field.

Yandrapalli, V. (2023). Revolutionizing Supply Chains Using Power of Generative AI. *International Journal of Research Publication and Reviews*, *4*(12), 1556–1562. DOI: 10.55248/gengpi.4.1223.123417

Zejjari, I., & Benhayoun, I. (2024). The Use of Artificial Intelligence to Advance Sustainable Supply Chain: Retrospective and Future Avenues Explored Through Bibliometric Analysis. *Discover Sustainability*, *5*(1), 174. Https://Doi.Org/10.1007/S43621-024-00364-6. DOI: 10.1007/s43621-024-00364-6

Chapter 2
AI for Green Logistics and Decarbonized Transport Performance of Transport Organizations in Morocco

Saida Ifiss
https://orcid.org/0009-0001-3491-0104
The National School of Business and Management, Morocco

ABSTRACT

This study addresses the climate emergency by assessing the environmental performance of 270 Moroccan transport organizations (2019–2023) using four machine learning algorithms: Random Forest, XGBoost, SVM, and PLS. Among them, XGBoost achieved the best results with an R^2 of 0.88 and F1-score of 0.83 in predicting fuel consumption and classifying environmental performance. The analysis identified route optimization and adoption of low-emission vehicles as key factors influencing environmental outcomes. The study also highlights differences in these factors' effectiveness across organization types (public vs. private) and sizes (SMEs vs. large firms). Ultimately, the research proposes a robust, data-driven framework to enhance environmental performance and guide public policies and managerial strategies towards greener logistics and decarbonized transport in Morocco.

INTRODUCTION

The world is currently experiencing an unprecedented climate emergency involving rising average temperatures, increased occurrence of extreme weather events, and global recognition that GHG emissions must drastically be curtailed (IPCC, 2006).

DOI: 10.4018/979-8-3373-3012-9.ch002

In this situation, the internationalization of firms, particularly SMEs, represents both value-added strategic threat and growth mechanism (OECD, 2022). However, many Moroccan SMEs see internationalization as more uncertainty, high logistic costs, and lack of technical support (Chamchati et al., 2021). The paradox can hold also to the transport sector, an industry with all the typical qualities which generates a considerable amount of GHG emissions, as organizations are being asked to change to sustainable practices (IEA., 2022).

National data suggests that the transport sector would contribute around 40% of total GHG emissions by 2022 with an emphasis on road transport (Climate Change Department, 2024). This indicates the key role that transport organizations must play in developing a viable path to sustainability. Transport organizations will include the Office National des Chemins de Fer, (ONCF), road freight and passenger transport companies, operators of ports such as Tanger Med and Casablanca, and airports such as Mohammed V Casablanca, and Marrakech-Ménara. Each of the above mentioned organizations are in a constant struggle to meet the expectations of their economic missions while adhering to demanded environmental requirements. Transport organizations particularly face challenges; the aging of their fleets, the lack of adapted infrastructures (electric charging stations, natural gas refuelling stations), and the burden of the initial costs associated with the ecological transition, as they very often operate in a economically constrained existence (AMEE, 2023; World Bank Group, 2022).

Furthermore, traditional approaches to assessment of the environment are showing their limitations due to the increasing complexity of transport systems. Traditional environmental audits using a finite number of static indicators and often based on descriptive analysis (averages, standard deviations) lack the ability to take into account the multifactorial and temporal dynamics contained within the environment (Patella et al. 2021). In addition, traditional assessments are incapable of effectively depicting the immense heterogeneous data space resulting from transport operations (e.g. telematics, GPS location, predictive maintenance, real-time consumption patterns) and are unable to model the complex non-linear relationships in the data space (Azad et al., 2024).

Given these constraints, machine learning (ML) is emerging as a new way to attain relevance that may overcome the shortcomings of traditional ways. ML can reveal patterns in data that were previously hidden, capture complex relationships, and provide robust predictions to provide strategic insight (Goodfellow et al., 2016 ; Tabachnick & Fidell, 2013). Algorithms like Random Forests (RF), XGBoost, Support Vector Machines (SVM) or Partial Least Squares (PLS) regression have shown their strength in predicting complex phenomena in many fields including finance, health or energy (Chen, & Guestrin, 2016 ; Breiman, 2001). However, while these algorithms have been applied extensively to many other areas, they are only just emerging in research in sustainable transport in Morocco and this provides an opportunity for further research.

This chapter will provide a rigorous and innovative examination of how AA applied in this domain, through Random Forests (RF), XGBoost, Support Vector Machine (SVM) and Partial Least Squares (PLS) regression, would be in order. This chapter will contribute to the literature by providing an empirical comparison of these methods in a particular domain: transport organizations in Morocco, including public and private entities.

The purpose of this chapter has three goals:

- **Explore the potential of machine learning** to assess and predict the performance of To examine the application of machine learning in evaluating and forecasting Moroccan transport organizations' performance regarding green logistics and decarbonized transport and to evidence the ability of any ML models to detect 'hidden' relationships in selected operational level and environmental data. We collected data over a five year period (2019-2023) from a representative sample of 270 Moroccan transport organizations to demonstrate capacity of ML models in identifying data trends; derive complex relationships between variables; and to provide accurate forecasts of performance. The analysis will consist of selected environmental performance indicators to assess Moroccan transport organizations' investment in, and commitment to, green logistics and decarbonized transport. These indicators included the following: (i) Specific fuel consumption in litres/100 km for road vehicles, converting litres to CO_2 equivalent using IPCC factors (Intergovernmental Panel on Climate Change) (IPCC, 2006) ; (ii) CO_2 emissions per tonne-kilometer transported (gCO_2/tkm) for other transport modes (rail, sea, air); (iii) a percentage for low-emission vehicles from total fleet and actual share of journeys performed by this vehicles, and (iv) utilization of sustainable practices, for instance, route optimization based on real-time traffic, employing intermodal, reverse logistics, or reusable packaging. Machine learning (ML) models will use this historical data for developing forecast on future environmental performance, and to identify the factors found to have the most significant impact on the indicators.
- **Compare the performance of AA algorithms** (RF, XGBoost, SVM and PLS) to determine which is best suited to predicting performance in green logistics and decarbonized transport. This comparison will be based on standard metrics rigorously applied to regression tasks (prediction of fuel consumption and emissions) and potentially classification tasks (e.g. classification of organizations according to their level of environmental performance: low, medium, high) (Geron, 2019). Metrics will include accuracy (for classi-

fication tasks, measuring the percentage of correct classifications), F1 score (particularly relevant where there is imbalance in performance classes), coefficient of determination (R2) (for regression tasks, indicating the proportion of variance explained by the model) and root mean square error (RMSE) (for regression tasks, measuring the mean square of prediction errors). Robust cross-validation techniques, such as k-fold validation, will be used to obtain a reliable and generalizable estimate of each model's performance on unseen data, and to minimize the risk of overlearning.

- **Identify the key factors influencing performance** in green logistics and decarbonized transport, for instance, fuel consumption, route optimization, low-emission vehicle use, and the rate at which green logistics practices are adopted (Gonzalez et al., 2010). Some AA models (especially RF and XGBoost) result in feature importance, which allows us to ask and determine the best action levers to implement, for transport organizations in Morocco, that will enhance their environmental performance. This can help us develop realistic recommendations for decision-makers and managers in the transport sector (Hansen et al., 2010; Staudigel et al., 2011). For instance, the research could quantitatively show whether working to improve route optimization has a greater impact on GHG reductions than the incremental implementation of electric vehicles, or vice-versa, in the business context of the firms studied. In this way, it will be possible to better steer investments and policies.

1. LITERATURE REVIEW

This section will comprehensively review key concepts and existing work. It will first define the key concepts of the study, including green logistics, decarbonized transport, environmental performance, and relevant key performance indicators (KPIs). A thorough description of the machine learning algorithms explored (Random Forests, XGBoost, Support Vector Machines and Partial Least Squares regression) will be given, describing their underlying features, advantages and disadvantages (Goodfellow et al., 2016). Finally, existing work on the application of AA in green logistics and decarbonized transport will be reviewed, highlighting gaps in the knowledge and establishing the relevance of this study within the Moroccan context.

1.1 Definitions of Key Concepts

Green logistics: Green logistics, also referred to as sustainable logistics, refers to the measures made by companies and organizations to assess environmental damage and to ultimately reduce the harmful effects of logistics activities along the entire supply chain. Green logistics considers emissions of greenhouse gases (GHG) through transportation, storage and handling; improves the utilization of natural resources requiring energy, water, and packaging supply; reduces and handles waste; promotes sustainable practices: responsible sourcing, eco-design (design products to recycle). With indirect transport examples of green logistics can be called diversifying modes of transport (using rail, river, electric vehicles, etc. when possible); improving transportation efficiency by making better route and load decisions to reduce empty mileage, and reduce fuel; raising vehicle energy efficiency through maintenance and driving behavior (eco-efficient driving); using sustainable alternative fuels (second and third generation biofuels as well as green hydrogen) (McKinnon, 2018).

Decarbonized transport: Decarbonized transport is best defined as a massive transformation toward a transport system that releases no or little carbon dioxide (CO2) and other GHGs affecting the climate. This lofty goal signifies a dramatic shift in propulsion technologies, energy sources, and the arrangement of transport flows. The strategies for achieving decarbonization of transport include enabling mass uptake of electric vehicles, powered by renewable energy; developing and rolling out hydrogen vehicles; utilizing sustainable biofuels that do not compete with food supply; and supporting active modes of transport (walking, cycling), and electrified public transport. Decarbonization is a critical part of the national commitments to reduce emissions via the Nationally Determined Contributions (NDCs) on a national and synced internationally by the Paris Agreement and NDCs (IEA., 2022).

Environmental Performance: Transport organizations assess their environmental performance by a variety of quantitative and qualitative indicators of their impact on the environment. The indicators can include the total and specific volume (per unit of service) of GHG emissions (gCO_2/tonne-kilometre), energy consumption, consumption of natural resources (water, fuels, materials), what proportion of the energy consumed is renewable, waste management, noise pollution, local air pollution (NO_X, PM, SO_2), all resulting from transport actions, and how the transport infrastructure impacts biodiversity and ecosystems crossed by it. Good performance means reducing these impacts successively over time, while maintaining logistics efficiency (ISO 1403, 2013).

The Key Performance Indicators (KPI) that are used for performance and impact evaluation and monitoring in green logistics and decarbonized transport must be representative of the environmental impacts associated with transport operations. The most relevant KPIs are

- **specific consumption or energy** use per unit of service, expressed in litres per 100 kilometres (L/100 km) for road transport, kilowatt-hours per kilometre (kWh/km) for electric vehicles, and grams of CO_2 per tonne-kilometre

(gCO_2/t.km) for sea, rail and air modes (EEA, 2021 ; ISO 1403, 2013). This allows energy performance to be expressed on an energy basis that is standardized according to distance and payload.

- Another fundamental indicator is it **specific greenhouse gas (GHG)** emissions, which can again be accounted for in gCO_2/t.km - this allows for comparing the amount of carbon emitted by transporting one tonne of goods one km (IPCC, 2006; IEA., 2022). It's also important to track the use and penetration rates of low- or zero-emission vehicles, which can be electric, plug-in hybrid or natural gas vehicles. The use and penetration rate can be represented in terms of the point of low- or zero-emission vehicles as a proportion of the total fleet, or the proportion of kilometres driven by these vehicles (IEA, 2022 ; WHO, 2018).
- The **rate of empty kilometers** in freight transport is an indicator of logistical efficiency: a high rate indicates wasted fuel and time (McKinnon, 2018). Similarly, the adoption rate of route optimization tools - including planning and routing software, update frequency, and inclusion of environmental criteria - indicates an organization's ability to reduce emissions by better organization (Zhu & Wen 2021).
- **Investment in green technologies** (clean vehicles, charging stations or multimodal infrastructures) can be measured by the percentage of the budget spent on green technologies (UNCTAD, 2020; AMEE, 2023). Commitment to environmental training is another relevant indicator which can be defined by the proportion of staff that are trained in eco-driving, energy management or sustainable logistics (ISO 1403, 2013; EEA, 2021).
- Finally, other related indicators are the share of alternative and sustainable modes of transport (rail, river, combined), the **size of the waste generated and recycling rates** (ISO, 2015; WHO, 2018), and the **level of noise reduction** in urban areas related to the use of quiet vehicles and the optimized time slots (WHO, 2018 ; EEA, 2021).

1.2 Introducing Machine Learning Algorithms:

This study leverages four prominent supervised machine learning algorithms, each offering unique strengths for predicting and analyzing environmental performance.

Random Forests (RF) are ensemble learning methods based on aggregating predictions from multiple decision trees, each trained on a different subset of data and features (Breiman, 2001). This structure enhances robustness, reduces overfitting, and allows RFs to capture complex, non-linear interactions. Additionally, RFs provide feature importance measures. However, they can be less interpretable than single trees and computationally heavy on large datasets.

XGBoost (Extreme Gradient Boosting) is a high-performance implementation of gradient boosting, which builds trees sequentially to correct errors from previous iterations (Chen & Guestrin, 2016). It incorporates advanced regularization techniques (L1/L2), handles missing data efficiently, and consistently achieves high predictive accuracy. It also offers insights into feature importance based on gain. Despite its power, tuning XGBoost requires handling numerous hyperparameters, which can complicate optimization.

Support Vector Machines (SVMs) are primarily used for classification but also support regression via Support Vector Regression (SVR) (Azad et al., 2024). SVMs seek to define the optimal hyperplane that maximizes separation between classes or fits the data within an acceptable margin. Kernel functions (linear, polynomial, RBF, sigmoid) allow SVMs to handle non-linear relationships by projecting data into higher dimensions. While effective in high-dimensional spaces and resistant to overfitting, SVMs can be sensitive to kernel choice and computationally demanding for large datasets.

Partial Least Squares (PLS) regression identifies latent components that best explain the covariance between independent and dependent variables (Wold et al., 2001). PLS is especially suited to datasets with multicollinearity and numerous predictors. However, as a linear method, it may underperform in modeling non-linear interactions compared to RF, XGBoost, or SVM.

1.3 Review Of Existing Work

1.3.1 Applying AA to Green Transport/Logistics

In recent years, increasing attention has been given to the role of digital transformation and machine learning in improving sustainability and performance in logistics and transport systems. While most studies focus on developed countries, important advances have been made in the Moroccan context, especially in port logistics and service supply chains.

Bentalha et al. (2019) propose a prospective framework for digitalizing service supply chains, emphasizing how anticipatory digital tools could enhance strategic agility. Similarly, Hmioui et al. (2020) explore sustainable supply chain management for service companies using a prospective methodology, shedding light on how digital and sustainable management intersect to boost global performance. These works are crucial for understanding digitalization beyond manufacturing logistics and include Moroccan service sectors, particularly relevant for public and private transport operators.

In the port logistics sphere, Bouklata & Bensfia (2020) highlight the impact of digitalization on goods transit times, a key performance factor in green logistics. Their research shows how dematerialized processes can reduce inefficiencies in port passage procedures. This is reinforced by Bensfia & Bouklata (2023), who evaluate the information system for electronic payment of port dues, confirming that digital integration can enhance transparency and operational fluidity, thus contributing indirectly to decarbonized transport objectives.

The concept of collective intelligence in port ecosystems has also emerged as a novel performance lever. Bensfia (2025) demonstrates how stakeholder coordination and information sharing improve responsiveness and sustainability of port territories. In parallel, the socio-technical approach developed by Bensfia & Aidi (2025) analyzes the effect of Port Community Systems (PCS) on organizational performance, providing a theoretical lens to assess digital platform adoption within logistics communities.

Moreover, Alla et al. (2022) propose a supply chain performance model adapted to post-COVID dynamics, emphasizing resilience, adaptability, and digital monitoring systems as critical dimensions. This aligns with Badr (2022), who envisions smart technologies as enablers of sustainable service supply chains through predictive analytics and anticipatory decision-making.

These contributions bridge local realities and global performance requirements, offering a valuable framework for our study. Despite these advancements, comparative evaluation of machine learning models remains largely unexplored in Morocco's transport sector, especially using large-scale empirical datasets. The present research extends existing work by applying and comparing ML algorithms (RF, XGBoost, SVM, PLS) to assess and predict the environmental performance of transport organizations. It also builds on the understanding that digital tools, stakeholder cooperation, and sustainable practices are deeply interconnected (Bensfia & Bouklata, 2022; Badr, 2022).

1.3.2 Gaps in Literature

While there is an increasing volume of international research on the use of machine learning (ML) in green logistics and decarbonized transport, the literature has multiple important gaps - particularly relating to the Moroccan context with its unique institutional, economic, and technological context.

a) Geographical and Contextual Bias

The first, and most important gap, the lack of empirical studies focused on Morocco, or countries with similar MENA characteristics. Much research has been conducted in developed economies (e.g., EU countries, North America, East Asia), which have benefited from more data, advanced technological infrastructure, and policies that challenge and advance the integration of AI (Yalçın, 2025). Many studies assume access to fully developed sensor networks, harmonized emissions, and rich regulatory environments-things that simply are not systematically present in Morocco, which means these studies cannot be generalized or transposed into Morocco's framework, where there is profound variability in the transport systems, more reliance on fossil fuels, old vehicle fleets, and regional variations in infrastructure and digitalization. In addition, AI-based frameworks rarely borrow from national strategies such as the Stratégie Nationale de Développement Durable 2030 (SNDD), so need situates the framework in a conceptually disconnected and operationally separated context from national policy instruments (World Bank, 2015).

b) **Fragmented Analytical Scope**

Secondly, there is a limited literature that deals with isolated aspects of environmental performance while neglecting the fact that sustainability in transport is multi-faceted and systemic. Numerous studies, (e.g. Paul, 2025) are typically on route optimization alone, or predicting energy consumption based on AI. These studies are still useful since they provide value, but they generally use single-target modeling and ignore other important aspects including fleet composition, green practices development, other modes of transportation integration, and/or training of the operator. The multiplicity of dimensions presents a challenge in terms of the practical real-life application of their suggestions and does not show the latter-synergetic effects from a number of different sustainability levers that could help firms' sustainability transition. Just as green logistics is multi-faceted by its very definition, the literature still lacks a holistic analytical framework to systematically consider both the operational, technological and human elements of green logistics, especially for an emerging economy like Morocco.

c) Lack of Benchmarking Methodology Across ML Models

For example, studies like Gong, et al. (2021) and Dong et al. (2025) use Random Forest or XGBoost for specific tasks (e.g. fuel prediction in diesel trucks or energy management for electric buses) but do not compare their results against SVM, PLS, or another regression/classification method using a common compensation, meaning that researchers and practitioners do not know the relative advantages and disadvantages of each model, especially in contexts that lack data or heterogeneity (as is the case in Morocco). And, advanced investigations, such as van Twiller et al. (2025), are also beginning to contribute the area incorporating deep reinforcement learning or hybrid neural networks, although the application is typically to a very specialized applications (e.g. the logistics of a shipping container, smart ports, etc.) and rarely includes a cross-algorithm comparison or reproducibility across various transport sectors (e.g., road, rail, maritime, urban logistics etc.).

d)) Little Integration with Local Policy and Stakeholder Contexts

Lastly, an important gap in the literature is insufficient alignment with local governance framework, stakeholder needs, and organizational constraints. In Morocco, transport organizations (either public or private) operate in a policy context defined by national aspirations (e.g., SNDD 2030, sectoral decarbonization plans, green procurement options) and regional priorities. There is less recognition of the contextualizing of model outputs to enable actionable insights for local managers or policymakers in the existing ML-based research. There is a lack of participatory approaches to incorporate the contributions of fleet operators, municipal authorities, or logistics planners to guide the design, interpretation, and implementation of AI solutions.

1.4 Relevance and Novelty Of The Study:

This research addresses these research gaps by examining how performance is assessed in the real-world context of transport organizations in Morocco using a multidimensional approach and assesses their performance in green logistics and decarbonized transport. In this research, a primary dataset was collected during a time-sensitive survey from a representative sample of Moroccan organizations, explicitly comparing the performance of four common AA algorithms. This research helped provide an understanding of what is the latent causes for environmental performance in this Moroccan context. Our findings will provide evidence-based insights for policy makers and managers within Morocco's transport sector as they seek to be deliberate in their use of variables and levers for action toward more sustainable practices and the national development goals. The type of comparative methodological approach to AA algorithms will also relevantly contribute to the literature through identifying the promising methods to analyze sustainable transport. More recently, for instance, Chegri et al. (2025). further underlined the importance of integrating and developing the role of artificial intelligence, toward catalyzing the sustainable mobility transformation in Morocco while recognising the barriers of such a transition. Likewise, Qbouche & Rhoulami (2024) demonstrated how machine learning algorithms such as those as AA in this analysis, understand, predict and enhance our understanding of transport modes in the Rabat region, offering practical applications to inform good urban planning and enhanced sustainable mobility.

2. METHODOLOGY

This section outlines the thorough methodological process used to accomplish the purpose of this study. It provides features of the data sample, an overview of the data sources and variables examined, a review of the machine learning methods used, model parameterization and adjustment to the Moroccan context, and model evaluation metrics to compare performance..

2.1 Sample Data

The study is based on a representative sample of 270 transport organizations in Morocco. The sample covers the five-year period (2019-2023), involving public and private transport organizations. The sample was proportionally stratified based on the structure of the transport industry in Morocco. The following percentages were used:

- Road haulage represented 42% of revenue.
- Passenger road transport represented 28% of revenue.
- Rail transport (including subsidiaries of ONCF), 16% of revenue.
- Maritime transport, including the key magnate greenhouse, Tanger Med, 9% of revenue.
- Urban delivery services, 5%.

The ratio of public (45%) to private (55%) contribution reflects the structure of the industry, although there are also representatives from urban transit authorities and private organizations.

In determining the sample size, we used the 2022 census of the Ministry of Transport and Logistics (2023b), where data are recorded for 1540 formal organizations. A sample of 270 units gives us a 95% confidence level with a margin of error of ±5%, using Cochran's formula for finite populations (Bartlett et al., 2001), where our calculation parameters were:

- Estimated proportion (p): 0.5 (maximizing variance)
- Confidence level (Z): 1.96
- Margin of error (e): 0.05

The stratified random sampling technique was used based on the following strata:

- Sub-sector of activity (refer back to the breakdown provided above)
- Size of organization (SME: <250 employees; large companies: ≥250 employees)
- Geographic area (North, Central, South Regions)

This sampling technique was chosen, in part, because it allows for balanced representation of the vast diversity within the sector, better precision of estimates and rigorous comparison of sub-groups. It reduces the variance of the total estimate, prevents over- or under-representation, and enables the analysis of differential effects by organization type or region (Lohr, 2010 ; OECD, 2022). It is particularly recommended in sector-based research focused on sustainable development and public policy.

The 2019-2023 period was chosen to examine the changes in practices prior to and following the COVID-19 pandemic. Specific methodological changes were made for the 2020-2021 data:

- Except for times of strict confinement (March-June 2020)
- Weighting of indicators by actual business volume

This approach enables the isolation of structural trends from cyclical elements, whilst maintaining the representative nature of four indicators. Data after 2021 show a full recovery in volumes and promote adding this phase to the characteristic timeframe for examining sustainable practices.

2.2 Data Sources

The data required, which are rich and reliable was collected from several complementary sources:

Direct and structured surveys: The environmental and operational managers of the 270 selected organisations were invited to complete a detailed questionnaire. The questionnaire was available in online and paper form and covered the following topics: fuel consumption (in litres/100 km or gCO_2/tkm depending on the mode of transport) (IPCC, 2006), the level of low-emission vehicles (percentage in the fleet and distance travelled) (IEA., 2022), use of route optimization (use of planning software, frequency of updates to the plan, number of empty runs reduced) (Zhu & Wen 2021), use of green logistics (intermodality, reverse logistics, groupage, etc.) (McKinnon, 2018), and staff environmental training (eco-driving, energy management, sustainable logistics) (ISO 1403, 2013). The questionnaire was also designed to question organisations about environmental monitoring tools, investment in clean technologies, and environmental performance indicators monitored by organisations

(EEA, 2021 ; AMEE, 2023). The overall response rate from the survey was 87%. To increase the number of responses received, this study used reminders and verification telephone interviews to ensure the datasets received were complete and reliable.

Secondary data: The study also drew the information from annual activity reports, environmental balance statements, and the internal management systems of the participating organizations (aggregated data fleet tracking systems, reports on use of route planning software). The confidentiality agreements and data protection policies of these organizations granted access to the information. Limitations experienced with the secondary data included the diversity of reporting formats and differences in available level of detail.

Aggregated data and official statistics: The study consulted economic and sector data from Moroccan government bodies, including the Ministry of Transport and Logistics (2023a) and the Moroccan Agency for Energy Efficiency (AMEE, 2023). yielding current information on the structure of the sector, energy transition policies, and geographical distribution of the operators. The study also gathered data from industry associations, in particular, the Confédération Générale des Entreprises du Maroc (CGEM, 2023). and the Fédération du Transport et de la Logistique (2023), part of the CGEM. These data provided background to the survey results, corroborated the reliability of the sector strata selected, and provided evidence of the representativeness of the sample relative to actual structure of the Moroccan transport market.

To ensure the integrity and quality of the analysis, we adopted a stringent data cleaning and pre-processing process. In cleaning our data in a manner sensitive to missing values, we took an approach based on the nature and distribution of the variables in question, either through statistical imputation - the mean for symmetric distributions and the median for asymmetric ones, or in the case of missing data for determining variables of concern, a logically reasoned exclusion of those observations (Little, & Rubin, 2019). We identified outliers and data entry errors using typical exploratory statistical methods (i.e. interquartile range, z-scores, and Mahalanobis for multivariate data) (Tabachnick & Fidell, 2013). We also supplemented this review phase with visual inspections (e.g. boxplots, histograms, scatterplots). Any cases in question were validated against their sources or removed, where an acceptable validation could not be performed.

In addition, one harmonization task dealt with inter-source harmonization, where attempts were made to unify the inconsistent and non-comparable data formats. All data included in the study included issues of measurement units, different observation dates and times, and inconsistent industry sector classifications; which are best practices in a multidimensional database management system (Hair et al., 2021). These operations were crucial to achieving comparability rigorously among the different datasets employed in this study.

The data collection faced special challenges related to the sensitivity of the data on costs and emissions, the variety of monitoring and reporting systems undertaken by the different organizations, and the trust relationship that had to be established between participants in order to receive complete and reliable data. The confidentiality of the data collected and the anonymity of respondents was protected in every way possible.

2.3 Studied Variables

The key variables examined in this study, which were chosen for their direct relevance in assessing performance in the area of green logistics and decarbonized transport, were:

Fuel consumption: Measured in liters/100 km for road transport, fuel consumption is converted to a CO_2 equivalent covering all production and end-of-life stages using the IPCC (IPCC, 2006) recommended emission factors. For sea, rail and air modes, consumption is in grams of CO_2/tonne-kilometre (gCO_2/tkm) transported (which is our aim in order to allow comparison across mode) (EEA, 2021; Facanha & Horvath, 2007).

Route optimization: This aspect of studies is analysed through a composite score covering elements such as existing planning software, how often they update their routes, how much empty mileage they have reduced, and whether they include environmental criteria in route planning (Zhu & Wen 2021 ; Paul, 2025 ; Bektas & Laporte 2011). A high score that reflects the frequent consideration of route optimization shows a higher level of rationalization of transport flows.

Use of low-emission vehicles: This variable was captured using two standalone indicators: (i) percentage of low-emission vehicle capacity (electric, plug-in hybrid or natural gas) to total fleet, and (ii) the percentage of kilometers traveled in them. These dimensions developed allow for differentiation between capacity of equipment and used capacity (IEA., 2022; Muehlegger & Rapson, 2022; WHO, 2018).

Adoption of green logistics practices: This variable was measured with an overall score based on various sustainable practices adopted (use of intermodal logistics, consolidation of deliveries, reverse logistics (returns), environmentally-friendly packaging, responsible purchasing, eco-driving training). Each practice adopted adds to the overall green logistics score and reflects an organization's desire to be sustainable (McKinnon, 2018; Patella et al. 2021; UNCTAD, 2020).

Staff environmental training: This specific variable provides information on the percentage of staff members that have undergone training for eco-driving, energy management, or sustainable logistics. It allows the user to evaluate how environmental issues are embedded into the organizational culture (ISO 1403, 2013 ; EEA, 2021; Gatta et al., 2019).

All these variables (and measures) will be subject to one form or another of statistical pre-processing (normalization or standardization based on the algorithm input requirements) to ensure compatibility and respond to inconsistencies in scale of measurements and the associated validity tests that best strengthen prediction robustness..

2.4 Machine Learning Techniques:

Four supervised machine learning algorithms were applied: Random Forests and XGBoost for their robustness and ability to model non-linear relationships; Support Vector Machines (SVM) using an RBF kernel; and Partial Least Squares (PLS) as a linear baseline. Hyperparameters were optimized via cross-validation (k=10), using grid or Bayesian search. Data preprocessing included normalization and standardization as required. These algorithms were selected for their proven performance in prediction tasks and their capacity to highlight key variable importance, especially in complex, heterogeneous transport data. PLS served as a reference to compare non-linear models against a traditional linear approach.

2.5 Parameterization and Context-Adaptation:

The data that has been collected will pass through several pre-processing steps before training the models. Categorical variables (e.g. transport sub-sector) will be treated using a one-hot encoding scheme so that it is appropriate for implementation with the machine learning algorithms. Numerical variables will be either normalized (scaled between 0 and 1) or standardized (scaled with mean = 0 and standard deviation = 1). The choice depend upon the sensitivity of the algorithm to scale. For example, SVMs, are sensitive to scale..

The data will be split into three parts - a training set (70% of the data) to use for learning the models, a validation set (15% of the data) which will be used for changing all the hyperparameters to select the best model for each algorithm, and a test set (15% of the data) to evaluate the performance of the selected models in an unbiased manner. The training set will undergo the k-fold cross-validation procedure (k=10) for model performance estimation during hyperparameter tuning. k-fold cross-validation is the procedure in which we divide the training set into k, subsets, or folds. We will train our model k times by training it using k-1 folds and then validating it against 1 fold. Finally all k validation estimates will be averaged to obtain a better estimate of model performance.

The interest in AA algorithms is warranted by their relative success in difficult prediction and classification challenges, their inherent capability to manage different types of data and crucially, their capability to provide information on feature importance (for RF and XGBoost), which is important for highlighting action levers to enhance environmental performance . PLS is included as a reference linear regression method to measure performance in the context of non-linear models.

2.6 Evaluation Metrics

Model performance will be evaluated using appropriate metrics based on task type. For classification, accuracy measures correct predictions, while F1-score balances precision and recall, especially for imbalanced classes. For regression, R^2 assesses explained variance, and MSE/RMSE estimate prediction errors, with RMSE being more interpretable. Models will be tested on a reserved dataset, and the best-performing model will be selected based on the highest R^2 and lowest RMSE (for regression), or the highest accuracy and F1-score (for classification), ensuring reliable prediction in the study's context.

3. RESULTS AND DISCUSSION

This section presents, in detail and systematically the findings produced after applying machine learning (ML) models on the dataset collected from transport organizations in Morocco. The aims are to determine the comparative performance of the algorithms and to determine the vital chain effect causing performance in green logistics and decarbonised transport, as well as discuss possible implications of findings in the particular Moroccan context.

3.1 Comparative Performance of Machine Learning Algorithms

We have analyzed the performance of four supervised learning algorithms - Random Forests (RF), XGBoost, Support Vector Machines (SVM) and Least Squares Partial Regression (PLS) - on the test set based on two tasks - (i) regression for predicting fuel consumption (stated in liters/100 km CO2 equivalents) and (ii) classification of overall environmental performance level (categories: low, medium and high) - and summarized all evaluation metrics from the test set for each model in Table 1.

Table 1. Comparative performance of Machine Learning Models on the Test Set

Model	Accuracy (Class.)	F1-Score (Class.)	R(2) (Reg.)	MSE (Reg.)	RMSE (Reg. in L/100km)
Random Forests	0.78	0.75	0.82	0.051	0.226
XGBoost	**0.85**	**0.83**	**0.88**	**0.039**	**0.197**
SVM	0.75	0.72	0.79	0.058	0.241
PLS	N/A	N/A	0.65	0.085	0.292

Note: Values in bold indicate the model with the best performance for the corresponding metric. N/A indicates that the metric is not applicable for this specific model/task (PLS is linear regression).

Table 1 clearly shows that the XGBoost model demonstrates the best performance for the two tasks examined. For the classification of environmental performance level, the XGBoost was more accurate (0.85) than Random Forests (0.78) and SVMs (0.75) where the F1-score for classification was also higher for XGBoost (0.83) than Random Forests (0.75) and SVMs (0.72). For predicting fuel consumption (regression) XGBoost had the highest coefficient of determination (R2) (0.88) and therefore explains 88% of the variance in fuel consumption. Not all regression algorithms were able to provide individual predictions of fuel consumption on a continuous scale but of those with multiple predictions, the XGBoost exhibited the lowest prediction errors (0.039 MSE and 0.197 L/100 km equivalent RMSE).

Random Forests also perform well and comes in at second-best in many instances, especially for the resilience shown to incomplete datasets. SVMs also have decent results, but are not as good as the tree-based approaches. As expected, Least Squares Partial Regression (PLS) performed least well on the regression task (R2=0.65, RMSE = 0.292) because it is a linear model. Importantly, PLS is linear by design, and as such cannot be used on classification tasks, which is why there was "N/A" in the precision and F1-score columns in Table 1. This stark contrast in performance between the linear models (PLS) and non-linear models (RF, XGBoost, SVM with RBF kernel) clearly confirms the presence of complex (non-linear) relationships among the factors that were modeled. Essentially, linear models cannot capture complex, non-linear relationships among variables that were modeled but non-linear algorithms can.

These results suggest tree-based models, especially XGBoost, are appropriate for modelling complexity due to their potential to approximate complex relationships and their sequential optimisation. They are also best suited for modelling various factors that influence performance in the context of green logistics and decarbonized transport in Morocco. Their robustness and ability to extract patterns from varied operational and organizational data likely explains their performance superiority across task types.

3.2 Determining the Key Influencing Factors

In determining which action levers could lead to the greatest improvements in environmental performance, we evaluated the relative importance of the explanatory variables using the XGBoost model, which produced the best result in the analysis. Here we showcase the importance of the variables in predicting future fuel consumption since this indicator constitutes a crucial KPI for green logistics and decarbonized transport. The importance of a variable is calculated based on its average contribution to lowering the prediction error across the model trees. Scores are provided in normalized format to improve readability.

Our analysis of the importance of variables for the XGBoost model in predicting future fuel consumption produced the following outcomes:

Figure 1. Relative importance of explanatory variables in predicting future fuel consumption using the XGBoost model

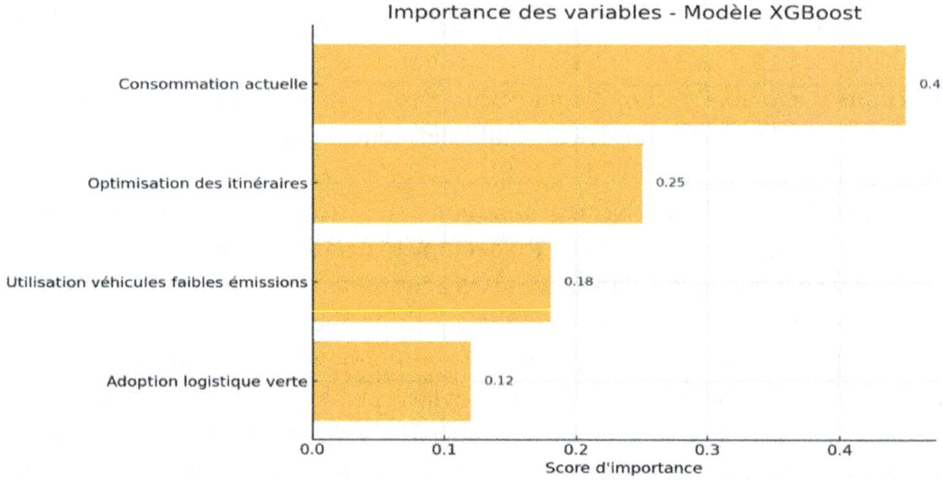

Figure 1 shows the structure of influence of the variables on fuel consumption. The outcome of interested focus is not surprisingly **past fuel consumption** is the strongest predictive variable (0.45), which reflects how past operational patterns and fleet characteristics are consistent from one time period to another.

Of course, organizations can only act on the levers, and, specifically, the **Route Optimization** Index is the most important (0.25). This indicates planning journeys that do not empty mileage and take real traffic conditions into account has a group and direct impact on mitigating consumption.

The adoption of **low-emission vehicles** is rated third (0.18). While essential in the longer-term decarbonization of transport, it is of slightly less importance than route optimization in this analysis, because of the current levels of market penetration of electric or low-emission vehicles in the aggregate fleet of organizations included in this analysis over the 2019-2023 period, which may still be relatively low. Of course, the impact of this option should be logically increasing as the transition accelerates.

Finally, the Index of Adoption of Green Logistics Practices in the broadest sense, including intermodality, consolidation, etc., is rated the least strongly (0.12) for its direct prediction of fuel consumption for the near future / medium future than operational and technological offerings (infrastructure and fuel) options. This does not de-emphasize its overall strategic significance, but points to its impact on fuel consumption potentially being more indirect, or manifested over a longer time-frame.

3.3 Performance and Importance Analysis By Organization Sub-Group

To drill down on the multi-faceted insights we gleaned from our machine learning applications and better reflect the diversity of Morocco's transport ecosystem, we undertook a disaggregated performance analysis of two key structural variables: the type of organization (public vs private) and size of organization (SME vs large enterprise). This evaluation is stratified for a specific purpose in the Moroccan context as issues related to governance approaches, level of resource, and operational maturity can have a big impact on both the quality of the data and what is being done from an environmental practice perspective.

The analysis has been narrowed down to focus on the XGBoost algorithm as it typically outperformed all models across the regression and classification models. The models were evaluated both within the public and private organizations, which to some degree were highly variable. Table 2 captures a summary of the model performances as evaluated by the various organizational sub-groups.

Table 2. Predictive Performance by Organization Type and Size (XGBoost Model)

Subgroup criterion	Subgroup	R^2 (Conso Prediction)	Precision (Class. Perf. Env.)	F1-Score (Env. Perf. Class)
Type of organization	Public organizations	0.80	0.78	0.76
	Private Organizations	0.91	0.88	0.87
Organization size	SMES	0.88	0.83	0.81
	Large Companies	0.91	0.86	0.84

As shown in Table 2, predictability appears to be highest from the **private organizations** and **large companies** regarding regression (R^2) and classification (Precision and F1 score) models. There may be some reasons driven by the following factors that are interrelated:

- **Operational standardization**: Private organizations and large companies tend to follow more standardized operating procedures which both streamline operations and make reporting more systematic, systematic data is better for machine learning applications. Public organizations, on the other hand, could be experiencing poor data quality due to the ill-equipped government administrative system which can't provide proper operational information in disaggregated form between the local authority and department or office which can lead to different reporting methods, completely different logics regarding usage which add to the overall variability of the data and thus the predictive accuracy.
- **Technological infrastructure**: larger private organizations are also more likely to use advanced technology to provide information regarding environmental operations, for example telematics systems, GPS based fleet systems, and Enterprise Resource Planning (ERP) software. Instead of gathering more useful data, the technology gives more reliance or trust in the data that is gathered. Sophisticated data processing systems have more reliability which increase the predictive learning capability of the model.
- **Organizational agility and innovation adoption**: SMEs and public entities are likely to have more budget restrictions, human capital, and change processes and may find themselves constrained to the levels of innovation, speed of change, and scope of innovation, such as real-time route optimization or low emission vehicles, which have already been defined in the model as critical inputs, and will influence predictive performance.

Despite this equality of variability across groups, it can be noted that the XGBoost model is still performing relatively well in all sub-groups with R^2 values over 0.80 and F1 scores above 0.75. This indicates that the model is performing strongly in terms of generalizability even in a less than structured or potentially noisy data environment.

From a policy perspective, the sub-group analysis suggests a need for **targeted interventions**. Specifically:

- **public organizations** may identify a priority for the improvement of digital infrastructure and compatibility in reporting formats to more consistently enable environmental monitoring, which could enhance the application of artificial intelligence driven instruments.

- **SMEs** may use subsidized access to route optimization technologies and provide eco-driving education program to demonstrate relatively improved environmental performance, since small efficiency improvements lead to a greater percentage change.c

In conclusion, this disaggregated analysis highlights the need to place AI applications in context with the structural realities of the transport sector. Machine learning models like XGBoost not only provide power in a predictive sense, but also provide diagnostic utility when disaggregated by organizational characteristics, which makes them indispensable evidence-based policy design aids for sustainable transport.

3.4 Variation in the Importance of Key Factors by Sub-Group

To engage in finer-grained understanding of the problem, we re-assessed the importance of key variables for our entire dataset using the XGBoost model, this time stratifying the sample by organization type (public vs private) and by size (SME vs Large Enterprise). Analysis according to sub-groups allows us to expose particular dynamics and visualize alternatives for recommendations. Table 3 below shows the relative importance (normalised) of the main variables accounting for fuel consumption for each sub-group.

Table 3. Relative Importance of Key Factors (XGBoost) by Organization Subgroup

Key factor	SMES	Large Companies	Public organizations	Private Organizations
Route Optimization Index	**0.32**	0.24	0.23	**0.29**
Rate of use of low-emission vehicles	0.14	**0.21**	**0.20**	0.16
Green Logistics Practices Adoption Index	0.11	0.15	0.18	0.12
Past fuel consumption (for reference)	0.43	0.40	0.39	0.43

Note: Importance scores are normalized for each sub-model. Values in bold indicate the most important levers of action (excluding past consumption) for the sub-group concerned, or significant differences.

Examining importance of variables within subgroups exposes some interesting subtleties:

- **For SMEs (small and medium-sized enterprises):** The Route Optimization Index (relative importance of 0.32) is identified as the most important action lever - even more than for large companies (0.24). This may be related to the smaller financial space afforded to SMEs, where every better planned fuel saving resonates more to their profitability.

- For large organizations: Although route optimization continues to have a top priority (0.24), **Low Emission Vehicle Utilization Rate** (0.21) is gaining relative importance from SMEs (0.14). Large organizations typically have more resources for large initial investments, helping them to more actively explore fleet renewals towards clean technologies. Their sheer size also often allows them to better plan and manage the recharging or fuelling infrastructure required. Regrettably, the **Green Logistics Practices Adoption Index** (0.15) was also slightly more important than SMEs (0.11) which may reflect a more structured sustainability strategy with more resources dedicated to sustainable practices.
- **For Public Organizations**: The **Low Emission Vehicle Utilization Rate** (0.20) and the **Green Logistics Practices Adoption Index** (0.18) were shown to be considerable in importance. This is probably related to government regulations, national sustainable development targets (such as SNDD 2030), and increased sensitivity to public image and exemplarity. Policies around green public procurement can also push investment towards sustainable solutions.
- **For private organizations**: Just like with SMEs (mostly private), the **Route Optimization Index** (0.29) was the most important consideration after past consumption. The importance of operational efficiencies and costs makes investigation into direct savings in fuel consumption really worthwhile.

These differences emphasize that although the levers that we can take action on are basically the same, the relevance of such actions and the urgency required to implement them can be different. In this regard, it would be reasonable to customize approaches to much improvements in environmental performance according to the distinctive characteristics of each type and size of organization in the Moroccan transport sector.

3.5 Relationships Between Variables: Correlation Analysis

To complement the analysis of variable importance conducted through machine learning models, a correlation analysis provides a deeper understanding of the underlying interrelationships between the key environmental performance levers. This statistical procedure, which was based on the scipy library to calculate pearson correlation coefficients, allows for an organized presentation of the directional links (either positive or negative) and the strength of these links between the different key variables examined.

Table 4. Correlation matrix

	Conso_ fuel	Optim_ itinerary	Vehicles_ greens	Adoption practices	Training
Conso_fuel	1.00	-0.52	-0.44	-0.38	-0.35
Optim_itineraire	-0.52	1.00	0.48	0.42	0.40
Green_vehicles	-0.44	0.48	1.00	0.36	0.38
Adoption_practices	-0.38	0.42	0.36	1.00	0.45
Training	-0.35	0.40	0.38	0.45	1.00

Note: Pearson correlation coefficients are shown. Negative values indicate an inverse relationship, positive values a direct relationship. Higher absolute values indicate stronger correlation.

a) Negative relationships with fuel consumption

The matrix results show that significant negative correlation exist between fuel consumption and a number of variables of interest:

- **Route optimization (r = -0.52):** this result demonstrates that organizations that have invested in smart planning tools (a reduction in empty runs, accommodating real-time traffic, etc.) experienced a considerable decrease in energy consumption. This solid correlation indicates a direct relationship between operational logistics organization improvements and environmental performance.
- **Utilization of low emissions vehicles (r= -0.44):** The decision to use clean technologies (electric, hybrid or natural gas vehicles) involves an expected reduction in emissions and, consequently, reduced energy consumption.
- **Adoption of green logistics practices (r = -0.38) and environmental training of staff (r = -0.35):** Although these correlations are more moderate in nature, they do provide some indication that organizational sustainability efforts may have a role in reducing consumption, either directly or indirectly

b) Positive correlations between green logistics levers

The study also indicates several positive correlations between sustainability levers indicating some systematic dynamics:

- **Route rationalization and green vehicles (r = 0.48):** Organizations creating route rationalization strategies are likely also the organizations who are implementing clean vehicles into their fleet.

- **Adoption of green practices and staff training (r = 0.45):** This relationship specifically indicates that when organizations are investing in training staff on awareness and skill development in the area of sustainable development, they are more likely to be implementing green solutions (reverse logistics, intermodality, sustainable use of packaging, etc.) in a more wide scale fashion.
- **Green practices and route optimization (r = 0.42):** There is a synergistic effect between organizational projects focused on rationalizing flows and those which promote sustainable operations.

c) Transversal reading and executive recommendations

The correlation matrix outlines the systemic coherence of environmental approaches. Specifically, companies committed to one aspect (e.g. environmental training) are often participating in other areas of sustainability as well. This overlap indicates that the best environmental performance comes from an integrative methodology - combining technologies, organizational practices, and human skills.

For decision-makers, this analysis points to the value of coherent transformation programs that promote more than just one technical lever (e.g. replacing the fleet), but encompass an array of training, optimization tools, and practices related to green logistics.

4. KEY FINDINGS AND STRATEGIC IMPLICATIONS

Ultimately, the findings of this study provide strong evidence that machine learning models are able to model the complex dynamics governing the environmental performance of Moroccan transport organizations. XGboost outperformed each of the other algorithms exhibited, demonstrating or suggesting an accuracy and robustness beyond the level provided by any of those other models. This supports Chen and Guestrin's (2016) findings in which boosting is effective in conditions with extreme heterogeneity and non-linear interactions. This is particularly salient in Morocco, where structural variability and diversity exist between small and large companies, often with a distinction between public and private operators.

The strong emphasis on route optimization and low-emission vehicles aligns the conclusions of Zhu & Wen (2021) and Muehlegger & Rapson (2022). that major environmental gains are often related to organizational practices rather than technology; the analysis of each sub-group shows enhancements accrued more rapidly with operational modifications (primarily logistics optimization) for SMEs while large companies and public organizations leveraged their scale and resources to invest in clean fleets and long-term plans. These distinctions of action levers are necessary for targeted public policies adapted to the capacity and constraints from each category of the player (Gatta, et al., 2019)

Finally, the distinct correlations between environmental training indicators, green practice adoption and overall performance suggest a systemic approach is important: sustainable logistics transition depends not only on technologies but an organizational culture and proactive behaviors (ISO 1403, 2013 ; McKinnon, 2018). This suggests the need for a comprehensive strategy combining regulatory incentives, training and support for logistical innovation.

CONCLUSION

This study makes a valuable contribution to the field of sustainable logistics by comparing four machine learning (ML) algorithms—Random Forests, XGBoost, Support Vector Machines, and Partial Least Squares—in the specific context of Moroccan transport organizations. Drawing on data from 270 public and private entities over a five-year span, the research highlights the superior predictive power of non-linear models, particularly XGBoost, in assessing environmental performance.

XGBoost and Random Forests proved especially effective in modeling complex interactions and heterogeneous data—features that are common in Morocco's transport sector. Their ability to identify key drivers of fuel consumption and emissions makes them powerful tools for decision-making.

Beyond the technical results, the study also identifies key levers for improving environmental performance. Route optimization emerges as the most impactful short-term solution, particularly for SMEs and private companies. While low-emission vehicles represent a longer-term decarbonization strategy, they require significant upfront investment—more accessible to larger or state-supported entities. Additionally, combining green logistics practices with staff training fosters a culture of sustainability across the supply chain.

The analysis also confirms that a one-size-fits-all approach is ineffective. Policies should be tailored to the size and type of organization, accounting for their unique capacities and constraints. Public bodies might need regulatory incentives, while SMEs may benefit more from technical and financial support.

The findings offer valuable insights for Moroccan policymakers and suggest integrating ML-based diagnostics into national strategies like the SNDD 2030. Finally, the research opens the door to future studies involving broader geographic and sectoral applications, time-series modeling, advanced AI techniques, and causal analysis to better guide sustainable transformation in transport.

In essence, machine learning is not just a predictive tool—it's a catalyst for greener, smarter logistics systems.

REFERENCES

Alla, L., Bentalha, B., & Bouhtati, N. (2022, May). Assessing supply chain performance in the covid 19 context: a prospective model. In 2022 14th International Colloquium of Logistics and Supply Chain Management (LOGISTIQUA) (pp. 1-6). IEEE. DOI: 10.1109/LOGISTIQUA55056.2022.9938083

AMEE. (2023). Activity report 2022-2023. Moroccan Agency for Energy Efficiency. https://www.amee.ma/sites/default/files/inlinefiles/Rapport%20d'activité%20AMEE%202023-VF-VA-24-11-23.pdf

Azad, A. K., Atkison, T., & Shah, A. F. M. S. (2024). A Review on Machine Learning in Intelligent Transportation Systems Applications. *The Open Transplantation Journal*, *18*(1), e26671212330496. DOI: 10.2174/0126671212330496240821114216

Badr, B. (2022). Smart Technologies for a Sustainable Service Supply Chain: A Prospective Perspective. *Journal of Environmental Issues and Climate Change*, *1*(1), 77–89. DOI: 10.59110/jeicc.v1i1.81

Bartlett, J. E., Kotrlik, J. W., & Higgins, C. C. (2001). Organizational research: Determining appropriate sample size in survey research. *Information Technology, Learning and Performance Journal*, *19*(1), 43–50.

Bektas, T., & Laporte, G. (2011). The Pollution-Routing Problem. *Transportation Research Part B: Methodological*, *45*(8), 1232–1250. DOI: 10.1016/j.trb.2011.02.004

Bensfia, C. (2025). Collective Intelligence of Port Territory Stakeholders: Case of Moroccan Ports. In Utilizing Technology to Manage Territories (p. 335-370). IGI Global

Bensfia, C., & Aidi, S. (2025). Port Community Systems and Organizational Performance: A Socio-technical Approach to Evaluation. Industrial and Logistics Systems Design and Efficient Operation, 42-52. https://doi.org/DOI: 10.1007/978-3-031-87309-6_4

Bensfia, C., & Bouklata, A. (2022). Validity of the TAM Model in the context of electronic payment of port dues in Morocco. In *Proceedings of the 14th International Colloquium of Logistics and Supply Chain Management (LOGISTIQUA 2022)* (pp. 1–5)

Bensfia, C., & Bouklata, A. (2023). Evaluation of the electronic port right of way payment information system. *Management Systems in Production Engineering*, *31*(3), 254–262. DOI: 10.2478/mspe-2023-0028

Bentalha, B., Hmioui, A., & Alla, L. (2019). The digitalization of the supply chain management of service companies: a prospective approach. In *Proceedings of the 4th International Conference on Smart City Applications* (pp. 1-8) DOI: 10.1145/3368756.3369005

Bouklata, A., & Bensfia, C. (2020). Digitalization of port passage procedures: Focus on the transit time of goods. In Proceedings of the 13th International Colloquium of Logistics and Supply Chain Management (LOGISTIQUA 2020) (Article no. 9353931, pp. 1–6). IEEE. https://doi.org/DOI: 10.1109/LOGISTIQUA49782.2020.9353931

Breiman, L. (2001). Random forests. *Machine Learning*, *45*(1), 5–32. DOI: 10.1023/A:1010933404324

CGEM. (2023). *Livre blanc: Les orientations du secteur privé pour le développement de l'industrie nationale*. Confédération Générale des Entreprises du Maroc.

Chamchati, O., El Mabrouki, M. N., & Minialai, C. (2021). Internationalisation of Moroccan SMEs in Sub-Saharan Africa: an Analysis Based on the Uppsala Model. In Apitsa, S. M., & Milliot, E. (Eds.), *Doing Business in Africa* (pp. XX–XX). Palgrave Macmillan., DOI: 10.1007/978-3-030-50739-8_5

Chegri, B. E., ... (2025). Navigating the Future: Artificial Intelligence and Sustainable Transportation in Morocco. In Innovations in Smart Cities Applications Volume 8 (pp. 751-765). Springer. https://doi.org/DOI: 10.1007/978-3-031-88653-9_71

Chen, T., & Guestrin, C. (2016). XGBoost: A scalable tree boosting system. In *Proceedings of the 22nd ACM SIGKDD International Conference on Knowledge Discovery and Data Mining* (pp. 785-794). https://doi.org/DOI: 10.1145/2939672.2939785

Climate Change Department. (2024). National inventory of GHG emissions. Ministère de la Transition Énergétique et du Développement Durable. https://unfccc.int/sites/default/files/resource/Maroc%20-%20Rapport%20National%20d%27Inventaire%20-%20D%C3%A9cembre%202024.pdf

Dong, C., Xiong, Z., Li, N., Yu, X., Liang, M., Zhang, C., Li, Y., & Wang, H. (2025). A real-time prediction framework for energy consumption of electric buses using integrated Machine learning algorithms. *Transportation Research Part E, Logistics and Transportation Review*, *194*, 103884. DOI: 10.1016/j.tre.2024.103884

European Environment Agency (EEA). (2021). Monitoring CO_2 emissions from new passenger cars and vans in 2020.

Facanha, C., & Horvath, A. (2007). Evaluation of Life-Cycle Air Emission Factors of Freight Transportation. *Environmental Science & Technology*, *41*(20), 7138–7144. DOI: 10.1021/es070989q PMID: 17993160

Fédération du Transport et de la Logistique (2023). Note de conjoncture logistique.

Gatta, V., Marcucci, E., & Nigro, M. (2019). Environmental transport policy and stakeholders' preferences: An application to freight transport in Italy. *Research in Transportation Economics*, *73*, 1–10. DOI: 10.1016/j.retrec.2018.11.002

Geron, A. (2019). *Hands-On Machine Learning with Scikit-Learn, Keras, and TensorFlow: Concepts, Tools, and Techniques to Build Intelligent Systems* (2nd ed.). O'Reilly Media.

Gong, J., Shang, J., Li, L., Zhang, C., He, J., & Ma, J. (2021). A Comparative Study on Fuel Consumption Prediction Methods of Heavy-Duty Diesel Trucks Considering 21 Influencing Factors. *Energies*, *14*(23), 8106. DOI: 10.3390/en14238106

Goodfellow, I., Bengio, Y., & Courville, A. (2016). *Deep Learning*. MIT Press.

Hair, J. F.. (2021). *A Primer on Partial Least Squares Structural Equation Modeling (PLS-SEM)* (3rd ed.). Sage Publications. DOI: 10.1007/978-3-030-80519-7

Hmioui, A., Bentalha, B., & Alla, L. (2020, December). Service supply chain: A prospective analysis of sustainable management for global performance. In 2020 IEEE 13th International Colloquium of Logistics and Supply Chain Management (LOGISTIQUA) (pp. 1-7) DOI: 10.1109/LOGISTIQUA49782.2020.9353940

IEA. (2022). Transport sector CO2 emissions. International Energy Agency. https://www.iea.org/reports/transport

Intergovernmental Panel on Climate Change (IPCC). (2006). *2006 IPCC Guidelines for National Greenhouse Gas Inventories*. IGES.

International Organization for Standardization (ISO). (2015). *ISO 14001:2015 - Environmental management systems - Requirements with guidance for use*. ISO.

ISO 1403. (2013). Environmental management - Environmental performance evaluation - Guidelines. International Organization for Standardization Little, R. J., & Rubin, D. B. (2019). Statistical Analysis with Missing Data (Vol. 793). John Wiley & Sons. https://doi.org/DOI: 10.1002/9781119482260

Lohr, S. L. (2010). *Sampling: Design and Analysis* (2nd ed.). Cengage Learning.

McKinnon, A. (2018). *Decarbonizing Logistics: Distributing Goods in a Low Carbon World*. Kogan Page.

Ministry of Transport and Logistics (2023a). Key figures for the transport sector in Morocco.

Ministry of Transport and Logistics. (2023b). Annual report on the transport sector in Morocco. Kingdom of Morocco. https://www.transport.gov.ma/Transport-routier/Transport-routier-de-personnes/Pages/Chiffres-cles.aspx

Muehlegger, E., & Rapson, D. S. (2022). Subsidizing Low- and Zero-Emission Vehicles: Evidence from California's Clean Vehicle Rebate Project. *American Economic Journal. Economic Policy*, *14*(2), 81–115. DOI: 10.1257/pol.20190398

OECD. (2022). *SME and Entrepreneurship Outlook 2022*. OECD Publishing.

Patella, S. M., Grazieschi, G., Gatta, V., Marcucci, E., & Carrese, S. (2021). The Adoption of Green Vehicles in Last Mile Logistics: A Systematic Review. *Sustainability (Basel)*, *13*(1), 6. DOI: 10.3390/su13010006

Paul, J. (2025). AI-powered route optimization: Reducing costs and improving delivery efficiency. ResearchGate. https://www.researchgate.net/publication/389987796_AI-Powered_Route_Optimization_Reducing_Costs_and_Improving_Delivery_Efficiency

Qbouche, K., & Rhoulami, K. (2024). Predicting Transport Mode in the Rabat Region: A Machine Learning Approach. *Ingénierie des Systèmes d'Information*, *29*(2), 493–500. DOI: 10.18280/isi.290210

Tabachnick, B. G., & Fidell, L. S. (2013). *Using Multivariate Statistics* (6th ed.). Pearson Education.

United Nations Conference on Trade and Development (UNCTAD). (2020). Trade and Development Report 2020: From Global Pandemic to Prosperity for All - Avoiding Another Lost Decade. United Nations.

Van Twiller, J.. (2025). Navigating Demand Uncertainty in Container Shipping: Deep Reinforcement Learning for Enabling Adaptive and Feasible Master Stowage Planning. arXiv. https://arxiv.org/abs/2502.12756

Wold, H., Sjöström, M., & Eriksson, L. (2001). PLS-regression: A basic tool of chemometrics. *Chemometrics and Intelligent Laboratory Systems*, *58*(2), 109–130. DOI: 10.1016/S0169-7439(01)00155-1

World Bank. (2015). Green growth: Putting Morocco in the lead against climate change. World Bank.https://documents.worldbank.org/en/publication/documents-reports/documentdetail/368381638794966349/

World Bank Group. (2022). *Morocco Country Climate and Development Report*. World Bank Group.

World Health Organization (WHO). (2018). *Environmental Noise Guidelines for the European Region*. WHO Regional Office for Europe.

Yalçın, S. E. (2025). Estimation of CO_2 Emissions in Transportation Systems Using Artificial Neural Networks, Machine Learning, and Deep Learning: A Comprehensive Approach. *Systems*, *13*(3), 194. https://www.mdpi.com/2079-8954/13/3/194. DOI: 10.3390/systems13030194

Zhu, A., & Wen, Y. (2021). Green Logistics Location-Routing Optimization Solution Based on Improved Genetic Algorithm Considering Low-Carbon and Environmental Protection. *Journal of Mathematics*, *2021*, 1–16. DOI: 10.1155/2021/6101194

Chapter 3
Aligning Sustainable Procurement with Broader Sustainability Goals:
The Role of Digital Innovations in Public Sector Accountability

Adam Sandow Saani
https://orcid.org/0009-0007-4286-6055
Tamale Technical University, Ghana

Lord Emmanuel Yamoah
Takoradi Technical University, Ghana

Abdul-Aziz Mustapha
Tamale Technical University, Ghana

Mohammed Awal Zingnaa Iddrisu
Tamale Technical University, Ghana

ABSTRACT

The abstract discusses sustainable procurement, which aligns purchasing with sustainable development principles to support goals like the UN SDGs. It examines strategies to integrate sustainable procurement with sustainability frameworks, emphasizing economic, social, and environmental impacts. Benefits include advancing SDGs related to poverty, health, education, and responsible consumption, while challenges involve policy support, supply chain collaboration, and performance

DOI: 10.4018/979-8-3373-3012-9.ch003

measurement. Case studies highlight successful strategies, and the paper calls for ongoing dialogue among stakeholders and further research to maximize sustainable procurement's impact.

1. INTRODUCTION

Sustainable procurement, defined as the strategic acquisition of goods, services, and works to maximize economic, social, and environmental benefits, has emerged as a pivotal mechanism for advancing global sustainability agendas (Institute for Supply Management [ISM], 2020). This strategic function increasingly intersects with service supply chain management, where digitalization creates new accountability paradigms requiring specialized frameworks for sustainable global performance (Hmioui, Bentalha, & Alla, 2020).

In the public sector, where procurement constitutes 12–15% of gross domestic product (GDP) in developed economies and up to 20% in developing ones, it wields unparalleled influence over markets, supply chains, and societal outcomes (Organisation for Economic Co-operation and Development [OECD], 2022). The United Nations Sustainable Development Goals (SDGs), adopted in 2015 as part of the 2030 Agenda, provide a universal framework to address pressing challenges such as poverty (SDG 1), inequality (SDG 10), climate change (SDG 13), and responsible consumption (SDG 12) (United Nations, 2015). Public procurement's scale positions it as a transformative tool to operationalize these goals, yet its alignment with such frameworks remains inconsistent due to systemic barriers like fragmented policies, cost-driven decision-making, and limited transparency (European Commission, 2021).

This chapter examines how sustainable procurement can be strategically aligned with the SDGs, with a particular emphasis on the role of digital innovations in enhancing accountability and efficiency within local government procurement. Accountability, in this context, encompasses the obligation of public institutions to demonstrate transparent, responsible, and impactful use of resources while meeting stakeholder expectations (Bovens, 2010). Local governments, as stewards of community resources, face unique pressures to balance fiscal prudence with social and environmental priorities, making them ideal contexts for exploring procurement's potential. Digital technologies—such as e-procurement platforms, blockchain, artificial intelligence (AI), and open data systems—offer solutions to longstanding challenges by improving data access, streamlining processes, and fostering public trust (Smith et al., 2025).

The chapter's purpose is fourfold:

1. To define sustainable procurement comprehensively, tracing its evolution and principles to establish a foundation for its application in public sector contexts.
2. To analyze the synergies between sustainable procurement and the SDGs, identifying specific goals and targets where procurement can have the greatest impact.
3. To explore the challenges and opportunities of integrating sustainable procurement with broader sustainability frameworks, focusing on local government dynamics.
4. To evaluate how digital innovations can enhance accountability, transparency, and performance measurement, supported by empirical case studies and actionable recommendations.

The chapter analyzes sustainable procurement through institutional theory (DiMaggio & Powell, 1983), examining how coercive regulations, normative standards, and mimetic isomorphism shape policy adoption. Digital tools are evaluated using the Technology Acceptance Model (Davis, 1989), addressing perceived usefulness and ease-of-use barriers in public sector contexts.

While drawing on global examples, special attention is given to sub-Saharan Africa—where procurement spending averages 17% of GDP yet faces acute digital divides (OECD, 2023). Ghana's experience illustrates these dynamics, as explored in Section 6.

This work contributes to the literature by addressing a critical gap: the under-explored intersection of digital technologies and public sector accountability in achieving SDG-aligned procurement outcomes. Unlike prior studies that focus on private sector practices or general sustainability principles, this chapter centers on local governments, offering practical insights for practitioners and theoretical advancements for researchers. It is directed toward a diverse audience, including municipal officials seeking to modernize procurement, policymakers crafting sustainability regulations, practitioners navigating complex supply chains, and academics studying governance and technology.

The chapter is organized as follows: Section 2 defines sustainable procurement and its theoretical underpinnings; Section 3 maps its alignment with the SDGs; Section 4 details challenges; Section 5 examines digital innovations; Section 6 presents case studies; Section 7 discusses measurement approaches; and Section 8 concludes with recommendations and future research directions. By weaving together theory, evidence, and practice, this chapter aims to inspire transformative change in how local governments leverage procurement for a sustainable future.

2. UNDERSTANDING SUSTAINABLE PROCUREMENT

2.1. Definition and Evolution

Sustainable procurement involves purchasing decisions that balance economic viability, social equity, and environmental stewardship, often framed as the triple bottom line (TBL) of people, planet, and profit (Elkington, 1997). The Institute for Supply Management defines it as "a spending and investment activity that maximizes positive environmental, social, and economic impacts while minimizing negative ones" (ISM, 2020, p. 3). This definition reflects a shift from traditional procurement's focus on cost and efficiency to a holistic approach that considers long-term value creation.

The concept has evolved significantly since the 1990s, when environmental concerns first prompted green procurement initiatives (Preuss, 2009). Early efforts focused on reducing waste and emissions, driven by regulations like the European Union's 1994 Packaging Directive (European Commission, 1994). By the 2000s, social dimensions—such as fair labor practices and supplier diversity—gained prominence, influenced by global frameworks like the International Labour Organization's conventions (ILO, 2020). The adoption of the SDGs in 2015 marked a turning point, positioning procurement as a strategic tool for broader sustainability goals (United Nations, 2015). Today, sustainable procurement encompasses lifecycle costing, ethical sourcing, and stakeholder engagement, reflecting a multidimensional approach to value creation.

In the public sector, sustainable procurement is both a policy instrument and a governance mechanism. Governments use it to influence markets, support local economies, and meet regulatory mandates (European Commission, 2021). For example, procuring energy-efficient buildings reduces operational costs and carbon footprints, while prioritizing minority-owned suppliers fosters social inclusion. However, public procurement's complexity—spanning diverse sectors like healthcare, education, and infrastructure—requires robust frameworks to ensure consistency and impact.

2.2. Principles and Frameworks

Sustainable procurement rests on several core principles, each grounded in international standards and best practices:

- **Lifecycle Thinking:** Assessing a product's impacts from raw material extraction to disposal, as outlined in ISO 20400: Sustainable Procurement (International Organization for Standardization [ISO], 2017). For instance, choosing durable furniture over cheaper alternatives reduces replacement costs and waste.

- **Inclusivity:** Ensuring opportunities for marginalized groups, such as women-owned or indigenous businesses, to promote equity (Walker & Brammer, 2012). This aligns with SDG 10 (Reduced Inequalities).
- **Transparency:** Providing clear, accessible information on procurement decisions to stakeholders, enhancing accountability (Bovens, 2010). Open tender processes, for example, deter corruption and build trust.
- **Ethical Sourcing:** Prioritizing suppliers who adhere to fair labor, human rights, and environmental standards (United Nations Environment Programme [UNEP], 2019). This includes excluding vendors linked to forced labor, supporting SDG 8 (Decent Work).
- **Risk Management:** Identifying and mitigating risks in supply chains, such as environmental disasters or labor violations, to ensure resilience (Preuss, 2009).

These principles are operationalized through frameworks like the European Commission's Green Public Procurement (GPP) guidelines, which provide criteria for eco-friendly purchasing (European Commission, 2021), and the UNEP's Sustainable Public Procurement Initiative, which promotes global best practices (UNEP, 2019). Such frameworks help standardize approaches, but their adoption varies due to regional differences in capacity and priorities.

2.3 Institutional Theory Framework

Public procurement operates within institutional fields where three pressures drive SDG adoption:

Coercive: International mandates (e.g., Ghana's *Public Procurement Act 663* requiring local content preferences)

Normative: Professional networks like Africa's Public Procurement Network

Mimetic: Replication of successful models (e.g., Rwanda's e-GP system)

Resistance to change often stems from misaligned institutional logics between fiscal austerity and sustainability goals (Thornton et al., 2012).

2.4. Public Sector Context

In local governments, sustainable procurement is shaped by unique dynamics. Municipalities manage diverse needs—school supplies, public transport, waste management—under tight budgets and intense scrutiny (OECD, 2022). Procurement decisions must balance immediate service delivery with long-term sustainability, often requiring trade-offs. For example, a city might choose costlier solar panels to meet climate goals, facing criticism from taxpayers prioritizing short-term savings (Walker et al., 2012).

Public procurement also serves as a market signal, encouraging private sector innovation. By demanding sustainable products, governments stimulate supply chains to adopt greener practices, as seen in the EU's push for circular economy solutions (European Environment Agency, 2021). However, implementation is hindered by bureaucratic inertia, limited expertise, and legal constraints favoring lowest-cost bids (Preuss, 2009). These barriers underscore the need for tools that simplify sustainable decision-making, such as digital platforms discussed later.

2.5. Benefits and Challenges

The benefits of sustainable procurement are manifold:

- **Economic:** Lifecycle costing reduces long-term expenses, while local sourcing boosts regional economies (OECD, 2022).
- **Social:** Inclusive procurement empowers marginalized communities, reducing inequality (Walker & Brammer, 2012).
- **Environmental:** Green purchasing cuts emissions and waste, supporting climate goals (European Environment Agency, 2021).

Yet, challenges persist:

- **Knowledge Gaps:** Procurement officers often lack training in sustainability metrics, leading to inconsistent practices (UNEP, 2019).
- **Cultural Resistance:** Stakeholders accustomed to traditional procurement may resist change, viewing sustainability as a cost burden (Preuss, 2009).
- **Data Limitations:** Assessing supplier sustainability requires robust data, which is often incomplete or unreliable (Walker et al., 2012).

Prospective modeling approaches like those developed by Alla, Bentalha, and Bouhtati (2022) demonstrate how predictive analytics can strengthen contingency planning, particularly for essential service supply chains facing systemic shocks.

Addressing these challenges requires integrating sustainability into procurement culture, supported by technology and policy reforms.

3. ALIGNMENT WITH THE SUSTAINABLE DEVELOPMENT GOALS (SDGs)

3.1. Overview of the SDGs

The SDGs, launched under the UN's 2030 Agenda, comprise 17 goals and 169 targets designed to address global challenges holistically (United Nations, 2015). They balance economic prosperity, social inclusion, and environmental protection, emphasizing interconnectedness. For instance, reducing poverty (SDG 1) requires access to health (SDG 3) and education (SDG 4), which depend on sustainable resource use (SDG 12). Public procurement's scale—\$10 trillion annually globally—makes it a linchpin for SDG progress (OECD, 2022).

The SDGs call for multi-stakeholder action, with governments, businesses, and civil society sharing responsibility. Local governments, as frontline service providers, are critical actors, using procurement to translate global goals into tangible outcomes (UN-Habitat, 2022). For example, procuring clean energy supports SDG 7 (Affordable and Clean Energy) while reducing costs for public facilities.

3.2. Direct Contributions to Specific SDGs

Sustainable procurement directly advances several SDGs:

- **SDG 12 (Responsible Consumption and Production):** By specifying eco-friendly products, governments promote sustainable supply chains. For instance, requiring recyclable packaging in tenders aligns with SDG 12's target of reducing waste by 2030 (UNEP, 2019). Circular procurement—reusing or leasing goods—further supports this goal, as seen in Amsterdam's lease-based furniture contracts (European Commission, 2021).
- **SDG 8 (Decent Work and Economic Growth):** Procurement policies mandating fair wages or excluding suppliers with labor violations uphold decent work standards. The ILO estimates that public procurement could influence 20% of global labor markets if aligned with SDG 8 (ILO, 2020). For example, South Africa's Black Economic Empowerment program prioritizes historically disadvantaged suppliers, boosting employment (South African Government, 2023).
- **SDG 13 (Climate Action):** Green procurement, such as purchasing electric buses, reduces emissions, supporting SDG 13's mitigation targets. The European Environment Agency (2021) reports that GPP in EU countries cut CO_2 emissions by 5% in transport sectors between 2015 and 2020.

- **SDG 11 (Sustainable Cities and Communities):** Local sourcing strengthens urban economies and reduces transport emissions. Bogotá's procurement of locally grown food for schools supports farmers and cuts carbon footprints, aligning with SDG 11 (UN-Habitat, 2022).

3.3. Indirect and Systemic Impacts

Beyond direct contributions, procurement creates ripple effects across SDGs:

- **SDG 2 (Zero Hunger):** Procuring nutritious food for public programs, like Brazil's school feeding initiative, improves health and agricultural livelihoods (World Food Programme, 2020).
- **SDG 5 (Gender Equality):** Prioritizing women-owned businesses fosters economic empowerment, as seen in Canada's supplier diversity programs (Government of Canada, 2023).
- **SDG 16 (Peace, Justice, and Strong Institutions):** Transparent procurement processes combat corruption, strengthening governance (United Nations, 2015).

These systemic impacts highlight procurement's role as a catalyst for integrated development, addressing multiple goals simultaneously. For instance, a single contract for eco-friendly school supplies can support SDG 4 (education), SDG 12 (sustainable production), and SDG 13 (climate action).

3.4. Private vs. Public Sector Adoption

The private sector has embraced SDGs for sustainability reporting, with 72% of OECD companies referencing them in 2021, up from 33% in 2016 (Pricewaterhouse-Coopers [PwC], 2021). Firms like Unilever use procurement to meet SDG targets, sourcing sustainable palm oil to support SDG 15 (Life on Land) (Unilever, 2023). Public sector adoption lags, with only 40% of municipalities in OECD countries integrating SDGs into procurement strategies (OECD, 2022). Barriers include bureaucratic silos, where procurement and sustainability teams operate independently, and a lack of standardized guidelines (Walker et al., 2012).

3.5. Opportunities for Alignment

Aligning procurement with SDGs requires:

- **Policy Integration:** Embedding SDG targets in tender criteria, as Sweden does with carbon-neutral requirements (European Environment Agency, 2021).
- **Stakeholder Collaboration:** Engaging suppliers, NGOs, and citizens to co-design sustainable solutions (UN-Habitat, 2022).
- **Capacity Building:** Training procurement officers to evaluate SDG-related metrics, such as social impact scores (UNEP, 2019).

Digital tools, discussed later, can accelerate alignment by providing data and streamlining processes, ensuring procurement delivers measurable SDG outcomes.

4. CHALLENGES IN ALIGNING SUSTAINABLE PROCUREMENT WITH SUSTAINABILITY GOALS

4.1. Policy Fragmentation

Procurement regulations often prioritize compliance over strategic outcomes, creating barriers to sustainability. Many jurisdictions mandate lowest-cost bidding, which conflicts with lifecycle costing needed for green products (Preuss, 2009). For example, a UK study found that 60% of local councils avoided sustainable options due to budget rules favoring upfront savings (Walker & Brammer, 2012). National policies, like the EU's GPP framework, provide guidance, but implementation varies, with smaller municipalities struggling to adapt (European Commission, 2021).

Fragmentation also arises from competing priorities. Environmental goals may clash with social ones, such as when eco-certified suppliers exclude local businesses (UNEP, 2019). Harmonizing policies requires flexible frameworks that balance multiple objectives, supported by clear incentives.

4.2. Cost-Centric Mindsets

Public sector budgets are constrained, leading to a focus on short-term savings over long-term value. Sustainable products, like energy-efficient appliances, often have higher upfront costs but lower lifecycle expenses (European Environment Agency, 2021). Convincing stakeholders to prioritize value requires demonstrating return on investment, which is challenging without robust data (OECD, 2022). For instance, a Canadian municipality faced backlash for choosing costlier recycled materials, despite projected savings over 10 years (Government of Canada, 2023).

Cultural attitudes exacerbate this issue. Procurement officers, trained in cost minimization, may view sustainability as a luxury, particularly in under-resourced settings (Preuss, 2009). Shifting mindsets demands education and leadership commitment to long-term goals.

4.3. Data Gaps

Evaluating supplier sustainability requires comprehensive data on environmental impact, labor practices, and economic contributions. However, many governments lack access to such data, relying on self-reported supplier claims that may be inaccurate (UNEP, 2019). For example, a 2022 audit in Australia found 30% of suppliers overstated their eco-credentials, undermining trust (Australian Government, 2023). Data gaps also hinder impact assessment, making it hard to quantify SDG contributions (Walker et al., 2012).

Standardized metrics, like those from the Global Reporting Initiative (GRI, 2023), can help, but adoption is uneven. Smaller municipalities often lack the infrastructure to collect or analyze data, necessitating technological solutions.

4.4. Capacity Constraints

Procurement officers frequently lack expertise in sustainability metrics, such as carbon footprint analysis or social impact evaluation (OECD, 2022). Training programs are limited, particularly in developing countries, where only 25% of municipalities offer sustainability-focused procurement courses (UN-Habitat, 2022). This gap leads to inconsistent practices, with some officers defaulting to familiar cost-based criteria (Preuss, 2009).

Capacity constraints also affect suppliers, especially SMEs, which may struggle to meet complex sustainability requirements without support. For instance, a Kenyan study found that 40% of SMEs were excluded from tenders due to unclear eco-certification processes (Nairobi City County, 2023).

4.5. Stakeholder Resistance

Balancing stakeholder demands—taxpayers, suppliers, regulators—creates tension. Citizens may oppose sustainable procurement if it raises costs, while suppliers resist stricter criteria that threaten profits (Walker et al., 2012). For example, a German municipality faced protests from local vendors when prioritizing eco-certified suppliers, perceived as favoring large firms (European Commission, 2021).

Public sector accountability adds complexity. Procurement decisions are subject to audits, media scrutiny, and political pressures, making innovation risky without clear evidence of impact (Bovens, 2010). Building consensus requires transparent communication and stakeholder engagement, supported by data-driven tools.

4.6. Systemic Barriers

Broader systemic issues, like digital divides and regulatory inertia, compound these challenges. Rural municipalities often lack internet access for e-procurement, limiting efficiency (UN-Habitat, 2022). Legal frameworks, designed for traditional procurement, may not accommodate modern sustainability practices, such as dynamic purchasing systems (OECD, 2022). Addressing these barriers demands a multi-pronged approach, integrating policy reform, capacity building, and technology.

In Ghana, only 45% of municipal procurement offices have broadband access—falling to 18% in Northern regions (GSS, 2023). This digital divide excludes SMEs like Kumasi-based textile cooperatives from e-tendering, contradicting SDG 9 (Industry, Innovation) and SDG 10 (Reduced Inequalities). Mobile-first solutions (e.g., Ghana's *GHANEPS Lite*) show promise but require offline capabilities.

5. THE ROLE OF DIGITAL INNOVATIONS IN SUSTAINABLE PROCUREMENT

5.1. Overview of Digital Transformation

Digital technologies are reshaping procurement by enabling data-driven decisions, enhancing transparency, and reducing administrative burdens. In the public sector, where accountability is paramount, these tools bridge gaps between policy goals and practical outcomes (Bovens, 2010). The global e-procurement market is projected to reach $50 billion by 2030, reflecting growing adoption (Smith et al., 2025). For sustainable procurement, digitalization offers solutions to challenges like data gaps, capacity constraints, and stakeholder distrust.

5.2. E-Procurement Platforms

E-procurement platforms streamline bidding, contract management, and supplier evaluation, embedding sustainability criteria into processes. The EU's Tenders Electronic Daily (TED) platform allows governments to specify requirements like energy efficiency or fair trade certification, increasing access for sustainable suppliers (European Commission, 2023). In Portugal, TED adoption reduced tender processing times by 30% and boosted SME participation by 15%, supporting SDG 8 (OECD, 2022).

User adoption of digital procurement platforms depends critically on perceived usefulness and ease of use. The Technology Acceptance Model (TAM) validation in electronic payment systems, as demonstrated in Moroccan port due payments by Bensfia and Bouklata (2022), provides transferable insights for designing public procurement interfaces that overcome resistance to technological change. Ghana's GHANEPS implementation showed 42% higher adoption after incorporating TAM-based design principles

These platforms enhance accountability by creating auditable records of decisions, deterring corruption (Bovens, 2010). They also democratize access, enabling smaller firms to compete, as seen in India's Government e-Marketplace, which awarded 20% of contracts to micro-enterprises in 2023 (Government of India, 2023). Challenges include high setup costs and the need for user training, particularly in resource-constrained settings.

5.3. Blockchain for Supply Chain Transparency

Blockchain creates immutable, decentralized records, ensuring traceability for sustainability metrics like ethical sourcing or carbon emissions. A Swedish pilot used blockchain to verify sustainable timber procurement, reducing illegal logging by 25% and aligning with SDG 12 (Nilsson & Berg, 2024). Each transaction—raw material extraction, manufacturing, delivery—is logged, allowing auditors to verify compliance in real time.

Blockchain's potential for enhancing supply chain transparency must be balanced against its operational requirements. Lessons from electronic port payment systems in Morocco (Bensfia & Bouklata, 2023) reveal that successful implementation requires both technical infrastructure and stakeholder training programs to achieve meaningful accountability gains.

Blockchain also fosters trust by making data accessible to stakeholders. In Australia, a blockchain-based system for food procurement disclosed supplier labor practices, boosting consumer confidence (Australian Government, 2023). However, scalability is limited by high energy consumption and technical complexity, requiring simplified solutions for widespread adoption (Smith et al., 2025).

5.4. Data Analytics and AI

AI and analytics optimize procurement by assessing supplier performance, predicting risks, and calculating lifecycle costs. In the Netherlands, AI models analyzed emissions data to prioritize low-carbon vendors, cutting transport-related CO2 by 8% (European Environment Agency, 2021). Predictive analytics can flag non-compliant suppliers, as seen in Canada's use of AI to detect labor violations, supporting SDG 8 (Government of Canada, 2023).

These tools also personalize procurement strategies. For example, machine learning algorithms in Singapore tailored tender criteria to local sustainability priorities, enhancing SDG 11 outcomes (Singapore Government, 2023). Challenges include data privacy concerns and the need for robust datasets, which may exclude smaller governments without digital infrastructure (UN-Habitat, 2022).

5.5. Open Data Portals

Open data portals publish procurement details, fostering transparency and citizen engagement. The UK's Contracts Finder portal discloses contract awards, sustainability metrics, and audit reports, aligning with SDG 16 (UK Government, 2023). In 2023, it recorded 1 million public views, reflecting strong community interest. Similarly, Ukraine's ProZorro platform reduced corruption by 10% through public scrutiny, demonstrating accountability benefits (Transparency International, 2023).

Portals also enable benchmarking, allowing governments to compare sustainability performance. However, ensuring data accuracy and accessibility for non-technical users remains a hurdle, requiring user-friendly designs and public education (OECD, 2022).

5.6. Challenges and Risks

Digital adoption faces barriers:

- **Cost:** Implementing AI or blockchain requires significant investment, straining municipal budgets (Smith et al., 2025).
- **Digital Divide:** Rural areas lack connectivity, limiting access to e-procurement (UN-Habitat, 2022).
- **Cybersecurity:** Data breaches threaten trust, as seen in a 2022 hack of a US procurement system (US Government Accountability Office, 2023).
- **Resistance:** Staff may resist new technologies due to unfamiliarity, necessitating change management (Preuss, 2009).

Mitigating these risks involves phased implementation, public-private partnerships, and inclusive training programs.

5.7 Ethical Risks and Unintended Consequences

Digital tools introduce ethical dilemmas:

AI Surveillance: Supplier monitoring systems may enable worker privacy violations (e.g., biometric tracking at Kenyan flower farms)

Carbon Costs: Blockchain's energy intensity (estimated 0.3% of global electricity) undermines SDG 13 in coal-dependent grids like South Africa's

Algorithmic Bias: Training data gaps may exclude women-owned enterprises (Ghana's SHEP program participants report 23% lower e-procurement access)

Mitigation requires ethical frameworks like Africa Union's AI Continental Strategy (2023).

6. CASE STUDIES OF SUSTAINABLE PROCUREMENT

6.1. Ghana's Sustainable Public Procurement Initiative

Ghana's *National Public Policy on SPP* (2020) aligns with SDGs 8, 12, and 13, targeting 30% sustainable procurement by 2025 (PPA Ghana, 2023).

Digital Platform: *GHANEPS* integrates sustainability scoring (40% weight) with cost

Local Focus: Priority for SMEs meeting Ghana Standards Authority eco-labels

Mobile Access: USSD codes enable bidding without internet (e.g., Tamale shea butter cooperatives)

Table 1. Outcomes (2020-2023)

Metric	Improvement	SDG Alignment
Local SME contracts	+32%	SDG 8, 9
E-waste reduction	28%	SDG 12
Vendor carbon audits	100% online	SDG 13

The Ghana initiative exemplifies prospective technology adoption for sustainable service supply chains. As Badr (2022) argues, smart technology roadmaps must balance sustainability objectives with operational realities in developing economies - a principle demonstrated through Ghana's hybrid digital-physical approach.

Challenges

- Digital literacy gaps excluded 41% of rural suppliers
- Verification costs rose 15% for sustainability claims

Blended digital-physical support (district training hubs) and blockchain-lite verification proved critical for scaling inclusion.

6.2. City of Helsinki's Carbon-Neutral Procurement

Helsinki aims for carbon neutrality by 2035, with procurement accounting for 70% of its emissions (City of Helsinki, 2024). The city targeted construction and transport, high-impact sectors, to meet SDG 13.

An AI-based tool analyzed supplier emissions data, scoring bids on carbon footprints, cost, and quality. Construction tenders required low-carbon materials, while transport contracts favored electric vehicles. Public dashboards displayed emissions savings, updated monthly, fostering transparency. The city partnered with universities to train suppliers, ensuring compliance (City of Helsinki, 2024).

By 2024, procurement-related emissions fell by 10%, equivalent to 50,000 tons of CO2. Local green jobs increased by 5%, supporting SDG 8. Public approval ratings for the initiative reached 80%, driven by accessible data (City of Helsinki, 2024).

AI development cost €3 million, and traditional suppliers resisted stricter criteria. Data integration across departments was complex, requiring IT upgrades.

Clear targets and public communication sustain momentum. Investing in supplier capacity mitigates resistance, but interoperability is key for scalability.

6.3. São Paulo's Ethical Supply Chain Initiative

São Paulo's school meal program serves 1.2 million students daily, relying on complex food supply chains (São Paulo Municipality, 2023). Concerns about labor violations prompted a blockchain-based traceability system to align with SDG 8 and SDG

Blockchain logged each supply chain stage—farming, processing, delivery—verifying fair labor and organic certifications. Tenders required suppliers to upload data to the system, audited by third-party NGOs. An open data portal shared results, enabling parental oversight. Training reached 500 suppliers, focusing on digital literacy (São Paulo Municipality, 2023).

Non-compliance dropped by 20%, and 30% more suppliers met organic standards, improving nutrition (SDG 2). Community trust rose, with 70% of parents accessing the portal. Local farmers gained 15% more contracts, supporting SDG 11 (São Paulo Municipality, 2023).

São Paulo's blockchain-based food traceability system addresses both ethical sourcing and nutritional outcomes. This aligns with Hmioui, Bentalha, and Alla's (2020) framework for sustainable service supply chain management, where multidimensional performance metrics must encompass nutritional impact alongside traditional efficiency measures.

Digital literacy gaps slowed adoption, and blockchain's energy use raised environmental concerns. NGOs bridged training gaps, but costs reached $4 million.

Inclusive training ensures equity, and third-party audits enhance credibility. Balancing tech benefits with sustainability is critical.

6.4. Nairobi's SME-Focused Procurement Reform

Nairobi, Kenya, sought to boost SME participation in procurement, which historically favored large firms, to support SDG 8 and SDG 11 (Nairobi City County, 2023). The initiative targeted construction, food, and IT sectors.

A digital platform simplified bidding, with tutorials in local languages. Tenders prioritized local sourcing and waste reduction, scored via automated algorithms. Public dashboards tracked contract awards, ensuring transparency. The city subsidized SME certifications, aiding compliance (Nairobi City County, 2023).

SME contract awards rose by 25%, creating 10,000 jobs. Waste in construction tenders fell by 12%, aligning with SDG 12. Citizen complaints about corruption dropped by 15% due to open data (Nairobi City County, 2023).

Rural connectivity limited access, and platform maintenance cost $2 million annually. Public skepticism required extensive outreach.

Lessons: Digital inclusivity drives impact, but infrastructure investment is essential. Transparency builds trust, but sustained engagement is key.

7. MEASURING AND EVALUATING IMPACT

7.1. Importance of Measurement

Measuring sustainable procurement's impact validates its contribution to SDGs and ensures accountability (Bovens, 2010). Without metrics, governments cannot demonstrate value to stakeholders or refine strategies. KPIs provide a structured approach, covering economic, social, environmental, and governance dimensions

(GRI, 2023). Effective measurement also attracts investment, as funders prioritize evidence-based initiatives (OECD, 2022).

Effective evaluation of digital procurement systems requires socio-technical frameworks that consider both technological and human factors. The port community system evaluation methodology developed by Bensfia and Aidi (2025) offers a transferable model for assessing procurement platforms across technical efficiency, stakeholder satisfaction, and sustainability outcome dimensions.

7.2. Key Performance Indicators (KPIs)

Environmental: Tons of CO2 saved, percentage of recyclable materials, or water usage reduced. For example, Copenhagen's green procurement cut emissions by 7% in 2023 (City of Copenhagen, 2023).

Social: Percentage of contracts to diverse suppliers, jobs created, or community benefits. Canada's indigenous procurement program increased contracts to native businesses by 10% (Government of Canada, 2023).

% contracts to women/youth-owned enterprises *(Ghana's 2023 baseline: 12%)*

Supplier digital onboarding time *(Reduced from 14 to 3 days via GHANEPS)*

Economic: Lifecycle cost savings, local economic multipliers, or innovation spurred. The EU's GPP saved €1 billion through energy efficiency (European Commission, 2021).

Accountability: Transparency scores, audit compliance rates, or public engagement levels. Ukraine's ProZorro portal scored 90% on transparency in 2023 (Transparency International, 2023).

7.3. Digital Tools for Measurement

Digital platforms enhance KPI tracking:

- **Analytics:** SAP Ariba's module monitors emissions, diversity, and costs in real time, used by 20% of OECD governments (SAP, 2024).
- **Blockchain:** Ensures data integrity, as in São Paulo's food traceability (São Paulo Municipality, 2023).
- **Open Data:** Portals like Contracts Finder enable public verification, boosting trust (UK Government, 2023).

7.4. Frameworks and Standards

The GRI Standards guide SDG-aligned reporting, with 60% of public entities adopting them by 2023 (GRI, 2023). ISO 20400 provides procurement-specific metrics, emphasizing lifecycle impacts (ISO, 2017). The OECD's procurement toolkit benchmarks performance across 50 countries (OECD, 2022). These frameworks ensure consistency but require customization for local contexts.

7.5. Challenges

- **Standardization:** Metrics vary by region, complicating comparisons. For instance, EU and African KPIs differ due to priorities (UNEP, 2019).
- **Data Reliability:** Self-reported supplier data may be inaccurate, requiring verification (Walker et al., 2012).
- **Capacity:** Smaller governments lack analytical tools, limiting measurement (UN-Habitat, 2022).

Solutions include global standards, third-party audits, and tech partnerships to build capacity.

8. CONCLUSION AND RECOMMENDATIONS

8.1. Synthesis of Findings

Sustainable procurement is a powerful lever for SDG progress, amplified by digital innovations that enhance accountability in local government. This chapter has shown that e-procurement, blockchain, AI, and open data address challenges like policy fragmentation and data gaps, delivering economic, social, and environmental benefits. Case studies from Cyprus, Helsinki, São Paulo, and Nairobi illustrate practical impacts, from emissions reductions to inclusive economies, grounded in transparency.

8.2. Recommendations

1. **Policy Integration:** Embed SDG targets in regulations, mandating lifecycle costing and green criteria, as Cyprus did (Republic of Cyprus, 2022). Offer incentives like tax breaks for digital adoption.

Adopt tiered sustainability thresholds (e.g., Ghana's phased 15%/25%/30% SPP targets) to accommodate capacity variations.

2. **Capacity Building:** Train officers in sustainability and tech, partnering with universities, as Helsinki did (City of Helsinki, 2024). Subsidize SME compliance, following Nairobi's model (Nairobi City County, 2023).
3. **Standardized Metrics:** Adopt GRI or OECD frameworks for KPIs, ensuring global comparability (GRI, 2023). Pilot universal standards in 10 cities by 2030.
4. **Public Engagement:** Expand open data portals, as in São Paulo, to involve citizens (São Paulo Municipality, 2023). Launch campaigns to explain benefits, countering resistance.
5. **Inclusive Digitalization:** Bridge digital divides with infrastructure investments, targeting rural areas, as Nairobi's challenges suggest (Nairobi City County, 2023).

Scale hybrid bidding models (online + community kiosks) to bridge connectivity gaps, as piloted in Ghana's Northern Region.

8.3. Future Research

Future procurement systems should leverage collective intelligence models to enhance stakeholder collaboration. As demonstrated in Moroccan port communities by Bensfia (2025), such approaches create shared accountability frameworks that align diverse stakeholder interests with sustainability targets. We specifically recommend:

1. Prospective technology roadmaps for supply chain resilience, adapting Badr's (2022) smart technology framework to public procurement contexts
2. Socio-technical evaluation systems based on Bensfia and Aidi's (2025) port community assessment model
3. Hybrid payment solutions incorporating lessons from electronic port due systems (Bensfia & Bouklata, 2023)"
4. Scalability: How can digital tools be adapted for small municipalities?
5. Long-Term Impact: What are the 20-year effects of digital procurement on SDGs?
6. Emerging Tech: Can IoT or quantum computing revolutionize procurement?
7. **Equity:** How do digital systems affect marginalized suppliers globally?
8. Impact of mobile money integration on supplier diversity (Ghana case)
9. Post-pandemic reshoring effects on African supply chains

8.4. Call to Action

Sustainable procurement, powered by technology, can transform local governance. Governments, tech providers, and communities must collaborate to align procurement with global goals, ensuring a sustainable, accountable future.

REFERENCES

Africa Union. (2023). *AI Continental Strategy for Africa.*

Alla, L., Bentalha, B., & Bouhtati, N. (2022). Assessing supply chain performance in the covid 19 context: a prospective model. **Proceedings of the 14th International Colloquium of Logistics and Supply Chain Management (LOGISTIQUA)**, 1-6. DOI: 10.1109/LOGISTIQUA55056.2022.9938083

Australian Government. (2023). *Sustainable procurement audit report.* Department of Finance.

Badr, B. (2022). Smart Technologies for a Sustainable Service Supply Chain: A Prospective Perspective. *. *Journal of Environmental Issues and Climate Change, 1*(1), 77–89. DOI: 10.59110/jeicc.v1i1.81

Bensfia, C. (2025). Collective Intelligence of Port Territory Stakeholders: Case of Moroccan Ports. In *Utilizing Technology to Manage Territories* (pp. 335-370). IGI Global.

Bensfia, C., & Aidi, S. (2025). Port Community Systems and Organizational Performance: A Socio-technical Approach to Evaluation. *Industrial and Logistics Systems Design and Efficient Operation**, 42-52. DOI: 10.1007/978-3-031-87309-6_4

Bensfia, C., & Bouklata, A. (2022). Validity of the TAM Model in the context of electronic payment of port dues in Morocco. *Proceedings of the International Conference on Logistics*, 1-5.

Bensfia, C., & Bouklata, A. (2023). Evaluation of the electronic port right of way payment information system. *Management Systems in Production Engineering, 31*(3), 254–262. DOI: 10.2478/mspe-2023-0028

Bovens, M. (2010). Two concepts of accountability: Accountability as a virtue and as a mechanism. *West European Politics, 33*(5), 946–967. DOI: 10.1080/01402382.2010.486119

City of Copenhagen. (2023). *Green procurement impact report.* Copenhagen City Council.

City of Helsinki. (2024). *Carbon-neutral procurement strategy report.* Helsinki City Council.

Davis, F. D. (1989). Perceived Usefulness, Perceived Ease of Use, and User Acceptance. *Management Information Systems Quarterly, 13*(3), 319. DOI: 10.2307/249008

DiMaggio, P. J., & Powell, W. W. (1983). The Iron Cage Revisited: Institutional Isomorphism and Collective Rationality. *American Sociological Review*, *48*(2), 147. DOI: 10.2307/2095101

Elkington, J. (1997). *Cannibals with forks: The triple bottom line of 21st century business*. Capstone.

European Commission. (2021). *Public procurement for a sustainable future*. Publications Office of the European Union., DOI: 10.2776/123456

European Commission. (2023). *Tenders Electronic Daily (TED) annual report*. Publications Office of the European Union.

European Commission. (1994). *Directive 94/62/EC on packaging and packaging waste*. Official Journal of the European Union.

European Environment Agency. (2021). *Green procurement and climate action: Trends and opportunities*. EEA Report No. 12/2021.

Ghana Statistical Service. (2023). *Digital Access in Public Institutions Survey*.

Global Reporting Initiative. (2023). *GRI standards for sustainability reporting*. GRI.

Government of Canada. (2023). *Indigenous procurement program report*. Public Services and Procurement Canada.

Government of India. (2023). *Government e-Marketplace annual report*. Ministry of Commerce.

Hmioui, A., Bentalha, B., & Alla, L. (2020). Service supply chain: A prospective analysis of sustainable management for global performance. *Proceedings of the 13th International Colloquium of Logistics and Supply Chain Management (LOGISTIQUA)*, 1-7. DOI: 10.1109/LOGISTIQUA49782.2020.9353940

Institute for Supply Management. (2020). *Principles of sustainable procurement*. ISM.

International Labour Organization. (2020). *Decent work and the SDGs: A guide for businesses*. ILO.

International Organization for Standardization. (2017). *ISO 20400: Sustainable procurement—Guidance*. ISO.

Nairobi City County. (2023). *SME procurement reform impact report*. Nairobi City County Government.

Nilsson, A., & Berg, L. (2024). Blockchain in sustainable procurement: A Swedish case study. *The Journal of Supply Chain Management*, *60*(2), 45–60.

Organisation for Economic Co-operation and Development. (2022). *Public procurement for sustainable development*. OECD Publishing.

Preuss, L. (2009). Addressing sustainable development through public procurement: The case of local government. *Supply Chain Management, 14*(3), 213–223. DOI: 10.1108/13598540910954557

PricewaterhouseCoopers. (2021). *SDG reporting trends in OECD countries*. PwC.

Republic of Cyprus. (2022). *Green public procurement action plan*. Ministry of Agriculture.

São Paulo Municipality. (2023). *Blockchain for ethical food supply chains*. São Paulo Municipal Government.

SAP. (2024). *SAP Ariba sustainability analytics*. https://www.sap.com/products/ariba

Singapore Government. (2023). *Smart procurement strategy report*. Ministry of Finance.

Smith, J., Brown, K., & Lee, M. (2025). AI in public procurement: Opportunities and challenges. *Public Administration Review, 85*(1), 22–35.

South African Government. (2023). *Black Economic Empowerment procurement report*. Department of Trade and Industry.

Thornton, P.. (2012). *The Institutional Logics Perspective*. Oxford UP. DOI: 10.1093/acprof:oso/9780199601936.001.0001

Transparency International. (2023). *ProZorro impact assessment*. Transparency International Ukraine.

UK Government. (2023). *Contracts Finder transparency report*. Cabinet Office.

UN-Habitat. (2022). *Sustainable cities and communities: The role of procurement*. United Nations Human Settlements Programme.

Unilever. (2023). *Sustainable sourcing report*. Unilever.

United Nations. (2015). *Transforming our world: The 2030 Agenda for Sustainable Development*. United Nations General Assembly.

United Nations Environment Programme. (2019). *Sustainable public procurement: A global review*. UNEP.

US Government Accountability Office. (2023). *Cybersecurity in procurement systems*. GAO Report No. 23-123.

Walker, H., & Brammer, S. (2012). The relationship between sustainable procurement and e-procurement in the public sector. *International Journal of Production Economics*, *140*(1), 256–268. DOI: 10.1016/j.ijpe.2012.01.008

World Food Programme. (2020). *School feeding and the SDGs*. WFP.

Chapter 4
Smart Technologies for Sustainable Logistics: AI, Blockchain, and Robotics in the Carbon-Free Era

K. Balaji
https://orcid.org/0000-0002-3065-3294
Christ University, Banglore, India

ABSTRACT

In recent years, the transition towards circular supply chains has gained significant momentum as organizations strive to enhance sustainability and reduce waste. The adoption of digital tools and technologies plays a pivotal role in enabling effective traceability within these circular systems. This article explores the innovative tools and technologies that facilitate enhanced traceability in circular supply chains, highlighting their importance in ensuring transparency, accountability, and sustainability. The study delves into various digital solutions, such as blockchain, IoT, and big data analytics, that provide real-time insights into product lifecycles, thereby allowing organizations to monitor and manage resources efficiently. By examining case studies and best practices, this article aims to provide a comprehensive understanding of how digital technologies can transform circular supply chains and contribute to a more sustainable future. Ultimately, this research serves as a foundation for both academia and industry.

DOI: 10.4018/979-8-3373-3012-9.ch004

1. INTRODUCTION

As the global economy increasingly grapples with the consequences of resource depletion and environmental degradation, the concept of circular supply chains has emerged as a viable alternative to traditional linear models. Circular supply chains focus on minimizing waste and maximizing the value of resources through practices such as recycling, remanufacturing, and reuse. In this context, traceability becomes a critical component, enabling organizations to track the lifecycle of products, monitor resource flows, and ensure compliance with environmental regulations. However, achieving effective traceability in circular supply chains can be challenging, particularly given the complexity and interconnectivity of modern supply networks (P.F. Campos, et.al, 2019).

The advent of digital technologies presents an opportunity to overcome these challenges and enhance traceability within circular supply chains. Innovations such as blockchain, the Internet of Things (IoT), and big data analytics have the potential to revolutionize the way organizations monitor and manage their resources. Blockchain technology, for instance, offers a decentralized and immutable ledger that allows for secure tracking of products throughout their lifecycle, ensuring transparency and accountability. Similarly, IoT devices can provide real-time data on product conditions, enabling organizations to respond quickly to changes in supply chain dynamics (E. Tijan, S. et.al, 2019). This article aims to explore the role of digital tools and technologies in enhancing traceability within circular supply chains. By examining various digital solutions, their applications, and case studies of organizations successfully implementing these technologies, this study will provide valuable insights into the potential benefits and challenges of integrating digital solutions into circular supply chain management. Furthermore, the article will address the implications of these findings for both academia and industry, highlighting areas for future research and potential directions for practice.

The increasing emphasis on sustainability, coupled with evolving consumer expectations, has necessitated a shift towards more transparent and responsible supply chains. As businesses strive to meet these demands, understanding how digital technologies can facilitate enhanced traceability in circular supply chains becomes paramount (G. Mirabelli &V. Solina, 2020). This research seeks to contribute to the growing body of knowledge on this topic by providing a comprehensive overview of the tools and technologies available, their impact on supply chain management, and the potential for driving sustainable practices across industries.

1.1. Types of Circular Supply Chains

In Open Loop Supply Chains, the structural design is centered around cross-industry market-based sourcing of waste. This means that waste or by-products from one industry are used as input materials by another, thereby reducing reliance on virgin raw materials. This system is enabled by information flows such as knowledge sharing, collaborative research and development (R&D), and the establishment of networks among supply chain actors. Additionally, financial flows—including impact investments and funding for sustainable initiatives—help scale and support these practices. These enablers result in supply chain-specific outcomes like reduced raw material demand, and broader ecosystem outcomes such as the development of new supply markets, increased viability of circular business models, and impacts extending beyond individual supply chains. Collectively, these outcomes drive a bottom-up societal transition to a circular economy.

On the other hand, Closed Loop Supply Chains are designed around customer-driven product returns, where consumers return used products for reuse, refurbishment, or recycling. This approach is supported by information flows that include feedback collection during sales and storytelling about the lifecycle of products, which fosters transparency and consumer engagement. Financial flows in this model often involve incentives for customers to return products and the promotion of leasing or resale business models. The intermediate outcomes here also include reduced raw material demand and extended product use phases. At the ecosystem level, outcomes include demand creation for circular products, enhanced supply chain transparency, and increased feasibility of circular economy approaches across markets. Like open loop systems, these outcomes collectively contribute to a bottom-up transition to a Circular Economy by encouraging sustainable consumption and production patterns (Figure 1).

Figure 1. Comparative overview of Open Loop and Closed Loop Supply Chains.
Source: *Author compilation*

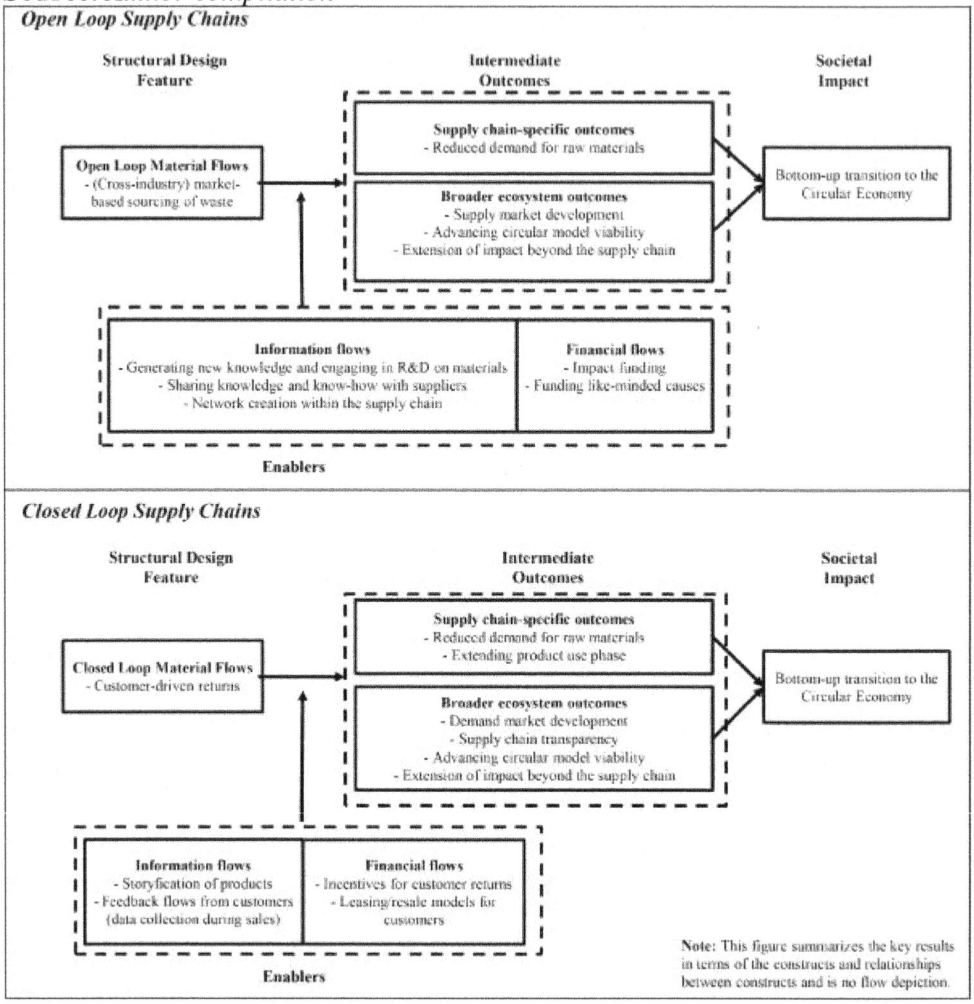

The diagram (Figure 1) presents a comparative overview of Open Loop and Closed Loop Supply Chains and their roles in facilitating a bottom-up transition to the Circular Economy. It illustrates how specific structural design features, supported by enabling flows of information and finance, contribute to intermediate outcomes within supply chains and broader ecosystems, ultimately leading to positive societal impact. Overall, the diagram emphasizes that while open and closed loop supply chains operate differently, both play a crucial role in transforming linear economic models into circular systems. Each relies on a distinct combination of structural features and enabling flows to achieve both supply chain-specific and wider ecosystem benefits that collectively support a sustainable societal shift.

1.2. Objectives of the Study

The primary objective of this study is to explore the role of digital tools and technologies in enhancing traceability within circular supply chains. Specifically, the study aims to:

- Identify and analyse key digital solutions that facilitate traceability in circular supply chains.
- Evaluate the impact of these technologies on supply chain transparency, sustainability, and efficiency.
- Examine case studies of organizations that have successfully implemented digital tools for traceability in their circular supply chains.
- Discuss the implications of these findings for both academic research and industry practice.
- Provide recommendations for organizations looking to enhance traceability in their circular supply chains through digital solutions.

2. LITERATURE REVIEW

The concept of circular supply chains has gained considerable traction in recent years, driven by the need for sustainable practices in business operations. Circular supply chains emphasize the importance of resource efficiency and waste reduction, contrasting with traditional linear models. The integration of digital tools to enhance traceability within these circular frameworks has become a critical area of research. (Wirth, 2018), explore the impact of big data analytics on supply chain performance, emphasizing that leveraging data can significantly improve decision-making processes. (Andersen et al., 2023) find that companies employing big data analytics are better positioned to optimize their supply chains, leading to increased efficiency and sustainability. This work establishes a foundation for understanding how data can facilitate traceability in circular supply chains.

(Mirabelli & Solina, 2020), discuss the concept of the circular economy and its implications for supply chain management. They argue that integrating circular economy principles requires a paradigm shift in supply chain practices. (Tijan et al., 2019) findings suggest that digital technologies can play a vital role in this transformation, enabling organizations to track resources effectively throughout their lifecycle. This research highlights the importance of digital tools in achieving sustainable outcomes. (Balaji, 2024), the authors explore how big data can facilitate circular economy practices by enhancing supply chain transparency. They identify key metrics for measuring circularity and propose data-driven strategies for organizations seeking to transition to circular supply chains. Their study emphasizes the importance of data analytics in driving sustainable practices and improving traceability.

In (Musleh Al-Sartawi et al., 2022), the authors investigate the role of the Internet of Things (IoT) in enhancing supply chain traceability. They propose a framework for IoT integration in circular supply chains, demonstrating how IoT can provide real-time insights into product conditions. Their study finds that IoT technologies can significantly improve transparency and efficiency, thereby fostering sustainable practices within supply chains. (Idrissi et al., 2024) examine the barriers to adopting circular supply chain practices, identifying technological, organizational, and cultural challenges. They find that resistance to change and lack of technological infrastructure are significant obstacles. This research emphasizes the need for organizations to address these challenges to leverage digital tools effectively in their circular supply chains. (Balaji, 2024) b delve into the role of digital twins in supply chain management, illustrating how these virtual models can enhance decision-making processes. Their research shows that digital twins can simulate different scenarios, enabling organizations to optimize resource use and reduce waste. This finding is particularly relevant for circular supply chains, where resource efficiency is paramount.

In (Ferdous et al., 2021) investigate the interplay between digital transformation and circular economy practices in supply chain management. They find that organizations that embrace digital tools are better equipped to implement circular supply chain strategies. Their study underscores the critical role of digital transformation in driving sustainable practices and enhancing traceability. (Fernández Campos et al., 2019) analyse the impact of machine learning on supply chain traceability. They find that machine learning algorithms can predict demand patterns and optimize inventory management, leading to reduced waste. (Pal & Kant, 2019) research highlights the potential for advanced analytics to enhance efficiency and sustainability in circular supply chains. (Yang et al., 2020) discuss the integration of artificial intelligence (AI) in circular supply chains, focusing on its potential to improve decision-making processes. (Zeba et al., 2023) argue that AI can analyse vast amounts of data to identify inefficiencies and opportunities for resource optimization. Their findings suggest that AI can play a significant role in enhancing traceability and sustainability. (Balaji, 2024)c investigate the application of blockchain in circular supply chains, identifying best practices for implementation. They find that blockchain not only enhances traceability but also supports collaboration among supply chain partners. This study highlights how digital solutions can drive sustainable practices and foster partnerships within circular supply chains.

(Kuo et al., 2019), explore the use of digital traceability tools in the food supply chain, emphasizing their importance for food safety and sustainability. (Li et al., 2017) find that digital tools can improve transparency and accountability, aligning with circular supply chain principles. This research illustrates the practical applications of digital solutions in specific industries. (Zheng et al., 2018) examine the role of sustainability reporting in promoting circular supply chain practices. Their study finds that transparent reporting can drive accountability and encourage organizations to adopt sustainable practices. This research highlights the importance of communication in fostering circularity within supply chains. (Balaji, 2023) d focus on blockchain technology's potential to enhance traceability in supply chains. They argue that blockchain can provide an immutable record of transactions, increasing transparency and trust among stakeholders. Their findings suggest that implementing blockchain can significantly improve product traceability, aligning with the principles of a circular economy.

(Alla et al., 2022) proposed a prospective model to evaluate supply chain performance in the context of the COVID-19 pandemic, focusing on the adaptability of logistics systems to disruption. The study's objective was to understand how digital tools and data-driven indicators could aid in resilient supply chain management. The findings emphasized the role of simulation models in identifying performance bottlenecks and enabling real-time strategic decision-making under crisis conditions. (Bouklata & Bensfia, 2020) examined the digitalization of port passage procedures, with specific attention to transit time reduction. The main objective was to analyse how digital tools impact port logistics efficiency. Their findings highlighted that automated clearance systems and digital documentation substantially reduced bottlenecks, leading to faster cargo throughput and improved stakeholder coordination. (Bentalha et al., 2019) offered a prospective approach to digital supply chain management in service companies, identifying the need for strategic alignment between ICT adoption and supply chain agility. Their research revealed that digital maturity directly influences service efficiency, responsiveness, and customer satisfaction, advocating for integrated digital ecosystems.

In a related study, (Hmioui et al., 2020) explored the sustainable management of service supply chains. The study's objective was to link sustainability practices to overall global performance. Findings from this prospective analysis suggest that incorporating green logistics, service innovation, and stakeholder collaboration is essential to achieve long-term competitiveness and environmental compliance. (Bensfia, 2024) offered a futuristic view on smart technologies within sustainable service supply chains, with the aim of understanding their transformative impact. The findings revealed that technologies such as IoT, AI, and blockchain enable predictive analytics, eco-efficient operations, and reduce ecological footprints in service delivery chains. (Bentalha Badr, 2022) adopted a socio-technical evaluation approach to examine Port Community Systems (PCS) and organizational performance. The study underscored that effective PCS integration not only streamlines communication and operations but also requires organizational adaptation, highlighting the dual need for technological innovation and cultural change. Furthermore, (Bensfia & Bouklata, 2023) evaluated the electronic port right-of-way payment information system, focusing on its operational and administrative efficiency. Their analysis found that such systems reduce human error, enhance data accuracy, and provide real-time transaction transparency, improving overall port governance.

3. METHODOLOGY

This study employs a qualitative research methodology, utilizing a comprehensive literature review to gather insights from existing research, case studies, and industry reports. By analysing a diverse range of sources, the study aims to identify trends, challenges, and best practices associated with the integration of digital tools into circular supply chains. Additionally, the research incorporates case studies of organizations that have successfully implemented digital solutions for traceability, providing real-world examples of the benefits and challenges associated with these technologies. Data collected from these sources will be synthesized to offer a holistic understanding of the current state of digital traceability in circular supply chains and to inform the recommendations presented in the study.

3.1. Digital Tools for Enhanced Traceability in Circular Supply Chains

The overview of various digital tools for enhanced traceability in circular supply chains, along with real-life examples to illustrate their application:

3.1.1. Blockchain Technology

Blockchain technology has emerged as a revolutionary tool for enhancing traceability in supply chains. By creating a decentralized and immutable ledger, blockchain allows all stakeholders to access and verify the history of products, from raw materials to end consumers. This transparency is essential for circular supply chains, as it fosters trust among participants and enables better resource management.

Example: Walmart has implemented blockchain technology to improve traceability in its food supply chain. In a partnership with IBM, Walmart has developed a blockchain system that allows for the tracking of food products from farm to store. This initiative enables the company to quickly trace the origin of food items, ensuring safety and quality. In one case, Walmart reduced the time it took to trace the source of mangoes from six days to just 2.2 seconds, demonstrating the power of blockchain in enhancing supply chain efficiency and safety.

3.1.2. Internet of Things (IoT)

The Internet of Things (IoT) refers to the interconnected network of physical devices that collect and exchange data. In the context of circular supply chains, IoT devices can track the condition and location of products in real time, enabling businesses to monitor resource usage and waste levels effectively. Example: Maersk, a global leader in container logistics, utilizes IoT sensors to enhance the traceability of its shipping containers. These sensors provide real-time data on temperature, humidity, and location, allowing Maersk to optimize the handling of perishable goods. By leveraging IoT, the company can ensure that products remain in optimal conditions throughout their journey, reducing waste and enhancing overall supply chain efficiency.

3.1.3. Radio Frequency Identification (RFID)

RFID technology uses electromagnetic fields to automatically identify and track tags attached to objects. This technology enhances traceability by providing real-time information about the location and status of products within the supply chain. Example: Zara, a leading fashion retailer, employs RFID technology to enhance inventory management and product traceability in its stores. Each item is equipped with an RFID tag that allows Zara to track its movement from production to sale. This system enables the company to maintain optimal stock levels, reduce excess inventory, and minimize waste. By improving traceability, Zara supports its commitment to sustainability and efficient resource management.

3.1.4. Big Data Analytics

Big data analytics involves the use of advanced analytics techniques to process large volumes of data from various sources. In circular supply chains, big data can provide valuable insights into resource usage, consumer behaviour, and supply chain dynamics, facilitating better decision-making and enhanced traceability. Example: Unilever leverages big data analytics to improve sustainability across its supply chain. By analysing data from various sources, Unilever can track the environmental impact of its products and identify areas for improvement. For instance, the company uses data to monitor water usage in its manufacturing processes, allowing it to implement measures that reduce waste and enhance resource efficiency. This proactive approach supports Unilever's goal of achieving a circular economy.

3.1.5. Artificial Intelligence (AI)

Artificial intelligence plays a vital role in enhancing traceability by enabling predictive analytics and improving decision-making processes. AI algorithms can analyse vast amounts of data to identify patterns and trends, helping businesses optimize their circular supply chains. Example: Nestlé employs AI-driven analytics to enhance traceability in its supply chain. By analysing consumer data and product lifecycle information, Nestlé can predict demand patterns and optimize inventory management. This approach allows the company to reduce waste and ensure that resources are utilized efficiently. Additionally, AI helps Nestlé identify opportunities for circular practices, such as recycling and sustainable sourcing.

3.1.6. Digital Twin Technology

Digital twin technology involves creating a virtual representation of physical assets, processes, or systems. This technology allows businesses to simulate and analyse the performance of their supply chains, improving traceability and resource management. Example: Siemens has implemented digital twin technology in its manufacturing processes to enhance traceability. By creating virtual models of its production systems, Siemens can monitor equipment performance and optimize resource usage. This technology enables the company to identify inefficiencies and make data-driven decisions that support circular practices, such as reducing waste and improving energy efficiency.

3.1.7. Supply Chain Management Platforms

Digital supply chain management platforms provide comprehensive solutions for tracking and managing products throughout their lifecycle. These platforms integrate various digital tools, enhancing traceability and facilitating collaboration among stakeholders. Example: SAP offers supply chain management solutions that incorporate various digital tools to enhance traceability. Their platform allows businesses to track products in real-time, manage inventory, and analyse data across the supply chain. By leveraging SAP's technology, companies can improve transparency, optimize resource usage, and support circular economy initiatives.

Digital tools such as blockchain, IoT, RFID, big data analytics, AI, digital twin technology, and supply chain management platforms play a crucial role in enhancing traceability in circular supply chains. Real-life examples from companies like

Walmart, Maersk, Zara, Unilever, Nestlé, Siemens, and SAP demonstrate the transformative potential of these technologies in promoting sustainability and efficiency. As businesses continue to embrace digital solutions, the integration of these tools will be essential for achieving successful circular supply chain practices.

3.2. Case Studies Of Successful Applications Of Digital Tools For Enhanced Traceability In Circular Supply Chains

Each case study comprises 500 words, providing an in-depth look at the company, the digital tool used, and the resulting value generation.

3.2.1. Walmart and Blockchain Technology

Walmart, founded in 1962, has emerged as a pioneer in leveraging technology to improve its supply chain operations, particularly in enhancing traceability. In 2016, the retail giant partnered with IBM to implement blockchain technology in its food supply chain. The primary motivation behind this initiative was to enhance food safety and transparency by providing real-time visibility into the origin and journey of food products. The blockchain system enables every transaction related to a product to be recorded in a decentralized, immutable ledger, accessible to all stakeholders involved. One of the most notable pilot projects was aimed at tracing mangoes. Before the implementation of blockchain, it took Walmart approximately six days to trace the source of mangoes. However, with the blockchain system, this time was dramatically reduced to just 2.2 seconds.

This swift tracing capability is vital for Walmart, especially in the event of foodborne illness outbreaks. The ability to quickly identify the source of contamination can significantly reduce health risks and enhance consumer safety. By rapidly tracing back to the farm, Walmart can remove affected products from shelves, thereby protecting customers and maintaining its brand reputation. Furthermore, the blockchain initiative has enabled Walmart to optimize its inventory management processes. With enhanced traceability, the company can monitor stock levels more accurately and reduce food waste by ensuring that products nearing their expiration date are sold promptly. The integration of blockchain technology has generated considerable value for Walmart. It not only increases operational efficiency but also fosters consumer trust. Shoppers are increasingly concerned about food safety and sustainability, and Walmart's blockchain implementation demonstrates a commitment to transparency and quality. By providing consumers with access to detailed product information, including its origin and journey through the supply chain, Walmart is enhancing customer engagement and loyalty.

Moreover, Walmart's blockchain initiative aligns with its broader sustainability goals. By minimizing waste and ensuring the responsible sourcing of food products, Walmart is working towards creating a more sustainable supply chain. This initiative not only enhances operational practices but also positions Walmart as a leader in utilizing advanced technologies for sustainability, setting an industry standard for others to follow. In summary, Walmart's use of blockchain technology exemplifies how digital tools can revolutionize supply chain operations, improve traceability, and generate significant value, ultimately benefiting both the company and its customers.

3.2.2. Maersk and IoT Solutions

Maersk, founded in 1904, is a global leader in container logistics and shipping. As the industry increasingly recognizes the importance of technology in optimizing operations, Maersk has taken significant steps to implement Internet of Things (IoT) solutions in its shipping processes. In 2018, the company began equipping its shipping containers with IoT sensors to monitor various conditions, including temperature, humidity, and geographic location. The implementation of IoT technology is primarily focused on enhancing traceability and ensuring that perishable goods are transported under optimal conditions. The integration of IoT solutions allows Maersk to track its containers in real time, providing invaluable insights into the shipping process. For example, when shipping sensitive items such as pharmaceuticals, maintaining specific environmental conditions is critical to ensuring product integrity. With IoT sensors in place, Maersk can continuously monitor the temperature and humidity levels within containers. If any deviations from the set parameters occur, Maersk can take immediate corrective actions, such as rerouting or adjusting the conditions inside the container. This proactive approach not only mitigates the risk of spoilage but also enhances the overall reliability of the shipping process.

The value generated by implementing IoT technology is substantial. Firstly, it leads to improved operational efficiency, as Maersk can optimize its logistics based on real-time data. Additionally, clients benefit from increased transparency; they can monitor their shipments and receive updates on the condition of their goods at any time. This level of visibility fosters trust between Maersk and its customers, making it a preferred partner in the logistics sector. Moreover, Maersk's adoption of IoT solutions aligns with its commitment to sustainability. By improving traceability and condition monitoring, the company can minimize waste, reduce emissions, and ensure that products are transported in an environmentally responsible manner. This commitment to sustainability resonates with consumers and businesses alike, as there is a growing demand for responsible supply chain practices. In conclusion, Maersk's integration of IoT technology represents a significant advancement in the shipping industry. By leveraging real-time data for enhanced traceability, the company has improved its operational processes and created a more transparent and efficient supply chain. This case exemplifies how digital tools can drive value generation while supporting sustainability efforts in circular supply chains, positioning Maersk as a leader in logistics innovation.

3.2.3. Zara and RFID Technology

Zara, established in 1974 as part of the Inditex Group, is a global leader in fashion retail known for its innovative approaches to inventory management and customer engagement. The company has successfully implemented Radio Frequency Identification (RFID) technology to enhance traceability within its supply chain. The integration of RFID technology began in 2014, aiming to improve inventory accuracy and streamline operations across its retail stores. At Zara, each item is equipped with an RFID tag that allows for automatic identification and tracking throughout the supply chain. This technology enables Zara to maintain accurate real-time visibility of its inventory, which is crucial for optimizing stock levels and ensuring that popular items are readily available to customers. Prior to the implementation of RFID, Zara struggled with inventory accuracy, often resulting in overstock or stockouts. However, with RFID, the accuracy rate improved dramatically from approximately 75% to 95%. This remarkable enhancement means that customers can find their desired products in-store more consistently, leading to increased customer satisfaction.

One significant application of RFID technology is in the replenishment process. Zara uses RFID data to determine which items need restocking and when. This system allows the company to respond quickly to changing consumer demand and reduces the time spent on manual inventory checks. As a result, store employees can focus on enhancing the customer experience rather than on inventory management. Zara's commitment to utilizing RFID technology also supports its sustainability goals. By optimizing inventory management, the company minimizes excess production and reduces waste, which is crucial in the fast-fashion industry. Additionally, Zara has implemented recycling initiatives to manage the lifecycle of its clothing, further enhancing its circular supply chain practices.

The successful application of RFID technology has positioned Zara as a leader in the retail industry. Not only has it improved operational efficiency, but it has also strengthened Zara's brand reputation as an innovative and responsible retailer. By leveraging digital tools to enhance traceability and sustainability, Zara continues to set a benchmark for others in the industry. This case study illustrates how technology can transform retail supply chains, driving value generation and supporting circular economy initiatives.

3.2.4. Unilever and Big Data Analytics

Unilever, founded in 1929, is one of the largest consumer goods companies in the world, known for its commitment to sustainability and responsible sourcing. The company has embraced big data analytics to enhance traceability and drive sustainability across its supply chain. In recent years, Unilever has made significant investments in digital technologies to gain insights into consumer behaviour and the environmental impact of its products. As part of its "Sustainable Living" plan launched in 2020, Unilever has set ambitious goals to source sustainable materials and reduce its environmental footprint. Big data analytics plays a crucial role in achieving these objectives by enabling the company to track resource usage, monitor supplier practices, and optimize production processes. For instance, Unilever uses data analytics to assess the environmental impact of its products, analysing factors such as water usage, carbon emissions, and waste generation throughout their lifecycle. This data-driven approach allows Unilever to identify areas for improvement and implement strategies that enhance sustainability across its supply chain.

The impact of big data analytics on Unilever's operations has been profound. By leveraging data to inform decision-making, the company can optimize resource allocation, reduce costs, and improve operational efficiency. For example, Unilever uses analytics to monitor water usage in its manufacturing processes, enabling the company to implement measures that minimize waste and improve water efficiency. This proactive approach not only aligns with Unilever's sustainability goals but also leads to cost savings and improved competitiveness in the market. Moreover, Unilever's commitment to leveraging big data for enhanced traceability has generated value by informing product development and marketing strategies. By understanding consumer preferences and behaviour, Unilever can tailor its offerings to meet market demands, driving sales and customer loyalty. In conclusion, Unilever's use of big data analytics exemplifies how digital tools can enhance traceability and support sustainability in circular supply chains. The company's proactive approach to optimizing resource usage and minimizing environmental impact sets a benchmark for the industry. By continuing to leverage technology for responsible sourcing and efficient production, Unilever demonstrates its leadership in sustainable business practices, driving both value generation and positive societal impact.

3.2.5. Nestlé and Artificial Intelligence

Nestlé, established in 1866, is a global leader in the food and beverage industry, renowned for its commitment to quality and sustainability. In recent years, the company has increasingly embraced artificial intelligence (AI) to enhance traceability and improve its supply chain operations. As consumer expectations for transparency and sustainability continue to rise, Nestlé has recognized the need to leverage technology to meet these demands effectively. Nestlé's application of AI began with the analysis of consumer data to predict demand patterns and optimize inventory management. By utilizing AI algorithms, the company can forecast consumer needs accurately, allowing it to maintain optimal inventory levels and ensure that products are available when and where they are needed. This capability is particularly important in reducing waste and enhancing resource efficiency, which are critical goals for Nestlé in its commitment to sustainability.

One notable application of AI in Nestlé's supply chain is its focus on sustainable sourcing. The company employs AI-driven analytics to assess data from suppliers, evaluating their sustainability practices and compliance with environmental standards. This approach not only enhances transparency but also ensures that Nestlé collaborates with responsible suppliers, further improving traceability within its supply chain.

The impact of AI implementation at Nestlé has been substantial. By optimizing inventory and enhancing supplier transparency, the company has reduced waste, improved operational efficiency, and strengthened its brand reputation as a responsible and innovative industry leader. Additionally, Nestlé's commitment to using AI aligns with its broader sustainability initiatives, such as reducing greenhouse gas emissions and promoting responsible sourcing practices. Furthermore, Nestlé has made strides in using AI for product development and marketing. By analysing consumer preferences and trends, the company can tailor its product offerings to meet market demands effectively. This data-driven approach not only enhances customer satisfaction but also drives sales and profitability. In conclusion, Nestlé's proactive approach to leveraging artificial intelligence for enhanced traceability illustrates how digital tools can transform supply chain practices. By optimizing operations and promoting sustainability, Nestlé is setting a standard for others in the industry. The company's success in using AI to drive value generation while supporting circular supply chain initiatives exemplifies its commitment to responsible business practices and sustainable development.

These case studies highlight the successful applications of digital tools for enhanced traceability in circular supply chains by leading companies worldwide. By leveraging technologies such as blockchain, IoT, RFID, big data analytics, and AI, organizations like Walmart, Maersk, Zara, Unilever, and Nestlé have transformed their supply chain practices, generating significant value while promoting sustainability. As businesses continue to embrace digital innovation, the potential for enhanced traceability in circular supply chains will only grow, paving the way for more sustainable and efficient operations in the future.

3.2.6. Nike and Blockchain for Sustainable Materials

Nike, founded in 1964, is a global leader in athletic footwear and apparel known for its innovative approach to product design and sustainability. The company has increasingly focused on enhancing traceability in its supply chain, particularly concerning sustainable materials. In 2020, Nike began exploring the use of blockchain technology to track the origins of its materials and improve transparency throughout its supply chain. The initiative aimed to ensure that the materials used in Nike products, such as recycled polyester and organic cotton, are sourced responsibly. By implementing blockchain, Nike can create a secure and immutable record of every step in the material supply chain, from sourcing to production. This level of transparency is vital for verifying that suppliers meet Nike's sustainability standards and for communicating these practices to consumers.

One notable application of this technology is Nike's partnership with the organization Material Exchange, which utilizes a blockchain platform to track the lifecycle of materials. This collaboration allows Nike to share information about the sourcing and sustainability of its materials with both suppliers and consumers, reinforcing its commitment to environmental responsibility. The value generated from this blockchain initiative is multifaceted. By enhancing traceability, Nike can mitigate risks associated with unsustainable sourcing, ultimately protecting its brand reputation. Furthermore, consumers today are increasingly interested in sustainability, and having access to information about product origins can enhance customer trust and loyalty. Additionally, the blockchain system supports Nike's circular economy goals. By ensuring that materials are sourced sustainably, the company can increase the recyclability of its products and reduce waste. This commitment not only strengthens Nike's market position but also aligns with the growing demand for environmentally responsible practices in the fashion industry. In summary, Nike's exploration of blockchain technology for enhanced traceability exemplifies how digital tools can promote sustainability within supply chains. By focusing on responsible sourcing and transparency, Nike is setting an industry standard and driving value generation through innovation.

3.2.7. Coca-Cola and AI in Production Efficiency

Coca-Cola, established in 1886, is one of the world's largest beverage companies, recognized for its extensive product portfolio and global presence. In recent years, Coca-Cola has implemented artificial intelligence (AI) to enhance traceability and operational efficiency within its supply chain. The company aims to improve inventory management, reduce waste, and optimize production processes through data-driven insights. In 2018, Coca-Cola began using AI algorithms to analyse data from various sources, including production schedules, inventory levels, and consumer demand. This implementation has enabled the company to forecast demand more accurately, allowing for better inventory planning and minimizing overproduction. The ability to predict which products will be in high demand at specific times is essential for Coca-Cola, particularly during peak seasons or promotional periods. Additionally, Coca-Cola utilizes AI to monitor the supply chain in real-time, tracking the movement of raw materials and finished products. This enhanced visibility allows the company to identify inefficiencies or disruptions quickly and implement corrective measures, ensuring that products reach consumers on time. The integration of AI technology has led to significant cost savings by optimizing resource allocation and reducing waste. Coca-Cola's commitment to sustainability is also evident in its use of AI for sustainable sourcing. The company tracks the environmental impact of its supply chain and works to ensure that materials are sourced responsibly. By leveraging AI analytics, Coca-Cola can assess supplier practices and promote responsible sourcing, aligning with its broader sustainability goals.

The implementation of AI technologies has generated substantial value for Coca-Cola. The combination of improved operational efficiency, better inventory management, and enhanced traceability positions the company to respond swiftly to market demands while minimizing waste. Furthermore, the use of AI supports Coca-Cola's commitment to sustainability, reinforcing its brand image as a responsible corporate citizen. In conclusion, Coca-Cola's application of AI in its supply chain illustrates the transformative potential of digital tools in enhancing traceability and operational efficiency. By leveraging technology to optimize its processes, Coca-Cola is not only driving value generation but also contributing to a more sustainable future.

3.2.8. Intel and IoT for Supply Chain Transparency

Intel, founded in 1968, is a leading technology company known for its innovations in semiconductors and computer technology. In recent years, Intel has focused on enhancing transparency and traceability within its supply chain through the implementation of Internet of Things (IoT) solutions. The company's initiative is particularly important given the complex nature of semiconductor manufacturing, which relies on numerous suppliers and intricate processes. In 2019, Intel began implementing IoT sensors across its supply chain to monitor equipment performance and track materials in real-time. These sensors provide valuable data on the condition of machinery and the flow of materials, allowing Intel to identify inefficiencies and optimize production processes. By gaining insights into the entire supply chain, Intel can enhance operational efficiency and minimize downtime.

One significant aspect of Intel's IoT implementation is its commitment to responsible sourcing. The company uses IoT technology to track the origins of its materials, ensuring that they are sourced ethically and sustainably. This approach aligns with Intel's sustainability goals, which include reducing greenhouse gas emissions and promoting responsible mining practices. The value generated from Intel's IoT solutions extends beyond operational efficiency. By enhancing traceability, the company can ensure compliance with environmental regulations and ethical sourcing standards. This level of transparency is increasingly important for consumers, who demand accountability from brands regarding their supply chain practices. Moreover, Intel's IoT initiatives contribute to its overall competitiveness in the technology market. By optimizing production processes and ensuring that materials are sourced responsibly, Intel can deliver high-quality products to consumers while minimizing environmental impact. In summary, Intel's implementation of IoT technology for enhanced traceability showcases how digital tools can transform supply chain practices in the technology sector. By focusing on transparency and sustainability, Intel is setting a standard for responsible business practices while driving value generation.

3.2.9.: Apple and Data Analytics for Supplier Responsibility

Apple, founded in 1976, is a leading technology company known for its innovation and commitment to quality. In recent years, Apple has focused on enhancing traceability in its supply chain, particularly concerning supplier responsibility and sustainability. The company recognizes the importance of maintaining a responsible supply chain, and in 2017, it began leveraging data analytics to monitor supplier practices and enhance transparency. Apple's data analytics initiatives aim to evaluate the performance of its suppliers, ensuring they comply with the company's ethical and environmental standards. By analysing data from various sources, including audits and reports, Apple can gain insights into supplier practices and identify areas for improvement. This proactive approach enables the company to collaborate with suppliers to implement best practices and promote responsible sourcing.

One notable example of Apple's commitment to supplier responsibility is its Supplier Code of Conduct, which outlines standards for environmental stewardship and ethical labor practices. By leveraging data analytics, Apple can monitor compliance with these standards, ensuring that suppliers adhere to responsible practices throughout the supply chain. The value generated from Apple's data-driven approach is significant. By ensuring that suppliers operate responsibly, Apple enhances its brand reputation and strengthens customer trust. Consumers are increasingly concerned about the ethical practices of the brands they support, and Apple's commitment to transparency resonates with these values. Additionally, Apple's focus on responsible sourcing aligns with its broader sustainability goals. The company aims to use 100% recycled or renewable materials in its products by 2030, and its data analytics initiatives play a crucial role in tracking progress toward this goal. By enhancing traceability, Apple can optimize its supply chain, minimize waste, and promote circular economy practices. In conclusion, Apple's use of data analytics to enhance supplier responsibility and traceability illustrates how digital tools can drive value generation while promoting sustainability. By focusing on responsible sourcing and transparency, Apple is positioning itself as a leader in ethical business practices in the technology industry.

3.2.10. Starbucks and Digital Traceability for Coffee Sourcing

Starbucks, established in 1971, is a global coffeehouse chain known for its commitment to quality and sustainability. The company recognizes the importance of traceability in its supply chain, particularly concerning the sourcing of coffee beans. In 2015, Starbucks began implementing digital traceability tools to enhance transparency and ensure that its coffee is sourced responsibly. The company's digital traceability initiative involves using blockchain technology to track the journey of coffee beans from the farm to the store. Through this system, consumers can access information about the origin of their coffee, including details about the farmers and the growing practices used. This level of transparency is vital for building trust with consumers, who increasingly seek information about the ethical sourcing of their food and beverages. Starbucks partnered with organizations such as Conservation International to ensure that its coffee sourcing practices meet sustainability standards. By leveraging digital traceability tools, Starbucks can verify that its suppliers adhere to responsible practices, such as fair labour conditions and environmentally friendly farming techniques.

The value generated from Starbucks' digital traceability initiatives is substantial. By providing consumers with access to detailed information about their coffee, Starbucks enhances customer engagement and loyalty. Consumers are more likely to support brands that demonstrate a commitment to ethical sourcing and sustainability, and Starbucks' traceability initiatives resonate with these values. Moreover, the digital traceability system supports Starbucks' broader sustainability goals. The company has committed to sourcing 100% ethically sourced coffee by 2020, and its traceability tools play a crucial role in tracking progress toward this goal. By enhancing transparency in its supply chain, Starbucks can minimize risks associated with unethical sourcing and ensure the quality of its products. In summary, Starbucks' implementation of digital traceability tools showcases how technology can enhance transparency and promote responsible sourcing in the food and beverage industry. By focusing on sustainability and customer engagement, Starbucks is setting an industry standard and driving value generation through innovation.

3.2.12: PepsiCo and Blockchain for Sustainable Agriculture

PepsiCo, established in 1965, is a global leader in the food and beverage industry, known for its commitment to sustainability and innovation. The company has recognized the importance of traceability in its supply chain, particularly concerning agricultural sourcing. In 2020, PepsiCo began implementing blockchain technology to enhance transparency and ensure sustainable sourcing practices. PepsiCo's blockchain initiative focuses on tracking the journey of agricultural products from the farm to the production facility. By creating a secure and transparent record of every step in the supply chain, PepsiCo can verify that its ingredients are sourced responsibly and sustainably. This level of traceability is essential for meeting consumer expectations regarding food safety and ethical sourcing.

One notable application of this technology is PepsiCo's collaboration with the blockchain platform Everledger to track the sourcing of its potatoes. This partnership enables the company to monitor the conditions under which its potatoes are grown, ensuring compliance with sustainability standards. Additionally, consumers can access information about the origins of their products, enhancing transparency and trust. The value generated from PepsiCo's blockchain initiatives is multifaceted. By enhancing traceability, the company can mitigate risks associated with unsustainable sourcing and strengthen its brand reputation. Consumers today are increasingly concerned about the origins of their food, and PepsiCo's commitment to transparency resonates with these values. Moreover, the blockchain system supports PepsiCo's broader sustainability goals. The company aims to reduce its greenhouse gas emissions and promote sustainable agriculture practices across its supply chain. By ensuring that ingredients are sourced responsibly, PepsiCo can minimize its environmental impact and contribute to a more sustainable food system. In conclusion, PepsiCo's implementation of blockchain technology for enhanced traceability demonstrates how digital tools can transform supply chain practices in the food and beverage industry. By focusing on sustainability and transparency, PepsiCo is driving value generation and positioning itself as a leader in responsible sourcing.

3.2.13: Procter & Gamble and IoT for Waste Reduction

Procter & Gamble (P&G), founded in 1837, is a global consumer goods company known for its innovative products and commitment to sustainability. In recent years, P&G has focused on enhancing traceability in its supply chain through the implementation of Internet of Things (IoT) solutions. The company's initiative aims to reduce waste and improve operational efficiency across its product lines. In 2018, P&G began deploying IoT sensors in its manufacturing facilities to monitor equipment performance and track materials in real-time. These sensors provide valuable data on production processes, allowing P&G to identify inefficiencies and optimize resource allocation. By gaining insights into the supply chain, P&G can minimize waste and ensure that products are produced efficiently. One significant aspect of P&G's IoT implementation is its commitment to sustainable sourcing. The company tracks the environmental impact of its supply chain and works to ensure that materials are sourced responsibly. By leveraging IoT technology, P&G can assess supplier practices and promote responsible sourcing, aligning with its broader sustainability goals.

The value generated from P&G's IoT initiatives is substantial. By enhancing operational efficiency and reducing waste, P&G can lower production costs and improve product quality. Additionally, the focus on responsible sourcing strengthens P&G's brand reputation as a leader in sustainability. Moreover, P&G's commitment to sustainability is evident in its efforts to reduce plastic waste in its packaging. By using IoT to optimize packaging design and materials, the company can minimize its environmental impact and contribute to circular economy practices. In summary, P&G's application of IoT technology for enhanced traceability showcases how digital tools can drive value generation while promoting sustainability. By focusing on waste reduction and responsible sourcing, P&G is positioning itself as a leader in the consumer goods industry.

3.3. Integration of AI, Blockchain, and Robotics: Towards Intelligent Green Logistics

The convergence of Artificial Intelligence (AI), Blockchain, and Robotics represents a transformative shift in the logistics industry, especially in the pursuit of sustainability and carbon-neutral operations. While each of these technologies offers standalone benefits, their integration creates a robust ecosystem capable of delivering end-to-end visibility, operational efficiency, and ecological responsibility. AI enhances decision-making through predictive analytics and real-time optimization, blockchain ensures transparency and data integrity, and robotics automates repetitive, resource-intensive tasks. When orchestrated together, these technologies facilitate the evolution of logistics from a reactive, fragmented model to an intelligent, adaptive, and eco-conscious system.

3.4. Cross-Technology Synergies

Synergistic integration of AI, blockchain, and robotics leads to amplified outcomes that none of the technologies could achieve independently. For instance, AI algorithms can analyze logistics data generated by robotic systems, enabling real-time adjustments to delivery routes or warehouse operations for energy efficiency. Blockchain can act as a secure layer for capturing and verifying all interactions among smart devices, ensuring that AI-driven decisions are based on tamper-proof and trusted data. Meanwhile, robotics systems can execute AI-optimized tasks—such as loading and unloading goods, sorting packages, or autonomous deliveries—while continuously logging their activities on the blockchain for auditability and compliance. Such synergy not only reduces emissions and energy waste but also builds resilient supply chains that are self-correcting and intelligent.

3.5. Interoperability and Data Sharing Models

Achieving true integration requires seamless interoperability among diverse systems and platforms. This involves establishing common communication protocols, data formats, and standards that allow AI engines, blockchain ledgers, and robotic devices to exchange information in real time. Data sharing models must be secure, scalable, and governed by transparent rules to ensure trust among stakeholders. Distributed ledger technology (DLT) such as blockchain provides a decentralized and immutable infrastructure that facilitates trusted data exchange without requiring centralized intermediaries. AI models can be trained on shared datasets aggregated from multiple sources, thereby improving prediction accuracy and contextual intelligence.

3.6. Challenges and Ethical Considerations

While the integration of AI, blockchain, and robotics in sustainable logistics holds tremendous promise, it also presents a range of challenges and ethical considerations that must be addressed to ensure inclusive, responsible, and environmentally sound adoption. These challenges span technological, social, economic, and environmental domains, requiring stakeholders to proactively manage risks and devise strategic responses.

3.6.1. Data Privacy, Cybersecurity, and Digital Inequality

The use of smart technologies in logistics entails the collection and processing of vast amounts of data, including sensitive operational, consumer, and environmental information. This raises significant concerns about data privacy and the potential misuse of personal or proprietary data. Cybersecurity threats such as hacking, ransomware, and data breaches become more severe in interconnected logistics networks powered by AI and blockchain. Ensuring robust encryption, secure access controls, and compliance with legal frameworks like the General Data Protection Regulation (GDPR) and India's Digital Personal Data Protection Act becomes paramount. Moreover, digital inequality poses a critical ethical challenge. Small and medium-sized enterprises (SMEs) and developing economies may lack the infrastructure, technical skills, or financial capacity to adopt these technologies, resulting in an uneven playing field and the risk of technological exclusion.

3.6.2. Environmental Costs of Smart Technologies

Ironically, the very technologies meant to reduce carbon emissions also come with their own environmental footprint. AI model training and data centers require substantial energy, while blockchain systems—particularly those using proof-of-work mechanisms—can be energy-intensive and resource-heavy. Robotics manufacturing involves the use of rare-earth elements, plastics, and electronic waste, all of which contribute to environmental degradation if not managed responsibly.

3.6.3. Workforce Displacement and Reskilling Needs

Automation driven by robotics and AI threatens to displace a significant segment of the logistics workforce, particularly in roles involving manual labour, routine processes, and repetitive tasks. While these technologies improve efficiency, they can also lead to job redundancies if not accompanied by robust reskilling and upskilling initiatives. Ethical deployment demands a human-centred approach that prioritizes workforce transition, social protection, and inclusive growth. Governments, industry associations, and educational institutions must collaborate to develop training programs, certifications, and learning platforms that equip workers with future-ready skills in robotics maintenance, data analytics, AI systems management, and cybersecurity. Furthermore, new job roles emerging from the technology shift—such as drone operators, automation supervisors, and sustainability analysts—must be aligned with inclusive labour policies and equitable access.

3.7. Future Trends in Circular Supply Chains

The future of sustainable logistics is being shaped by a convergence of intelligent technologies, environmental priorities, and system-level innovations. As the logistics industry evolves toward carbon-neutrality, emerging trends point toward deeper integration, decentralization, and digitization of supply chain ecosystems. Research is increasingly focusing on tools and systems that not only boost efficiency but also minimize ecological impact. From digital simulations to energy-efficient algorithms, the next wave of transformation will be driven by intelligent, adaptive, and environmentally responsible systems.

3.7.1. Digital Twins and Predictive Analytics

Digital twins—virtual replicas of physical logistics systems—are poised to revolutionize how supply chains are designed, monitored, and optimized. These high-fidelity models enable real-time simulation of logistics operations, allowing companies to test different scenarios, forecast disruptions, and optimize resource allocation without incurring physical costs or emissions. When coupled with predictive analytics powered by AI, digital twins can anticipate maintenance needs, optimize delivery schedules, and manage inventory more efficiently, significantly reducing energy consumption and waste. Future research in this area will likely focus on multi-agent modelling, real-time sensor integration, and edge-AI analytics for hyper-local decision-making in complex logistics networks.

3.7.2. Green AI and Low-Power Robotics

With growing awareness of the carbon footprint associated with AI model training and robotics, the emergence of Green AI aims to develop more energy-efficient algorithms that perform complex tasks using minimal computational resources. Green AI focuses on balancing performance with sustainability, advocating for smaller models, optimized hardware utilization, and adaptive learning techniques. In parallel, the development of low-power robotics is becoming critical in achieving net-zero goals, particularly in last-mile deliveries and warehouse automation. These robots are designed with lightweight materials, energy-efficient motors, and intelligent task scheduling to reduce their environmental impact. Future research will likely explore the integration of renewable energy sources with robotic systems and the use of biodegradable or recyclable components in hardware manufacturing.

3.7.3. Decentralized Autonomous Logistics Systems (DALS)

One of the most transformative concepts emerging in logistics is the Decentralized Autonomous Logistics System (DALS) a blockchain-enabled, AI-driven framework where logistics operations are managed through autonomous agents without centralized control. In a DALS ecosystem, smart contracts can initiate and validate logistics transactions, AI agents can optimize routing and warehouse operations in real-time, and robotic systems can execute these tasks with minimal human intervention. This decentralization enhances transparency, resilience, and scalability while promoting sustainability by eliminating redundancies and inefficiencies. Research in this domain is expanding rapidly, focusing on agent-based modelling, consensus protocols suited for logistics, and the ethical governance of autonomous supply chains. The long-term vision is a self-regulating, energy-efficient logistics system that dynamically adapts to demand patterns, environmental changes, and market shifts in real time.

4. RESULTS AND DISCUSSION

4.1. Solutions And Recommendations

To enhance traceability in circular supply chains through digital tools, organizations should consider several key strategies. First, investing in blockchain technology can provide a secure and transparent method for tracking products throughout their lifecycle. By enabling stakeholders to access and verify information about product origins and handling, blockchain fosters trust and accountability within the supply chain. Second, organizations should leverage IoT devices to collect real-time data on product conditions, facilitating proactive decision-making and resource management. By integrating IoT solutions, companies can monitor factors such as temperature, humidity, and location, ensuring that products are handled appropriately throughout their lifecycle. Additionally, organizations should utilize big data analytics to gain insights into supply chain performance and identify areas for improvement. By analysing data from various sources, companies can optimize their resource management strategies, reduce waste, and enhance sustainability efforts. Lastly, fostering collaboration among supply chain partners is crucial for achieving effective traceability. By sharing data and insights, organizations can create a more interconnected and transparent supply chain, driving collective efforts towards sustainability.

4.2. Implications of the Study

This study holds significant implications for both academia and industry. For academia, the research contributes to the growing body of knowledge on circular supply chains and the role of digital technologies in enhancing traceability. It provides a framework for future research on the integration of digital tools in supply chain management, opening avenues for exploration in areas such as sustainability, innovation, and technology adoption.

For industry practitioners, the findings underscore the importance of leveraging digital solutions to enhance traceability and promote sustainable practices. As consumer demand for transparency and accountability continues to rise, organizations must adopt innovative approaches to supply chain management. By implementing the recommendations outlined in this study, companies can position themselves as leaders in sustainability and drive positive change within their industries.

4.3. Future Research Directions

Future research in the field of circular supply chains and digital traceability should focus on several key areas. First, longitudinal studies that assess the long-term impact of digital tools on supply chain performance and sustainability outcomes would provide valuable insights into the effectiveness of these technologies. Second, research exploring the barriers to technology adoption in circular supply chains could help identify strategies for overcoming challenges related to implementation and integration. Additionally, comparative studies examining the effectiveness of different digital solutions across various industries could yield insights into best practices and industry-specific challenges. Finally, exploring the role of emerging technologies, such as artificial intelligence and machine learning, in enhancing traceability within circular supply chains presents a promising avenue for future research. By investigating how these technologies can complement existing digital tools, researchers can further advance the understanding of supply chain innovation and sustainability.

5. CONCLUSION

In conclusion, the transition towards circular supply chains represents a critical shift in supply chain management, emphasizing sustainability and resource efficiency. Digital tools and technologies play a pivotal role in enhancing traceability within these systems, enabling organizations to monitor and manage resources effectively. By leveraging innovations such as blockchain, IoT, and big data analytics, companies can foster transparency, accountability, and collaboration among supply chain partners. This study provides valuable insights into the potential benefits and challenges of integrating digital solutions into circular supply chain management, offering recommendations for organizations looking to enhance traceability. As the demand for sustainable practices continues to grow, the role of digital tools in transforming circular supply chains will be essential in driving positive change across industries. Ultimately, the findings of this research contribute to the ongoing dialogue on sustainability and innovation, providing a foundation for future exploration in this vital area of study.

REFERENCES

Alla, L., Bentalha, B., & Bouhtati, N. (2022). Assessing Supply Chain Performance in the Covid 19 Context: A Prospective Model. *2022 14th International Colloquium of Logistics and Supply Chain Management (LOGISTIQUA)*, 1–6. https://doi.org/ DOI: 10.1109/LOGISTIQUA55056.2022.9938083

Andersen, A.-L., Brunoe, T. D., Bockholt, M. T., Napoleone, A., Hemdrup Kristensen, J., Colli, M., Vejrum Wæhrens, B., & Nielsen, K. (2023). Changeable closed-loop manufacturing systems: Challenges in product take-back and evaluation of reconfigurable solutions. *International Journal of Production Research*, *61*(3), 839–858. DOI: 10.1080/00207543.2021.2017504

Badr, B. (2022). Smart Technologies for a Sustainable Service Supply Chain: A Prospective Perspective. *Journal of Environmental Issues and Climate Change*, *1*(1), 77–89. DOI: 10.59110/jeicc.v1i1.81

Balaji, K. (2024), "Artificial Intelligence for Enhanced Anti-Money Laundering and Asset Recovery: A New Frontier in Financial Crime Prevention," *2024 Second International Conference on Intelligent Cyber Physical Systems and Internet of Things (ICoICI)*, Coimbatore, India, 2024, pp. 1010-1016, DOI: 10.1109/ICoICI62503.2024.10696100

Bensfia, C. (2024). Collective Intelligence of Port Territory Stakeholders: Case of Moroccan Ports. In Alla, L., Alj, B., & Bentalha, B. (Eds.), (pp. 335–370). Advances in Electronic Government, Digital Divide, and Regional Development. IGI Global., DOI: 10.4018/979-8-3693-6854-1.ch011

Bensfia, C., & Bouklata, A. (2023). Evaluation of the Electronic Port Right of Way Payment Information System. *Management Systems in Production Engineering*, *31*(3), 254–262. DOI: 10.2478/mspe-2023-0028

Bentalha, B., Hmioui, A., & Alla, L. (2019). The digitalization of the supply chain management of service companies: a prospective approach. *Proceedings of the 4th International Conference on Smart City Applications*, 1–8. https://doi.org/DOI: 10.1145/3368756.3369005

Bouklata, A., & Bensfia, C. (2020). Digitalization of port passage procedures: focus on the transit time of Goods. *2020 IEEE 13th International Colloquium of Logistics and Supply Chain Management (LOGISTIQUA)*, 1–6. https://doi.org/DOI: 10.1109/ LOGISTIQUA49782.2020.9353931

Ferdous, M. S., Chowdhury, M. J. M., & Hoque, M. A. (2021). A survey of consensus algorithms in public blockchain systems for crypto-currencies. *Journal of Network and Computer Applications*, *182*, 103035. DOI: 10.1016/j.jnca.2021.103035

Fernández Campos, P., Trucco, P., & Huaccho Huatuco, L. (2019). Managing structural and dynamic complexity in supply chains: Insights from four case studies. *Production Planning and Control*, *30*(8), 611–623. DOI: 10.1080/09537287.2018.1545952

Hmioui, A., Bentalha, B., & Alla, L. (2020). Service supply chain: A prospective analysis of sustainable management for global performance. *2020 IEEE 13th International Colloquium of Logistics and Supply Chain Management (LOGISTIQUA)*, 1–7. https://doi.org/DOI: 10.1109/LOGISTIQUA49782.2020.9353940

Idrissi, Z. K., Lachgar, M., & Hrimech, H. (2024). Blockchain, IoT and AI in logistics and transportation: A systematic review. *Transport Economics and Management*, *2*, 275–285. DOI: 10.1016/j.team.2024.09.002

Kuo, T.-T., Gabriel, R. A., & Ohno-Machado, L. (2019). Fair compute loads enabled by blockchain: Sharing models by alternating client and server roles. *Journal of the American Medical Informatics Association : JAMIA*, *26*(5), 392–403. DOI: 10.1093/jamia/ocy180 PMID: 30892656

Li, Z., Kang, J., Yu, R., Ye, D., Deng, Q., & Zhang, Y. (2017). Consortium Blockchain for Secure Energy Trading in Industrial Internet of Things. *IEEE Transactions on Industrial Informatics*, •••, 1–1. DOI: 10.1109/TII.2017.2786307

Mirabelli, G., & Solina, V. (2020). Blockchain and agricultural supply chains traceability: Research trends and future challenges. *Procedia Manufacturing*, *42*, 414–421. DOI: 10.1016/j.promfg.2020.02.054

Musleh Al-Sartawi, A. M. A., Hussainey, K., & Razzaque, A. (2022). The role of artificial intelligence in sustainable finance. *Journal of Sustainable Finance & Investment*, •••, 1–6. DOI: 10.1080/20430795.2022.2057405

Pal, A., & Kant, K. (2019). Using Blockchain for Provenance and Traceability in Internet of Things-Integrated Food Logistics. *Computer*, *52*(12), 94–98. DOI: 10.1109/MC.2019.2942111

Tijan, E., Aksentijević, S., Ivanić, K., & Jardas, M. (2019). Blockchain Technology Implementation in Logistics. *Sustainability (Basel)*, *11*(4), 1185. DOI: 10.3390/su11041185

Wirth, N. (2018). Hello marketing, what can artificial intelligence help you with? *International Journal of Market Research*, *60*(5), 435–438. DOI: 10.1177/1470785318776841

Yang, R., Wakefield, R., Lyu, S., Jayasuriya, S., Han, F., Yi, X., Yang, X., Amarasinghe, G., & Chen, S. (2020). Public and private blockchain in construction business process and information integration. *Automation in Construction, 118*, 103276. DOI: 10.1016/j.autcon.2020.103276

Zeba, S., Suman, P., & Tyagi, K. (2023). Types of blockchain. In *Distributed Computing to Blockchain* (pp. 55–68). Elsevier., DOI: 10.1016/B978-0-323-96146-2.00003-6

Zheng, Z., Xie, S., Dai, H. N., Chen, X., & Wang, H. (2018). Blockchain challenges and opportunities: A survey. *International Journal of Web and Grid Services, 14*(4), 352. DOI: 10.1504/IJWGS.2018.095647

Chapter 5
Towards Decarbonized Port Management:
Systematic Review

Chafik Bensfia
https://orcid.org/0000-0001-8991-7023
Mohammed First University, Morocco

Abdelkader Bensfia
Abdelmalek Essaâdi University, Morocco

ABSTRACT

The contribution of port operational activities to climate change has made it a research focus of increasing concern. To identify the news and characteristics of research on carbon emissions in the maritime transport sector, this study used the "Scopus" and "Web of Science" databases for the period 2020-2024. In addition, we identify the contributions of the most cited review articles. The results of the analysis highlight that port carbon emission reduction strategies involve operational optimization, equipment transformation and energy management and that the inherent gaps between ports present challenges in establishing a standard methodology at the global level. Future research avenues have been suggested, such as comprehensive assessment of emission sources, harmonization of estimation methods, equipment-integrated planning, port automation and electrification, and energy management. This study guides the progress of port carbon emission assessment and reduction, thus enabling port operators to manage production efficiently.

DOI: 10.4018/979-8-3373-3012-9.ch005

INTRODUCTION

The needs of international trade have led to a rapid and significant development of maritime transport in recent years. This development has prompted ports and terminals to improve their operations in order to become more efficient, more competitive and more sustainable.(Okşaş, 2023). Due to the increased capacity of ports and the volume of cargo they handle, sustainability has become even more crucial. In this regard, the link between port performance and sustainability needs to be explained more clearly. Nowadays, the environmental impact of a port is one of the most crucial aspects of its sustainability. Ports, which are one of the main cornerstones of the maritime sector, are responsible for a disproportionate share of the environmental damage caused by maritime trade and operations.(Bensfia, 2025; Bensfia & Aidi, 2025; Okşaş, 2023).

For port regions, the negative environmental effects of climate change provide significant obstacles. Three per cent of the world's greenhouse gas emissions from marine transportation come from ports.(Barkley et al., 2017; Misra et al., 2017; L. Wang & Li, 2023). According to the International Association of Ports and Harbors (IAPH), greenhouse gas emissions from ports are responsible for environmental pollution and global warming (IAPH, 2010). Ports mainly generate greenhouse gases such as carbon dioxide (CO2), nitrogen dioxide (N2O) and methane (CH4) during their daily operations.(Yang, 2017), One of the main contributors to the potential for global warming is CO2(Styhre et al., 2017; L. Wang & Li, 2023).According to Wright, As vital hubs in the global supply chain, ports are vital to a nation's external growth (Wright, 2013). This is why experts are interested in how carbon dioxide emissions affect the atmosphere. There's no denying the importance to mankind of CO2 emissions from port operations. Carbon emissions from ports have a negative impact on local air, water and soil quality, as well as on the quality of life of local residents (Eyring et al., 2010). One of these impacts is the propensity of CO2 emissions to increase temperature degrees (Wang et al., 2023).

CO2 emissions have adverse effects on human health and national economies, in addition to their direct impact on the environment (Wang et al., 2023). Extreme temperatures, pollution of the air, decreased agriculture, rising sea levels, and other significant global problems are some of these effects, according to Du et al. (Wang et al., 2023). All of these issues have the potential to immediately jeopardise human life if they are not addressed. Therefore, one of the most important aspects of encouraging sustainable socioeconomic growth at ports is lowering these environmental issues (IPCC, 2013, L. Wang & Li, 2023).

However, a thorough examination of carbon emissions in port regions is necessary for the literature now in publication (Wang et al., 2023). A more thorough investigation of carbon emissions from ports is necessary given the substantial influence of carbon dioxide on climate-related concerns (Wang et al., 2023). This paper analyses existing gaps, suggests future research topics, and thoroughly assesses the body of knowledge about carbon emissions in port regions. Research findings on calculating carbon emissions from boats and port places and views on port carbon emission reduction and the efficiency of these measures are its two primary research sections (Wang et al., 2023).

METHODOLOGY

Because it is a powerful research technique that is frequently utilised in the port industry, this study employed a systematic review of the literature to thoroughly gather pertinent research material from two well-known databases. An orderly and reproducible method for locating topic-specific knowledge and determining potential study paths is provided by the systematic literature review technique. (Stechemesser & Guenther, 2012; BENSFIA et al., 2024; L. Wang & Li, 2023). This study employed the Scopus and WOS databases to search the literature. Choosing the right keywords to find pertinent material was the first step. Using inclusion and exclusion criteria, the second stage examined and extracted data and search results. Using thorough data from the acquired literature, the final phase was analysing connected research issues. Snyder, 2019; B. Wang et al., 2023; Denyer & Tranfield, 2009).

The aim of this study is based on two types of analyses: bibliometric and systematic analysis.(Donthu et al., 2021). Bibliometric analysis uses visualization tools such as VOSviewer to aggregate massive bibliometric data to shed light on the current state of knowledge and emerging trends in a specific field of study. Systematic literature reviews help to summarize and synthesize the findings of all the literature developed on a specific topic or field of study (BENSFIA et al., 2024).

Search Tactics

This cutting-edge work was completed in compliance with the "PRISMA" (Preferred Reporting Items for Systematic review and Meta-Analysis) criteria. The following search equation was used to explore the WOS and databases Scopus: "carbon emissions and sea transport". We should assume that the search was stopped on September 08, 2024.

Inclusion Criteria

The following criteria were applied to target the documents that require analysis: Year: 2020 to 2024; Document type: Journal articles; Language: English.

We chose this particular time frame because we wanted to review the literature this year which is marked by an abundance of scientific production worldwide on the topics of sustainable ports.

In addition, we performed a title analysis and abstract to adopt only those publications that meet the objective already set. Only writings that simultaneously address both themes were considered. Using the bibliographic management tool Zotero, a total of 236 publications were added and cleaned.

Figure 1 summarizes the search steps with the number of publications included according to the PRISMA method.

Figure 1. PRISMA diagram of systematic literature review.

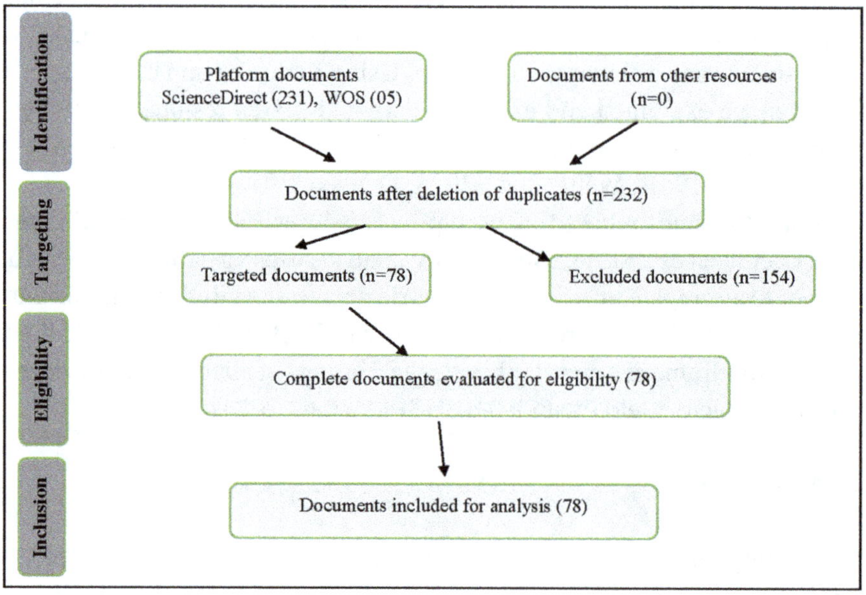

Analysis

Starting from an analysis which aims to give a general assessment of the articles included in this state of the art as well as their content via a method which describes the distribution of articles by year of publication and by journal using the VOSViewer software.

To guarantee relevance rather than quantity, the original sample was carefully screened and chosen. The particular literature selection procedure used for this investigation is depicted in Figure 1. The number of publications was initially constrained by the selection criteria shown in Figure 1. Reading the 236 publications' titles, abstracts, and keywords was the first step in the selection process. Articles about ports and carbon emissions were the main emphasis, with research from other disciplines, including atmospheric chemistry, environmental sciences, and marine engineering, being excluded (L. Wang & Li, 2023). 158 articles that didn't fit the inclusion criteria were therefore eliminated. The entire texts of the remaining 78 articles were then reviewed, paying particular attention to the conclusion and structure parts. The following criteria were considered for this stage: (1) Research on ship emissions at berths and CO_2 emissions in port areas was included. (2) Research on port resource management with an emphasis on emission reduction, such as equipment conversion and schedule optimization, was considered. (3) Studies directly addressing energy use and environmental pollution and providing guidance for further studies were also considered. (L. Wang & Li, 2023).

RESULTS AND DISCUSSIONS

The 78 articles that comprise this review have been examined and synthesized. The results are presented, discussed and their limitations are highlighted in this part. It also suggests potential future research topics and highlights theoretical and managerial ramifications.

Descriptive Analysis

The purpose of this subsection is to present a preliminary overview of the distribution of publications from two points of view, in particular by year and by type of document as well as the language of writing.

From the distribution of journal articles over time and according to Figure No. 2, it can be deduced that since 2020, there were not many publications included in the two scientific databases that address the topic of carbon emissions in ports.

Figure 2. Number of annual publications

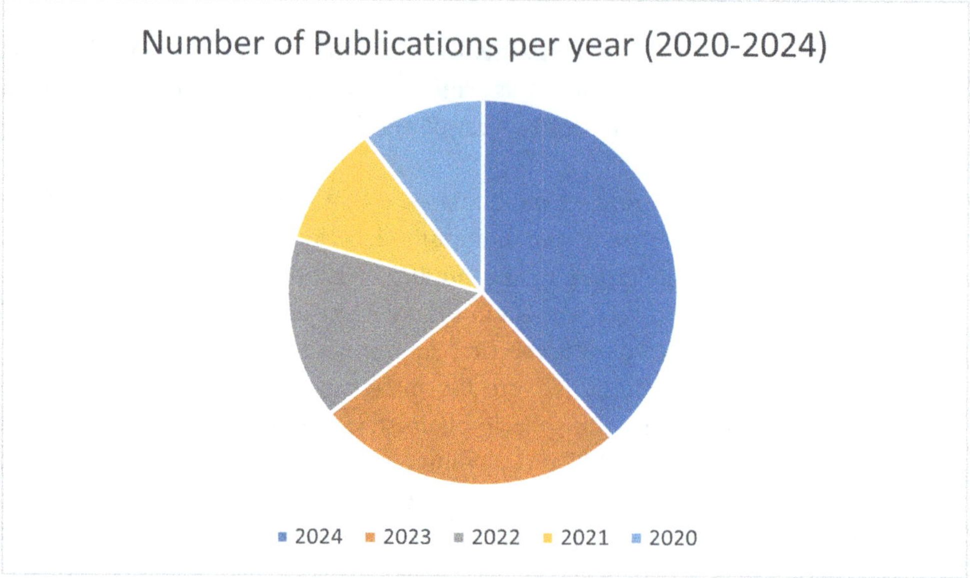

According to the analysis of the results obtained regarding the type of publication in the two databases, and according to Figure 3, we see that research articles represent the lion's share in our selected database.

Figure 3. Types of published articles

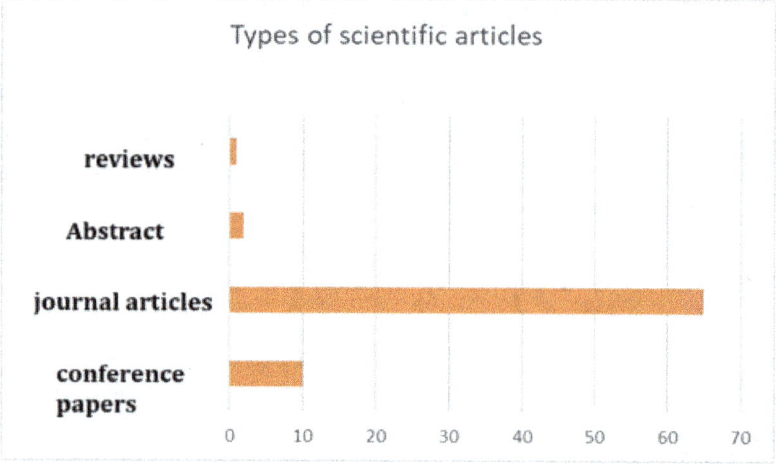

Cluster Analysis

We identified and categorized the most important research topics by performing a cluster analysis using VOSviewer software after descriptive analyses. VOSviewer is a decision-support tool.

Researchers can analyze study topics, explain co-occurrence networks and identify research clusters using this tool to represent bibliometric maps (Garrabou et al., 2022; Piao, 2024).VOSviewer software is a popular application because it is simple to use and offers a variety of capabilities (Fadiga et al., 2024).

Based on the abstracts and titles of the 124 articles, the program VOSviewer was used to do a co-word evaluation and generate a conceptual framework. for the rest of the study. The terms considered had to appear at least four times.

They were selected on the basis of their relevance to the research. Standardization of related ideas was necessary to produce a map of port decarbonization research groups.

The final map shown in Figure 4 comprises 04 objects and 2 main groupings.

Figure 4. The final map of the research groups.

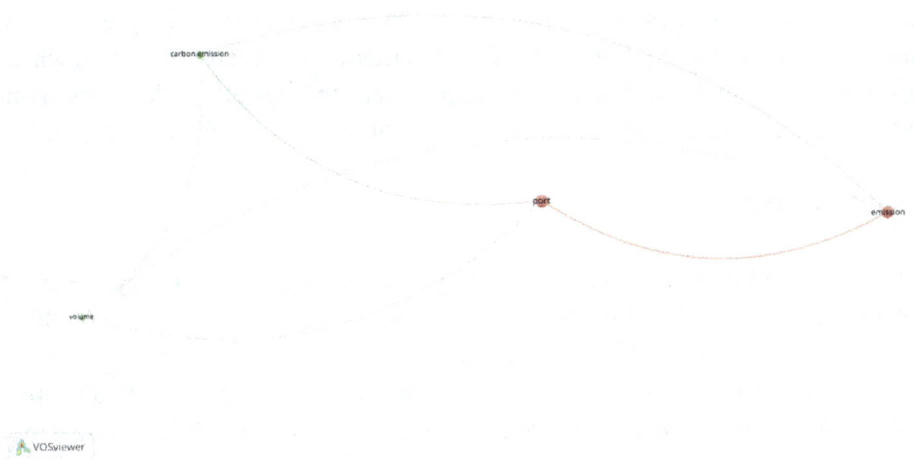

Estimating Carbon Emissions In Ports

In recent decades, a large number of ports worldwide have begun measuring and disclosing their carbon emissions. To gain a better knowledge of the environmental effect of port and marine operations, the initial step is to quantify emissions from port activities. (Sorte and others, 2021; L. Wang & Li, 2023). Port operators may visually monitor energy usage and carbon emissions in daily port operations, identify emission sources, and analyse carbon emissions from ongoing port operations by creating an emissions inventory. (L. Wang & Li, 2023). The estimation of carbon emissions serves as the foundation for additional studies aimed at lowering port emissions and enhancing port environmental performance. According to B. Wang et al. (2023), it also motivates international and national organisations to take steps towards a more environmentally friendly future for ports worldwide (L. Wang & Li, 2023). The researchers divided emissions from port operations and marine transportation into two categories to make estimating greenhouse gas emissions easier (Villalba & Gemechu, 2011). In addition to allowing the incorporation of emissions from port operations in urban emissions inventories, this separation has two benefits: it streamlines and increases the openness of the estimating process (L. Wang & Li, 2023). Estimating carbon emissions from port operations is the goal of this project. Therefore, emissions from docking ships and emissions from activities in the port area make up the two primary components of carbon emissions from ports (L. Wang & Li, 2023). The first step is to carefully choose pertinent and representative literature. A thorough comparative study of this literature is then conducted, which makes it easier to condense and synthesise research findings about the assessment of CO_2 in the environment of port areas and berthing ships (L. Wang & Li, 2023).

Docking of Ships

Over 80% of global trade is carried out by sea, making it the primary means of international trade (Deniz et al., 2010).International shipping produced 79.6 billion tonnes of CO_2 emissions in 2012, which accounted for 2.2% of all emissions in that year, according to research published by the International Maritime Organisation (IMO) (B. Wang et al., 2023). By 2050, shipping-related carbon dioxide emissions are predicted to reach 15% if pertinent maritime institutions do not step in (IMO, 2014). Ports are important locations where shipping routes meet, serving as centers for marine transportation. As a consequence, port communities deal with both ships berthing pollution and infrastructure (B. Wang et al., 2023).

Air quality in port cities is affected by over 70% of CO_2 originating from the maritime and road fleet, in addition to greenhouse gas emissions from ship berthing activities within port areas. According to research (Nunes et al., 2017). The inner harbour has been shown to be the area most affected by ship emissions, which have an even more serious environmental impact on port cities than air pollution produced on the open sea (Toz et al., 2021).To improve air quality management, precise calculations for monitoring and quantifying ships are needed, enabling an accurate assessment of the root causes that release more gas. It is therefore essential to draw up a complete inventory of carbon emissions from ships (B. Wang et al., 2023).

Port Areas

Determining the magnitude of emissions associated with port activities and energy use is essential to assessing carbon emissions from port area operations. Three sources of carbon emissions have been identified by the World Ports Climate Initiative (WPCI) to determine the extent to which ports are permitted to emit greenhouse gases (WPCI, 2010). The categorization of the range of CO_2 emissions coming from port platforms sites is illustrated in Figure 4 (B. Wang et al., 2023).

Figure 5. **Ports' carbon emission sources**

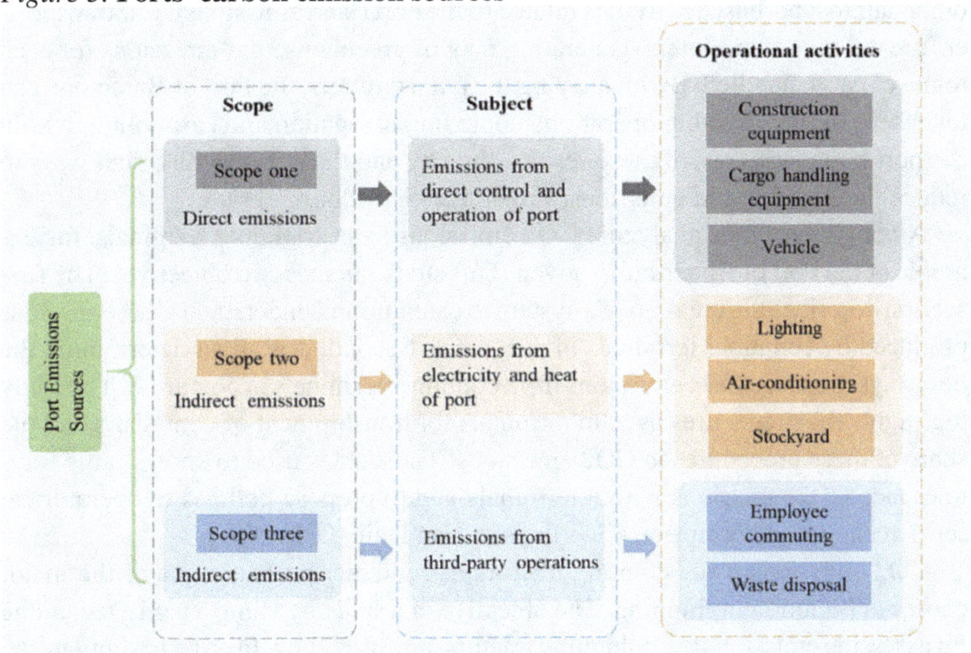

(B. Wang et al., 2023)

Direct emission sources, or CO_2 emissions produced by the management and how the port authority operates, are represented in Figure 4 by Scope one. Scope two covers indirect emissions. Direct emission sources are carbon emissions from the port's energy use. The main features of scope 2 are the tank farm, air conditioning and lighting. The third scope is the additional indirect emissions produced by the port's activities which concerns the additional indirect emissions produced by parties not affiliated with the port authorities. The main sources of scope 3 are employee travel and the removal of garbage.(Alamoush et al., 2020; Azarkamand et al., 2020; B. Wang et al., 2023).

Recently, cross-border activities that are part of a city's urban metabolism and economy have been added to cities' greenhouse gas inventories.(B. Wang et al., 2023). In order to integrate seaport emissions into the city's inventory or to allow the port to monitor its technological and policy advances aimed at mitigating climate change, a study on the Port of Barcelona focuses on calculating seaport emissions. As useful tools for policy formulation and long-term monitoring of emissions reduction efforts, the study proposes the following indicators: GHG emissions per tonne of cargo handled or per passenger, and emissions per value of cargo handled. The study determined a total of 331,390 tonnes of GHG emissions (CO2 equivalents) for the year 2008 by adapting current techniques to the Port of Barcelona. Half of these emissions are attributable to ship movements (maritime emissions) and the other half to land-based activities related to the port (land emissions). Car transporters are the biggest polluters, generating 6 kg of greenhouse gas emissions for each tonne of cargo handled. By identifying the main emitters, the Port of Barcelona can take steps to improve ship operations, including navigation and carpooling, within the port boundaries. With these results, the city and the port can also find ways to reduce their land-based emissions(Villalba & Gemechu, 2011).

Among the studies analyzing CO2 emissions from container terminals, the example of the Port of Rotterdam is given. This study pursues two objectives. The first section describes the creation of a system to examine and understand CO2 emissions produced by container terminals in port areas. Second, it is about determining the best ways to reduce CO2 emissions from container terminals in port areas. The study highlights the procedures used in terminals for transhipment of containers and the share of these procedures in CO2 emissions. This data is used to know viable ways to reduce CO2 emissions within terminals and to propose policies to operators in service and for governments.(Geerlings & Van Duin, 2011).

CO2 emissions from shunting tractors in container ports are one of the major causes of pollution in shipping. The objective of a study by Hung Yu on a technique for assessing emissions from shunting tractors during loading. In order to simulate the behavior of tractor arrivals at each block when loading begins, an export container allocation model is first constructed. Then, the yard congestion is modeled using queuing theory. Then, the arrival projections are used to assess tractor emissions from the yard. Tractor emissions are divided into two categories: one for vehicles operating normally and the other for vehicles stuck in traffic or in a stop-and-go situation. Simulated terminals with varying calls and storage usage are used to demonstrate the effectiveness of the suggested approach and identify viable traffic management strategies to reduce transportation-related emissions. Emissions from shunting tractors have been shown to be directly related to the location of export containers, which should be a major operational concern for a busy terminal.(Yu et al., 2017).

Strategies for Reducing Carbon Emissions From Ports

The idea that civilizations must combat climate change is now widely recognized. Governments and businesses are increasingly being asked to propose efforts to minimize CO2 emissions, as reducing CO2 emissions, one of the main causes of global warming, has become a priority. Given that the transport sector's share of the CO2 footprint is increasing while that of other sectors is decreasing, this issue is particularly relevant for the transport industry. The main current objective is to determine the best ways to reduce CO2 emissions from container terminals in port areas. There is no doubt that changing the layout of terminals, as the Rotterdam Shortsea Terminal has done, is the most effective way to reduce CO2 emissions. This reduces CO2 emissions from current terminals by around 70%.(Geerlings & Van Duin, 2011). The other approach is to blend 30% biofuels with the diesel currently used. This reduces CO2 emissions by 13-26% per terminal and reduces emissions from the container industry as a whole by 21%. Concrete suggestions for reducing CO2 emissions in container ports are proposed in light of these results. In several ports around the world, especially in China, rubber-tired container gantry cranes (RTGs) have already been largely replaced by electric rubber-tired container gantry cranes (ERTGs). By switching from diesel to electricity (energy replacement), ERTGs can reduce carbon dioxide emissions. The conversion of an RTG to an ERTG takes about six to twelve months, and due to mechanical reformation, construction of the power supply system, etc., the modified yard cranes cannot operate during this period. Consequently, the network of yard cranes impacts the service level and creates congestion in the container traffic when energy is replenished (L. Wang & Li, 2023). This essay's primary focus is on the problem of distributing scarce resources for yard cranes in order to lower carbon emissions. The main questions are how to manage the high levels of uncertainty in the transport network around container terminals and how to manage the problem of energy replacement at the network level.

Compared to transport, the energy consumption and CO2 emissions of combined rail-road transport are 81.34% and 74.24% lower respectively.(Tao & Wu, 2021). Consequently, reducing vehicle energy consumption and switching from road to mixed rail transport for container traffic remain the most effective ways of saving energy and reducing CO_2 emissions. The realistic estimation and structural analysis of energy consumption and CO2 emissions under different scenarios was made possible by the updated ASIF approach.

The International Maritime Organization (IMO, 2018) suggested a plan to lower greenhouse gas emissions as well as a number of energy-saving initiatives. Additionally, the International Maritime Organization suggested as soon as feasible, measures to cut emissions must be put into place (Balsalobre-Lorente et al., 2023). It is essential to put policies into place at all ports, irrespective of their size or management style (Alamoush et al., 2022).

But there is still a lot of variation in the capacity for CO2 reduction of current methods (Bouman et al., 2017), and several elements and characteristics (cost, complexity, adaptability to analyze the potential of all prospective measures, factors such as ability, sustainability, and dependability must be taken into account (Loza & Veloso-Gomes, 2023).

- Alternative fuels: LNG

In order to achieve its decarbonization goals, the maritime sector is relying on innovative solutions. The adoption of alternative fuels is an effective strategy for the sector. These fuels are used to generate electricity and provide propulsion, while maintaining the speed of trade.(Decarbonization in maritime transport: towards a more sustainable industry – SupplyChainInfo, nd). Various ecological fuels are being developed with the aim of becoming the new references for this type of transport:

There are many light gases. There are LNG (liquefied natural gas) and bio-LNG, which are synthetic and renewable natural gases, hydrogen. There are also bio-based synthetic biofuels in liquid or gaseous form such as biodiesel and bio H2. As an alternative, alcohols such as methanol, ethanol or ammonia are proposed. Among other solutions for reducing carbon emissions from ports, energy solutions, including hybrid energy, electrification technology and alternative fuels, are very promising for reducing carbon emissions in the port industry.(Hoang et al., 2022; L. Wang & Li, 2023). Fuels that are clean or renewable can greatly lower carbon emissions at ports. Switching from fossil fuels to biofuels is an important step in energy conservation techniques. Major alternative energy methods studied in the literature include LNG. Ships increasingly use LNG as fuel to reduce carbon emissions (Styhre et al., 2017).

Another workable strategy to lower port emissions is to switch to biofuels from fossil fuels. In addition to biodiesel, It is possible to mix biomass fuels with different fuels that are fossilised (L. Wang & Li, 2023). When combined, they reduce carbon emissions (Misra et al., 2017).

The impact of emission reductions varies depending on the blending ratio used. For example, carbon emissions can be reduced by 30% by blending 30% biofuels with diesel (Geerlings and Van Duin, 2011).

In addition, the combination of fuel and electricity is promising and that conventional diesel-powered ships will be replaced by diesel-electric hybrid propulsion ships. Vessels using hybrid fuel can consume 12% less energy.(Wang et al., 2023;Geerlings and Van Duin, 2011).

- Joint optimization of ship routes and speeds

On longer trips, vessel route optimisation can cut emissions by up to 30% during strong storms (L. Wang & Li, 2023).However, all situations examined demonstrate a reduction in emissions when vessel routing optimization is applied. The predicted increase in extreme weather events, in terms of frequency, intensity and duration due to climate change, indicates a progressive benefit of applying vessel routing optimization in all types of journeys, regardless of distance.(Boren et al., 2022).

The sailing speed of ships has a direct impact on fuel consumption. In other words, increasing speed has led to increased costs and fuel consumption. On the other hand, reducing speed has had an impact on the ship's schedule. Figure 1 shows how fuel consumption increases steadily with speed and how this upward trend continues to increase (Lashgari et al., 2021).

Figure 6. Fuel consumption at varying speeds.

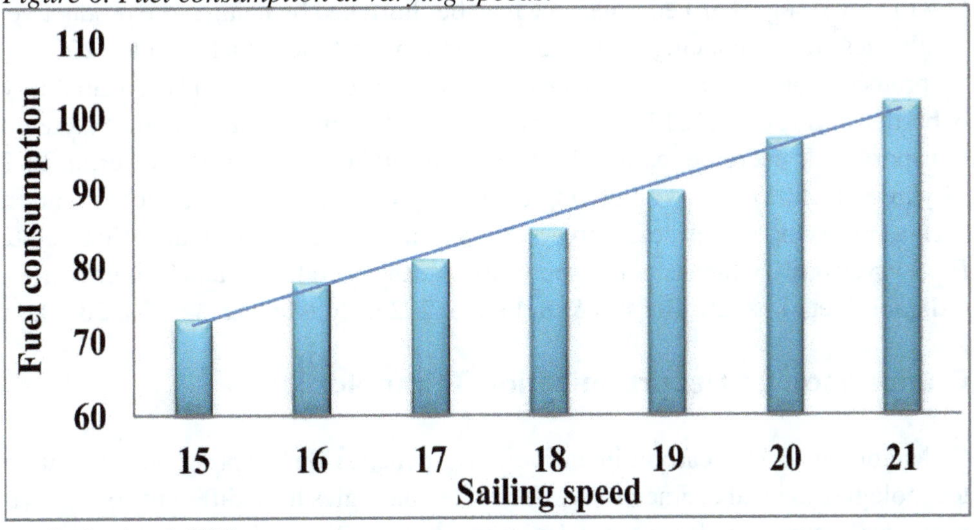

(Lashgari et al., 2021)

- Electrification of the ship fleet

The analysis of the fleet electrification potential revealed that it would reduce total CO2 emissions from 495.42 tonnes to 277.32 tonnes per year. The port of Koper, located in the Northern Adriatic region and with separation zones and many ports hosting the largest container ships, serves as a case study for the viability of adding electric tugs and a macroscopic overview of the traffic(Luin et al., 2024).

- Measures of the equipment

Equipment-related measures might include replacing outdated equipment or adapting to use cleaner and more energy-efficient

Additionally, proper equipment upkeep might lower excess emissions and save energy (Alamoush et al., 2020). To increase productivity and safeguard the environment, digitalization aids in the identification, tracking, and aggregation of data. Fuel consumption may be decreased by managing logistical flows with the use of remote sensing, cloud computing (Xia et al., 2021), and the Internet of Things. Automation of container terminals through the use of drones, autonomous guided boats, automated machinery, and automated guided vehicles (AGV)(Fadiga et al., 2024).

- Ship turnaround time

The overall emissions would be directly impacted by shorter turnaround times (TAT) for the ships berthed. The TAT may be shortened by being more productive, waiting less time, reducing traffic, and being more efficient(Fadiga et al., 2024).

protocols for clearance, the effectiveness of crane equipment, and the availability of berths (Styhre et al., 2017; Winnes et al., 2015). Shipping businesses may increase transport work, decrease sea speed, and expand port berth capacity by lowering TAT (Styhre et al., 2017). Ports could implement improved alternative service practices, such as reserving berths before to arrival, as ships typically berth on a first-come, first-served basis, which has been shown to raise the overall TAT and CO2 emissions (Alamoush et al., 2020; Bensfia & Bouklata, 2023; Bouklata & Bensfia, 2020).

Comparison Of Decarbonization Technologies

Migration to less carbon-intensive ports depends on adapting decarbonation technologies to local characteristics. These technologies have different costs, environmental returns and large-scale deployment possibilities. However, the choice of these solutions depends as much on available resources and the strategic priorities of port authorities as on technical and regulatory constraints. The following table compares the main technological options according to three criteria: implementation cost, performance in terms of CO_2 emissions reduction and scalability capabilities.

Table 1. Comparative table of decarbonization technologies (cost, efficacy, scalability).

Technolgy	cost	CO_2 reduction efficiency	scalability	Remarks
Alternatives fuels (LNG, H2, biofuels)	High	efficiency not exceeding 40%.	Moderate	Regulations and fuel availability determine potential.
Electrification of port equipement	Medium to high	efficiency not exceeding 60%.	High	Clean electrical sources are necessary for a sustainable solution.
Operational optimization (logistics)	Low	efficiency not exceeding 30%.	Very High	Low cost, quick deployment, and significant return on investment.
Digitalization and automation	Meduim	efficiency not exceeding 30%.	High	increases overall efficiency but necessitates digital infrastructure and expertise.

Source: (Villalba & Gemechu, 2011; L. Wang & Li, 2023) Table adapted and summarized by us

Obstacles to Decarbonization

According to Sorrell et al. the hurdles prevent investments in economically and ecologically effective technology and strategies. Although rigorous laws exist, their proper application is hampered by obstacles (L. Wang & Li, 2023). The majority of barriers, particularly to energy efficiency, have been studied from the point of view of classical economic theory, such as agency theory and contract theory, despite some studies from the point of view of energy efficiency in shipping, including decarbonization (L. Wang & Li, 2023).

Sorrell et al. developed a framework to analyze the following types of barriers: economic market failures (principal agent problem, split incentives, moral hazard and imperfect/asymmetric information); behavioral barriers (form of information, rationality limits, trust and transparency); and non-market barriers (market inhomogeneity, hidden costs) (L. Wang & Li, 2023). Consequently, this study examined barriers to port decarbonization using these categories in addition to a few others identified in the study(Bensfia & Bouklata, 2022; L. Wang & Li, 2023).

Theoretical and Managerial Implications

Contribution to the literature on port emissions: These studies provide good input for research, since they provide well-founded classifications of emission sources (berthing, port areas, scopes 1-2-3) and draw attention to persistent methodological shortcomings.

Integrated systemic approach: They also emphasize the port's role in an integrated vision of emissions in urban metabolism (Villalba & Gemechu), echoing city inventories and confirming the postulate that ports are not isolated.

Energy transition theory: both papers reflect on the conditions required to achieve decarbonized ports, based on the paradigms of sustainability, technology and environmental governance.

For port authorities: this research provides an operational framework for drawing up accurate emissions inventories, identifying the sources that should be prioritized for processing (such as ships and diesel engines in terminals), and building action plans that are as close as possible to their context.

For public decision-makers: they provide ideal ground for integrating ports into local and national climate policies, by emphasizing their role in determining GHG emissions in urban environments.

For companies operating in these ports: The results that sum up their interest are unequivocal, and involve both technological modernization (electrification of port tools, clean energies, virtuous equipment) and the need to adopt eco-efficient practices. The demand for greater transparency in environmental management is also a win-win situation.

For international cooperation: these results reinforce the case for more common standards (for ports) and shared mechanisms (WPCI or IPCC guidelines, which should aim to standardize practices to enable better inter-port benchmarking).

CONCLUSION

Key participants in the battle against global warming are ports, which are the real engines of the marine economy. In fact, they concentrate high-intensity activities that produce a lot of greenhouse gas emissions. As a result, decarbonising these infrastructures has gained international attention. Ports are critical nodes in the global transport system, whose activities emit large volumes of greenhouse gases, particularly CO_2. As the methods here are heterogeneous, all point in the direction of greater effort in taking into account all emissions from the ships that dock there, but also those from land port infrastructures. Too often underestimated, they represent a major strategic challenge for the environmental sustainability of urban and port areas.

With a view to rationalizing less virtuous practices, a number of levers can be identified: electrification of equipment, improved energy efficiency, automation, use of renewable energies, reconfiguration of logistics processes. But their implementation requires both heavy investment and multi-stakeholder coordination.

Methodologically, purely bottom-up approaches are favored for their accuracy, even if their implementation is hampered by the availability of real data. Standardization efforts, on the other hand, are severely limited by the known heterogeneity of ports (size, type of activity, climate, history of equipment, energy policy), making it difficult to develop universal models.

Quantitative assessments form the basis of the majority of studies on reducing carbon impact in ports. However, it is essential to understand that the port enclosure is characterized by a very remarkable complexity with the involvement of several stakeholders, including port authorities, shipping companies, freight forwarders with their fleets as well as tenants, who have to work together to manage complex port operations (Wang and Peng, 2023).

Therefore, combining carbon emissions assessment with a wider variety of environmental indicators and metrics, as well as ecological costs, is an alternative to consider in future research. Furthermore, In this context, it is essential that future studies consider energy costs and carbon emissions as key variables, which will facilitate comparative assessments between traditional and electric crane equipment, automation and electrification will influence Port operations' future, with implications for economic profitability and sustainability of the environment.Thorough economic and environmental analyses are essential to make informed decisions (L. Wang & Li, 2023).

The fact that several publications relevant to this work were not retained because of the keywords used could be one of the effects of these limitations. It may also result from the use of incorrect keywords or different names for the same idea. The subjectivity of editors and reviewers naturally affected the selection, inclusion, classification, and analysis of articles.

Key opportunities for future research include the need for more research in ports in Europe, Africa and South America prioritized by port authorities.

In conclusion, the summaries of existing literature presented convincing arguments for the adoption of new technologies to lower seaports' greenhouse gas emissions.

REFERENCES

Alamoush, A. S., Ballini, F., & Ölçer, A. I. (2020). Ports' technical and operational measures to reduce greenhouse gas emission and improve energy efficiency : A review. *Marine Pollution Bulletin, 160*, 111508. DOI: 10.1016/j.marpolbul.2020.111508 PMID: 32798921

Alamoush, A. S., Ölçer, A. I., & Ballini, F. (2022). Port greenhouse gas emission reduction : Port and public authorities' implementation schemes. *Research in Transportation Business & Management, 43*, 100708. DOI: 10.1016/j.rtbm.2021.100708

Azarkamand, S., Ferré, G., & Darbra, R. M. (2020). Calculating the Carbon Footprint in ports by using a standardized tool. *The Science of the Total Environment, 734*, 139407. DOI: 10.1016/j.scitotenv.2020.139407 PMID: 32460080

Balsalobre-Lorente, D., dos Santos Parente, C. C., Leitão, N. C., & Cantos-Cantos, J. M. (2023). The influence of economic complexity processes and renewable energy on CO2 emissions of BRICS. What about industry 4.0? *Resources Policy, 82*, 103547. DOI: 10.1016/j.resourpol.2023.103547

Barkley, M. P., González Abad, G., Kurosu, T. P., Spurr, R., Torbatian, S., & Lerot, C. (2017). OMI air-quality monitoring over the Middle East. *Atmospheric Chemistry and Physics, 17*(7), 4687–4709. DOI: 10.5194/acp-17-4687-2017

BENSFIA, C., Bouklata, A., & NEJJARI, R. (2024). Explorer la digitalisation pour l'économie circulaire : Une revue systématique. *Revue de Gestion et d'Économie, 12*(2), 305–321.

Bensfia, C. (2025). Collective Intelligence of Port Territory Stakeholders : Case of Moroccan Ports. In *Utilizing Technology to Manage Territories* (p. 335-370). IGI Global.

Bensfia, C., & Aidi, S. (2025). Port Community Systems and Organizational Performance : A Socio-technical Approach to Evaluation. *Industrial and Logistics Systems Design and Efficient Operation*, 42-52. DOI: 10.1007/978-3-031-87309-6_4

Bensfia, C., & Bouklata, A. (2022). *Validity of the TAM Model in the context of electronic payment ofport dues in Morocco*. 1-5.

Bensfia, C., & Bouklata, A. (2023). Evaluation of the electronic port right of way payment information system. *Management Systems in Production Engineering, 31*(3), 254–262. DOI: 10.2478/mspe-2023-0028

Boren, C., Castells-Sanabra, M., & Grifoll, M. (2022). Ship emissions reduction using weather ship routing optimisation. *Proceedings of the Institution of Mechanical Engineers. Proceedings Part M, Journal of Engineering for the Maritime Environment, 236*(4), 856–867. DOI: 10.1177/14750902221082901

Bouklata, A., & Bensfia, C. (2020). *Digitalization of port passage procedures : Focus on the transit time of Goods*. 1-6.

Bouman, E. A., Lindstad, E., Rialland, A. I., & Strømman, A. H. (2017). State-of-the-art technologies, measures, and potential for reducing GHG emissions from shipping–A review. *Transportation Research Part D, Transport and Environment, 52*, 408–421. DOI: 10.1016/j.trd.2017.03.022

Deniz, C., Kilic, A., & Cıvkaroglu, G. (2010). Estimation of shipping emissions in Candarli Gulf, Turkey. *Environmental Monitoring and Assessment, 171*(1-4), 219–228. DOI: 10.1007/s10661-009-1273-2 PMID: 20058072

Donthu, N., Kumar, S., Mukherjee, D., Pandey, N., & Lim, W. M. (2021). How to conduct a bibliometric analysis : An overview and guidelines. *Journal of Business Research, 133*, 285–296. DOI: 10.1016/j.jbusres.2021.04.070

Fadiga, A., Ferreira, L. M. D., & Bigotte, J. F. (2024). Decarbonising maritime ports : A systematic review of the literature and insights for new research opportunities. *Journal of Cleaner Production, 452*, 142209. DOI: 10.1016/j.jclepro.2024.142209

Garrabou, J., Gómez-Gras, D., Medrano, A., Cerrano, C., Ponti, M., Schlegel, R., Bensoussan, N., Turicchia, E., Sini, M., Gerovasileiou, V., Teixido, N., Mirasole, A., Tamburello, L., Cebrian, E., Rilov, G., Ledoux, J.-B., Souissi, J. B., Khamassi, F., Ghanem, R., & Harmelin, J.-G. (2022). Marine heatwaves drive recurrent mass mortalities in the Mediterranean Sea. *Global Change Biology, 28*(19), 5708–5725. DOI: 10.1111/gcb.16301 PMID: 35848527

Geerlings, H., & Van Duin, R. (2011). A new method for assessing CO2-emissions from container terminals : A promising approach applied in Rotterdam. *Journal of Cleaner Production, 19*(6-7), 657–666. DOI: 10.1016/j.jclepro.2010.10.012

Hoang, A. T., Foley, A. M., Nižetić, S., Huang, Z., Ong, H. C., Ölçer, A. I., Pham, V. V., & Nguyen, X. P. (2022). Energy-related approach for reduction of CO2 emissions : A critical strategy on the port-to-ship pathway. *Journal of Cleaner Production, 355*, 131772. DOI: 10.1016/j.jclepro.2022.131772

Lashgari, M., Akbari, A. A., & Nasersarraf, S. (2021). A new model for simultaneously optimizing ship route, sailing speed, and fuel consumption in a shipping problem under different price scenarios. *Applied Ocean Research, 113*, 102725. DOI: 10.1016/j.apor.2021.102725

Loza, P., & Veloso-Gomes, F. (2023). Literature review on incorporating climate change adaptation measures in the design of new ports and other maritime projects. *Sustainability (Basel), 15*(5), 4569. DOI: 10.3390/su15054569

Luin, B., Al-Mansour, F., & Perkovič, M. (2024). Optimization of shipping routes with AIS data. *Thermal Science and Engineering Progress, 56*, 103042. DOI: 10.1016/j.tsep.2024.103042

Misra, A., Venkataramani, G., Gowrishankar, S., Ayyasam, E., & Ramalingam, V. (2017). Renewable energy based smart microgrids—A pathway to green port development. *Strategic Planning for Energy and the Environment, 37*(2), 17–32. DOI: 10.1080/10485236.2017.11907880

Nunes, R., Alvim-Ferraz, M., Martins, F., & Sousa, S. (2017). The activity-based methodology to assess ship emissions-A review. *Environmental Pollution, 231*, 87–103. DOI: 10.1016/j.envpol.2017.07.099 PMID: 28793241

Okşaş, O. (2023). Carbon emission strategies for container handling equipment using the activity-based method : A case study of Ambarlı container port in Turkiye. *Marine Policy, 149*, 105480. DOI: 10.1016/j.marpol.2023.105480

Piao, S. (2024). Barriers toward circular economy transition : Exploring different stakeholders' perspectives. *Corporate Social Responsibility and Environmental Management, 31*(1), 153–168. DOI: 10.1002/csr.2558

Styhre, L., Winnes, H., Black, J., Lee, J., & Le-Griffin, H. (2017). Greenhouse gas emissions from ships in ports–Case studies in four continents. *Transportation Research Part D, Transport and Environment, 54*, 212–224. DOI: 10.1016/j.trd.2017.04.033

Tao, X., & Wu, Q. (2021). Energy consumption and CO2 emissions in hinterland container transport. *Journal of Cleaner Production, 279*, 123394. DOI: 10.1016/j.jclepro.2020.123394

Toz, A. C., Buber, M., Koseoglu, B., & Sakar, C. (2021). An estimation of shipping emissions to analysing air pollution density in the Izmir Bay. *Air Quality, Atmosphere & Health, 14*(1), 69–81. DOI: 10.1007/s11869-020-00914-7

Villalba, G., & Gemechu, E. D. (2011). Estimating GHG emissions of marine ports—The case of Barcelona. *Energy Policy, 39*(3), 1363–1368. DOI: 10.1016/j.enpol.2010.12.008

Wang, B., Liu, Q., Wang, L., Chen, Y., & Wang, J. (2023). A review of the port carbon emission sources and related emission reduction technical measures. *Environmental Pollution*, *320*, 121000. DOI: 10.1016/j.envpol.2023.121000 PMID: 36621720

Wang, L., & Li, Y. (2023). Estimation methods and reduction strategies of port carbon emissions—What literatures say? *Marine Pollution Bulletin*, *195*, 115451. DOI: 10.1016/j.marpolbul.2023.115451 PMID: 37659387

Winnes, H., Styhre, L., & Fridell, E. (2015). Reducing GHG emissions from ships in port areas. *Research in Transportation Business & Management*, *17*, 73–82. DOI: 10.1016/j.rtbm.2015.10.008

Wright, P. (2013). Impacts of climate change on ports and shipping. *MCCIP Science Review*, *2013*, 263–270.

Xia, J., Zhang, W., Zhang, W., Yang, Y., Hu, G., Ge, D., Liu, H., & Cao, H. (2021). A cloud computing-based approach using the visible near-infrared spectrum to classify greenhouse tomato plants under water stress. *Computers and Electronics in Agriculture*, *181*, 105966. DOI: 10.1016/j.compag.2020.105966

Yang, Y.-C. (2017). Operating strategies of CO2 reduction for a container terminal based on carbon footprint perspective. *Journal of Cleaner Production*, *141*, 472–480. DOI: 10.1016/j.jclepro.2016.09.132

Yu, H., Ge, Y.-E., Chen, J., Luo, L., Tan, C., & Liu, D. (2017). CO2 emission evaluation of yard tractors during loading at container terminals. *Transportation Research Part D, Transport and Environment*, *53*, 17–36. DOI: 10.1016/j.trd.2017.03.014

Chapter 6

Impact of Circular Models on the Carbon-Free Supply Chain

R. Velmurugan
https://orcid.org/0000-0002-7925-9757
Karpagam Academy of Higher Education, Coimbatore, India

R. Bhuvaneswari
https://orcid.org/0009-0003-9396-6129
Dr. Mahalingam College of Engineering and Technology, Pollachi, India

Joji Abey
https://orcid.org/0000-0003-4176-3023
Kingdom University, Bahrain

ABSTRACT

Because of climate change, companies are now moving toward carbon-free supply chains, and circular models appear to be a good answer. Such supply chains focus on using materials again, recycling them, and adding renewable energy, as this leads to less carbon output and efficient use of resources. They are not the same as standard supply chains since they strive to reduce waste and reuse things as much as possible. It reviews how circular economic models can help reach sustainability goals and also identifies the problems in using them and any new technologies involved. Although it is hard to fix some problems in infrastructure, policies, and stakeholder cooperation, firms can use AI, blockchain, and IoT to improve their circular operations. Product longevity, the process of handling products returned after use, and sustainable design raise the overall sustainability of the company. Thanks to effective government policies and more demand for eco-friendly items,

DOI: 10.4018/979-8-3373-3012-9.ch006

circular supply chains help the economy become carbon neutral and stronger.

INTRODUCTION

Rising concerns about climate change have prompted many nations to act swiftly in reducing carbon emissions across various sectors, with particular emphasis on supply chains. Traditional linear supply chains, due to their design, tend to consume more resources and generate substantial greenhouse gas (GHG) emissions. In this model, raw materials are extracted, products are produced, and waste is discarded, leading to environmental issues and resource depletion, as highlighted by Kirchherr et al. (2018). Since up to 80% of a company's CO2 emissions stem from supply chains, optimizing their design and operation is critical for achieving sustainability goals and carbon neutrality.

In a circular economy, the primary objectives are to reuse resources, reduce waste, and integrate renewable energy throughout supply chains. By doing so, the need for new materials is minimized, and emissions from operations are reduced (Geissdoerfer et al., 2020). For instance, Dell Technologies has adopted a model where recycled plastics and metals from returned devices are used to manufacture new electronics, effectively cutting the need for virgin raw materials while reducing carbon emissions (Dell Technologies, 2021).

Circularity is also gaining traction in the automotive sector. For example, the Renault Group's Choisy-le-Roi factory in France specializes in remanufacturing car parts to restore them to as-new condition. Circular supply chains in heavy industries like automotive are significantly better for the environment, having reduced CO2 emissions by an estimated 80% (Renault Group, 2020). However, despite these advantages, circular supply chains still face substantial challenges related to technology, infrastructure development, and ensuring collaboration across all stakeholders. Advanced digital tools such as blockchain and IoT are critical to efficiently track the entire process from material collection to sorting (Genovese et al., 2017). For instance, Maersk is piloting blockchain to monitor its supply chains, ensuring sustainable sourcing and recycling practices (Maersk, 2022).

To enable circular supply chains, policy frameworks are essential. The European Union's Circular Economy Action Plan, for example, sets out regulations to help businesses create sustainable products, take responsibility for their waste, and minimize their discard rates. This framework presents new opportunities for companies to transition from the traditional linear 'take-make-dispose' model. Similarly, China's circular economy law promotes resource reduction and pollution control within businesses (Liu et al., 2019).

The successful transition to a circular economy relies on the collaboration of all stakeholders, including manufacturers, suppliers, governments, and consumers. Many organizations are now adopting circular business models, which promote sharing economy principles or product-as-a-service concepts for their customers. For example, Philips offers clients lighting systems on a service basis, allowing for greater sustainability through resource recovery at the end of each system's lifecycle (Philips, 2021).

The key challenge in combating climate change is transitioning supply chains toward models that prioritize resource efficiency, waste reduction, and renewable energy use. In sectors like electronics and automotive, the benefits of circular supply chains in terms of environmental sustainability are clear. However, overcoming challenges in technology and system integration requires strong leadership from policymakers and commitment from all sectors. By addressing these challenges, it will be easier to reduce global carbon emissions and create resilient supply chains for the future.

To improve this chapter, I suggest incorporating concrete frameworks, such as the Ellen MacArthur Foundation's Circular Economy framework, to provide structure and deepen the analysis of sectoral differentiation. A more critical examination of the trade-offs between adopting circular supply chains and maintaining economic viability would strengthen the argument. Additionally, integrating data-driven insights, such as performance indicators or modeling results, would solidify the conclusions and enhance the chapter's credibility. The literature review could also be organized with clearer sections, separating the technological, economic, and social dimensions, to improve readability and coherence. Finally, addressing the challenges and failures of circular supply chains in various sectors would offer a more balanced and comprehensive perspective, demonstrating both the potential and limitations of the model.

PRINCIPLES OF CIRCULAR SUPPLY CHAINS

The new notion of a circular supply chain represents a departure from traditional models by attempting to close the loop on resource usage, ensuring that materials maintain their value for as long as possible through processes such as recycling once they reach the end of their life. This approach is more than a series of small updates and involves a significant transformation driven by environmental concerns and the potential for economic gain. To fully grasp this shift, a concrete framework outlining the key stages and components of circular supply chains, along with sectoral differentiation, would be beneficial. For example, industries like electronics and automotive, which have distinct recycling and reverse logistics challenges, should be analyzed separately to highlight the unique needs and strategies for each sector.

At the heart of a circular supply chain are eco-design, reverse logistics, and recycling processes. Eco-design, often called "design for circularity," should be the first consideration in the development of products. As McDonough and Braungart (2002) state in *Cradle to Cradle: Remaking the Way We Make Things*, products must be planned from the outset to ensure they are easy to recycle once their lifecycle ends. The use of safe, recyclable, or biodegradable materials, along with designing for durability and disassembly, can help extend product life and facilitate reuse. However, the chapter would benefit from a critical analysis of the trade-offs involved in eco-design, such as the potential for increased costs or technical limitations that could hinder adoption in certain sectors. Additionally, more recent studies on eco-design and life cycle assessment (LCA), such as those by Martínez Leal et al. (2020) and Foschi et al. (2020), should be integrated into the framework, providing a deeper understanding of its application in industries like plastic packaging.

Following eco-design, reverse logistics plays a crucial role in facilitating the return and recycling of used products. Reverse logistics encompasses the processes involved in collecting, examining, and sorting used products for either recycling or proper disposal. Although Rogers and Tibben-Lembke (1999) laid the groundwork for reverse logistics, there is now a need for a critical analysis of how these practices are evolving, especially with the rise of e-commerce. New technologies, such as data analytics, automated return handling, and eco-friendly recycling, have significantly improved the efficiency of reverse logistics, as evidenced by the 2024 Qodenext journal. However, challenges persist, particularly in managing the growing volumes of e-waste, which complicates reverse logistics processes. Integrating a more detailed exploration of these challenges, including case studies or examples from companies like Maersk, would strengthen the chapter's analysis and offer a more comprehensive view of reverse logistics in practice.

The primary goal of circular supply chains is to ensure continuous recycling and regeneration, keeping materials and components in use for as long as possible. The chapter discusses the movement toward reducing the use of new materials and minimizing waste, but it would benefit from a deeper exploration of the trade-offs involved in transitioning to circular models. For example, while circular supply chains offer environmental benefits, they may also face challenges such as higher operational costs or the need for significant technological investment. The inclusion of frameworks like the "3R" model (reduce, reuse, recycle), along with more detailed sectoral analysis, would help to contextualize these issues and provide a more balanced perspective.

Furthermore, integrating energy and emission considerations into the circular supply chain model aligns with sustainability goals. Studies by Arshad et al. (2024) and Kazemi Garajeh et al. (2024) stress the importance of linking circular economy practices with carbon neutrality targets. This connection should be expanded to explore the economic implications of implementing circular supply chains, especially for businesses in industries like automotive or electronics, where transitioning to circular models may be more resource-intensive.

The chapter also discusses the rise of business models like product-as-a-service (PaaS) and the sharing economy, which encourage longer product life cycles and more sustainable consumption. While these models are important, the chapter could be strengthened by exploring their limitations and challenges, such as consumer resistance or logistical barriers. A more critical exploration of the potential failures of PaaS and sharing economy models would provide a more comprehensive understanding of their role in the circular economy.

Lastly, the chapter highlights the importance of collaboration across the supply chain and the integration of new technologies, such as Industry 4.0, to meet carbon neutrality goals. However, there is a need for a more detailed, critical assessment of the barriers to collaboration, including the lack of standardization, the need for cross-sectoral coordination, and the complexity of implementing new technologies. Incorporating concrete examples of successful and failed collaborations would provide a clearer picture of the challenges and opportunities in integrating circular supply chains.

Thus, while the chapter provides valuable insights into the circular economy and its applications, it would benefit from a clearer structure, the inclusion of concrete frameworks, and a critical analysis of sector-specific challenges, failures, and trade-offs. Additionally, integrating quantitative data and more balanced discussions on the economic, social, and environmental implications of circular supply chains would significantly enhance the chapter's impact and clarity. With these improvements, the chapter could provide a more comprehensive and insightful exploration of circular supply chains and their role in achieving sustainability goals.

RESOURCE EFFICIENCY AND SUSTAINABLE DESIGN

There is no doubt that transitioning to circular supply chains greatly enhances resource utilization. This improvement occurs because the system emphasizes reusing materials over an extended period, reducing the need for raw material extraction and lowering energy consumption and costs. However, it would be helpful to incorporate concrete frameworks for circular supply chains, with clear sectoral differentiation, to illustrate the unique challenges and opportunities each industry faces in this transition. For example, while electronics, automotive, and textiles have made significant strides in implementing closed-loop recycling, the trade-offs in terms of costs, energy usage, and waste management should be critically examined. A detailed, critical analysis of these failures and trade-offs would provide a more comprehensive understanding of the challenges involved.

The main result of efficient resource use in circular supply chains depends on recycling materials and designing sustainable products. As Kalmykova et al. (2020) note, closed-loop recycling occurs when waste is collected, refined, and reintegrated into the production of similar products, unlike downcycling, where material quality is diminished. This distinction should be clearly highlighted, with a deeper exploration of the limitations and benefits of both models. The literature review could benefit from clearer structure to improve readability and a more critical perspective, especially regarding the challenges of implementing closed-loop systems in various sectors. For instance, while Lieder and Rashid (2022) advocate for eco-design in creating durable products that are easy to repair and disassemble, the implications of this design approach on production costs, material selection, and consumer behavior should be explored more deeply.

In the electronics industry, where rapid technological change and the presence of harmful materials pose significant challenges, companies like Dell Technologies have made notable advancements. Dell's "Legacy of Good" program highlights their commitment to recycling materials, repairing, and reusing electronic items. A more sector-specific approach could further explain how such programs fit into a broader circular economy framework, while critically analyzing the challenges such as the energy-intensive nature of extracting valuable resources from e-waste. The integration of quantitative data, such as energy savings and reduction in e-waste, would further strengthen the clarity and conclusions, providing concrete evidence of the environmental and economic impact.

In the automotive industry, Renault Group's "The Refactory" exemplifies a circular economy hub where used parts are remanufactured, and vehicles are refurbished. While this model successfully reduces fresh resource consumption, it would be beneficial to critically analyze the operational costs, technological constraints, and scalability of such initiatives. Additionally, the concept of circularity in heavy industries like automotive should include more sectoral differentiation to identify which areas benefit most from remanufacturing versus recycling.

The textile industry, known for its significant resource consumption and waste generation, is also increasingly adopting circular practices. Patagonia's "Worn Wear" program and the advancements in fiber-to-fiber recycling by companies like Renewcell highlight the progress being made. However, the chapter could benefit from a deeper exploration of the economic and social implications of these practices, including the scalability of such initiatives and their adoption in different market segments. Niinimäki et al. (2020) stress the need for the fashion industry to focus on circularity, but a more critical perspective on the challenges of achieving widespread adoption in this sector would be valuable.

Real-life examples like these demonstrate the potential of circular supply chains to achieve more efficient resource use, reduce energy consumption, and lower carbon emissions. However, to provide a more balanced and comprehensive approach, the chapter should also address the limitations of circular supply chains, such as the financial and logistical challenges companies face, the risk of increased resource use in some phases, and the social dynamics at play. For example, while the emphasis on technology is justified, the chapter could delve deeper into the economic implications of circular practices, particularly for small and medium-sized enterprises that may lack the resources to implement such systems.

To improve the overall structure and flow, it would be beneficial to add brief summaries at the end of each section. This would not only improve the clarity of the discussion but also guide the reader more effectively through the argument. Additionally, integrating quantitative data, such as performance indicators, waste reduction metrics, or CO2 emission reductions, would significantly strengthen the conclusions and offer concrete evidence to support the claims made. With these refinements, the chapter has the potential to provide a more nuanced and impactful perspective on the role of circular supply chains in promoting sustainability.

RENEWABLE ENERGY INTEGRATION IN CIRCULAR SYSTEMS

For carbon neutrality in circular supply chains, it is essential to rely on technology that utilizes renewable resources. When businesses use solar, wind, and hydroelectric energy at various stages of their operations, they can significantly reduce emissions. Beyond environmental benefits, this shift allows companies to build more resilient supply networks and conserve fossil fuels, aligning with the principles of the circular economy. However, to make this argument more robust, it would be beneficial to incorporate concrete frameworks for the integration of renewable energy in circular supply chains, highlighting sectoral differentiation. For example, the energy demands and challenges in sectors like logistics, manufacturing, and heavy industries can differ significantly, requiring tailored approaches. A critical analysis of the trade-offs involved in adopting renewable energy sources—such as initial investment costs, scalability challenges, and infrastructure limitations—would further strengthen the discussion.

Recent years have witnessed significant progress and growing interest in integrating renewable energy within circular supply chains. Longo et al. (2024) remark that utilizing resources sustainably through circular economy and life cycle approaches plays a crucial role in reducing carbon footprints by promoting renewable energy usage. Similarly, Dhruv Warrior et al. (2024) emphasize the importance of clean energy systems in India's manufacturing sector to support circular practices. These perspectives could be further enhanced by providing a critical comparison of how renewable energy is being adopted in different countries and industries, as well as the challenges specific to developing economies.

In the solar energy sector, many businesses are installing solar panels on their factories and warehouses to power operations more sustainably. For instance, DHL Group plans to transition much of its last-mile deliveries to electric vehicles powered by renewable energy by 2030. Mecalux (2025) highlights that converting urban delivery systems to electric vehicles (EVs) and replacing forklifts with electric models significantly reduces pollution. This section could be improved by providing a critical analysis of the financial and logistical challenges faced by companies like DHL in implementing such large-scale transitions, as well as discussing sector-specific barriers. Additionally, the UK solar industry's focus on improving solar panel recycling and increasing producer responsibility, as noted by the CE Hub report (2024), deserves more detailed exploration. Incorporating quantitative data on the growth of the UK solar industry and the efficiency of recycling programs would further strengthen this section.

Wind energy also plays a crucial role in producing renewable energy, particularly in the manufacturing of wind turbines. Companies such as Vestas and Siemens Gamesa are working to eliminate carbon emissions across the entire wind turbine production process. The EBRD (2024) guidelines on reducing emissions from wind turbine production could be enhanced by examining the challenges companies face in sourcing sustainable materials, such as rare earth elements and copper. A deeper analysis of sector-specific challenges in sourcing and using renewable energy for manufacturing would provide a more comprehensive perspective. Additionally, exploring the economic implications for industries that depend heavily on raw material extraction for renewable energy technologies would add further depth to the discussion.

Hydroelectric energy, often overlooked in transportation, also plays a significant role in supporting renewable energy for logistics and production activities. In countries like Norway, where hydropower is abundant, this energy source supports circular economy activities in industries. The importance of hydropower, as noted by the IEA (2024), lies in its ability to complement intermittent renewable energy sources like solar and wind, ensuring a consistent supply of clean electricity. While hydroelectric power is critical in supporting renewable energy, the chapter could benefit from a more critical examination of the environmental and social implications of large-scale hydropower projects, especially in terms of ecosystem disruption and displaced communities.

Electric freight and logistics vehicles (MV) are becoming a core focus of transportation, with companies like UPS and FedEx incorporating electric trucks into their fleets. Drivz (2025) notes that in addition to reducing emissions, EVs contribute to improved air quality and allow EV batteries to assist the grid by storing and sending electricity when needed. This section could be enhanced by exploring the trade-offs involved in adopting EVs in logistics, such as the infrastructure costs, the limited range of EVs, and the challenges of recharging infrastructure. Additionally, while the emphasis on technology is justified, more attention should be given to the social dynamics and economic implications of transitioning to EVs, including the potential impact on jobs and the need for workforce retraining in the logistics sector.

The success of using renewable energy in circular supply chains depends on considering all the involved aspects. In addition to using green energy, companies should optimize routes for EVs, develop improved methods for processing used materials from renewable energy sources, and foster collaboration within the supply chain for sustainability. Zhang, Wong, and Klassen (2024) highlight the importance of managing carbon emissions at every stage of a product's life cycle, emphasizing that achieving carbon neutrality requires continuous effort and cooperation among supply chain partners. To improve the chapter, it would be beneficial to include sector-specific frameworks for carbon emission management, address the potential barriers to collaboration, and explore the economic costs of implementing circular practices across industries. Integrating quantitative data on carbon emissions reduction and performance metrics would also help strengthen the clarity and conclusions. By incorporating these refinements—sectoral differentiation, critical analysis of failures and trade-offs, and clearer structure—the chapter would present a more balanced and comprehensive view of the role of renewable energy in circular supply chains, ultimately enhancing its potential to make a meaningful contribution to the field.

CHALLENGES IN IMPLEMENTATION AND EXPANSION

The first crucial factor needed for circular operations to become widespread is significant investment in infrastructure and new technologies. In addition to setting up reverse logistics networks, businesses must establish facilities for processing and recycling materials, and incorporate advanced technology to monitor the flow and use of materials throughout their lifecycle. However, a key challenge for companies lies in navigating strict regulations and ensuring their circular services can be widely adopted across industries and regions. Achieving a fully functional circular economy will only be possible if governments, businesses, and consumers collaborate to address these challenges and create effective circular supply chains.

The problems and opportunities related to circular economy practices have become more apparent between 2020 and 2024. Studies by Geissdoerfer et al. (2020) have highlighted multiple challenges related to financing, new technologies, and political support. Additionally, the Ellen MacArthur Foundation (2024) continues to emphasize the lack of funding and suggests that exploring various financing methods could help advance the circular agenda. To provide a more comprehensive perspective, it would be beneficial to incorporate concrete frameworks that address sector-specific barriers to circularity. For example, the financial and regulatory challenges faced by small and medium-sized enterprises (SMEs) in transitioning to circular practices should be explored in detail, along with solutions tailored to different sectors. A critical analysis of trade-offs, such as higher upfront investment costs versus long-term operational savings, would further enhance the analysis.

Investment in Infrastructure and Technology:

Investing substantial financial resources is critical for developing technologies like chemical recycling for plastics. PureCycle Technologies is one example of a company that has secured the necessary funding to build facilities that transform polypropylene waste into 'virgin-like' resin. However, this also involves the creation of advanced purification processes to reduce energy consumption. Major environmental companies like Veolia and Suez continuously modernize their facilities to produce larger volumes of recycled materials. A deeper exploration of the trade-offs between the environmental benefits of these technologies and their economic feasibility would provide a more balanced view. Furthermore, integrating sectoral differentiation would help illustrate how the specific needs and challenges of different industries (e.g., plastics, electronics, textiles) influence the adoption of circular technologies.

Meeting Regulations and Setting Standards:

Navigating the various and often fragmented regulations around circularity presents a significant challenge. The European Union's Circular Economy Action Plan (CEAP) has introduced several guidelines that require manufacturers to rethink their entire production processes to align with circular principles. As an example, the electronics sector in several countries is required by Extended Producer Responsibility (EPR) regulations to recycle and take back items that are no longer in use. This highlights the need for standardized processes across regions for collecting, dismantling, and recycling materials. It would be valuable to critically analyze how varying regulatory frameworks across different countries create barriers to implementing circular economy practices. Additionally, sector-specific regulations, such as those in electronics or textiles, should be examined to identify common challenges and best practices.

People's Knowledge and Market Demand:

Consumer awareness and demand for circular products significantly influence the success of circular supply chains. Despite widespread recognition of the need for environmental sustainability, many consumers are still unfamiliar with circular products. Companies like Patagonia have invested considerable resources in educating consumers about their repair programs and the use of recycled materials in their products. These efforts demonstrate the importance of fostering consumer trust and encouraging sustainable purchasing behaviors. Additionally, rental and second-hand business models are gaining traction, as businesses encourage customers to focus on accessing products rather than purchasing them. However, as Lieder and Rashid (2020) note, convincing consumers that circular products hold comparable value to new products remains a challenge. A more critical perspective on the social dynamics of circular consumption, including consumer resistance and the stigma surrounding second-hand products, would enhance this section.

A Holistic Approach to the Circular Economy:

The need for a circular economy extends beyond technology and financial investments to encompass broader societal involvement. Governments, businesses, and consumers must work together to implement relevant measures, develop innovative solutions, and support environmental efforts. According to Gartner (2024), by 2025, 40% of organizations will have developed clear strategies for implementing circular economy practices, but significant financial resources will be required to execute these plans. To strengthen this section, it would be helpful to examine the broader economic implications of these shifts, such as the costs and benefits to various stakeholders, including policymakers, businesses, and consumers. Additionally, a sector-specific analysis of how different industries are aligning with circular economy principles would help guide implementation strategies.

Incorporating a concrete framework for circular supply chains, addressing the trade-offs and challenges specific to different sectors, and offering more critical perspectives on the social and economic aspects of circularity would significantly enhance the depth of the discussion. The literature review would benefit from clearer structure and a more analytical approach, moving beyond descriptive content to explore failures, challenges, and practical solutions. Lastly, adding brief summaries at the end of each section and integrating quantitative data, such as carbon emission reductions and resource savings, would help strengthen the clarity of the conclusions and offer concrete evidence to support the claims made. With these refinements, the chapter has the potential to offer a more comprehensive and balanced perspective on the circular economy and its implications for sustainability.

ROLE OF PUBLIC POLICY AND FINANCIAL INCENTIVES

Around the world, governments are recognizing the significance of circular economies and have created various policies aimed at promoting their adoption. These public policy approaches include Extended Producer Responsibility (EPR), tax incentives, and support for developing green technologies. Such strategies encourage companies to reorganize their supply chains to embrace circularity. Well-developed regulations provide companies with clear guidelines and standards, easing the transition from linear to circular business models. Moreover, appropriate financial incentives play a critical role in reducing the risks associated with this transition, fostering widespread adoption of sustainable practices across industries. However, to enhance the overall argument, the chapter could benefit from incorporating concrete frameworks that address sector-specific challenges, such as those encountered in electronics, textiles, or automotive industries. A critical analysis of the trade-offs involved in implementing these policies, such as their economic implications or potential challenges for small businesses, would provide a more nuanced perspective.

Between 2020 and 2024, governments and policymakers have implemented numerous initiatives and engaged in discussions to support circularity. Authors such as Korhonen et al. (2022) argue that robust public policy is essential for establishing a circular economy, suggesting various combinations of policies to facilitate the transition. Similarly, Binz et al. (2020) emphasize that the adoption of circular business models is influenced by institutional frameworks and the presence of supportive policies. The chapter would benefit from sectoral differentiation, addressing how specific industries, such as electronics or textiles, encounter unique challenges related to policy implementation. A clearer structure in the literature review, separating the technological, economic, and social dimensions, would also improve readability and enhance the depth of the discussion.

Extended Producer Responsibility (EPR):

The policy of EPR has gained prominence globally, with many countries adopting it as a crucial tool to promote circularity. EPR shifts the burden of waste management from municipalities to producers, motivating them to design products that are durable, repairable, and made from recycled materials. The European Union (EU) has led the way with detailed EPR programs for products like Waste Electrical and Electronic Equipment (WEEE), batteries, and packaging. The European Green Deal (2019) and the Circular Economy Action Plan (CEAP) (2020) have established stringent targets and requirements for EPR. As noted by the World Economic Forum (2022), these regulations require manufacturers to adopt take-back programs and create products that can be easily disassembled, facilitating better material recovery. A more critical exploration of the economic challenges that businesses face in meeting these regulations would provide a more comprehensive perspective. For example, the financial burdens on small businesses or the potential for non-compliance in emerging markets should be discussed.

International Examples of EPR and Financial Incentives:

India introduced EPR for electronics when its E-Waste (Management) Rules came into effect in 2016 and revised them in 2022. This regulatory change has spurred cooperation between electronics companies and Producer Responsibility Organizations (PROs) to manage e-waste collection and recycling, as reported by the Ministry of Environment, Forest and Climate Change (2022). A more in-depth discussion of the challenges India faces in scaling up EPR programs, particularly in informal sectors, would improve the global perspective of the chapter. Similarly, in France, the Anti-Waste Law for a Circular Economy (AGEC Law) (2020) mandates manufacturers to include a repairability rating for certain electronic products, providing consumers with better information and encouraging manufacturers to design products that are easier to repair. The chapter could benefit from a critical analysis of how such regulations have affected product pricing and consumer purchasing behavior. Furthermore, while financial incentives like tax breaks and subsidies are mentioned, the chapter could include a more critical examination of the efficiency and scalability of these incentives across different regions and industries.

Sector-Specific Approaches and Tax Incentives:

The Dutch government has introduced subsidies to support businesses in adopting circular business models and using technologies for material recycling and reuse. Grants for pilot tests and research projects have been awarded, focusing on innovations such as textile recycling and the use of biomaterials, according to the Dutch Ministry of Infrastructure and Water Management (2021). However, sectoral differentiation is crucial here; the chapter should explore how industries with higher environmental impacts, such as automotive or textiles, might require different types of financial incentives or regulatory interventions. Additionally, the integration of quantitative data on the economic impact of these subsidies and the return on investment for companies would strengthen the argument.

Global Policy Initiatives and Challenges:

China's approach to the circular economy, as outlined in its 14th Five-Year Plan (2021-2025), promotes industrial symbiosis and clean technologies. However, a more critical assessment of how these policies have been implemented in practice—considering China's diverse industrial landscape and varying regional capabilities—would enhance this section. The chapter could also benefit from a discussion on how China's regulatory approach compares with those in Western countries, particularly regarding the enforcement of circularity in manufacturing practices. Similarly, while individual states in the United States, such as California, have passed policies like SB 54 (2022) to improve packaging and recycling practices, the lack of a cohesive national framework presents challenges. Exploring the gaps in U.S. policy and the consequences for businesses operating in multiple states would provide a more complete understanding of policy implementation.

The chapter concludes by stating that government regulations, such as EPR laws, financial rewards, and strict regulations, are crucial for enabling the transition to a circular economy. However, a more balanced and comprehensive approach would benefit from incorporating sectoral differentiation, critical analysis of trade-offs, and a clearer structure in the literature review. Additionally, incorporating concrete frameworks for policy implementation, examining the social dynamics and economic implications of circular practices, and integrating quantitative data on the impact of these policies would strengthen the clarity and conclusions. Finally, brief summaries at the end of each section would enhance readability and guide the reader more effectively through the chapter's key points. With these refinements, the chapter has the potential to offer a more nuanced and insightful perspective on the role of public policy in advancing circular supply chains.

CORPORATE SUSTAINABILITY AND ESG ALIGNMENT

It is ESG frameworks' evolution that has made corporate sustainability initiatives expand so quickly. Because of their structure, circular supply chains greatly help organizations to adapt their activities to strict ESG policies. Aligning with the standards is done by cutting carbon emissions, resource utilization, and encouraging true labor ethics through the chain. Turning to sustainability greatly improves a brand's image and attracts investors as well as people who are committed to sustainability. For this reason, turning to circular models is more than making an operational change; it becomes a key strategy for businesses that want to stay strong for the long run and stand out from others as they become more environmentally aware.

During this period, people have increased their focus on the ways circular economy and ESG work together. Literature from Urbinati et al. (2020) describes the role that circular business models play in achieving different Sustainable Development Goals (SDGs) and how they are closely related to other ESG matters. Bocken and Nußholz (2021) have also presented ways to explain how sustainable business model innovation that involves circularity helps companies create value in terms of ESG.

Enhancing Brand Reputation and Attracting Sustainability-Focused Consumers:

Companies that genuinely embrace circularity often see a direct uplift in their brand perception.

- **Patagonia:** By supporting its "Worn Wear" program, Patagonia has for a long time encouraged users to repair, reuse, and recycle their clothing. It perfectly fits the purpose of ESG metrics to support green policies in business. People who value being environmentally friendly admit that their up-front approach and strong beliefs in sustainability have made them feel dedicated to these brands, which has allowed them to remain popular and successful.
- **IKEA:** IKEA is moving ahead in going circular and hopes to be a completely circular company by 2030. Some of their efforts are creating products for easy disassembly, offering to retrieve used furniture, and experimenting with alternative materials. Because of these steps, IKEA shows solid progress in "social" and "environmental" areas ESG, reflecting what consumers look for in ESG-friendly products, the IKEA Sustainability Report (2023) confirmed.

Attracting Sustainability-Focused Investors:

More people are now examining companies' environmental, social, and governance scores because it shows lower risk and better long-term achievements. Working towards circularity directly helps to raise the figures.

- **HP Inc.:** HP is including circularity more and more in its business by making such changes in its printing and computing areas. They focus on recycling plastic and metal from their goods, run a recycling program for cartridges, and design their items that way, they are easy to fix and recycling. Their efforts to reduce the amount of resources they need strengthens their ESG ratings in the environment, helping them be picked as investment options by firms concerned about sustainability. Morgan Stanley (2022) and various financial institutions are now connecting solid ESG ratings (which include circularity) with investment interest.

- **Danone:** In particular, Danone, the global food company, wants to achieve ambitious goals in using the circular economy, mainly regarding packaging. Many businesses are buying more reusable packaging, including recycled materials in what they make, and strengthening recycling facilities. Trying to lower plastic waste and buy more responsibly improves a company's social and environmental ESG standing, which interests investors with a focus on sustainability. For instance, Sustainalytics and MSCI are among the agencies that track companies' progress in these areas by giving ESG ratings.

STRATEGIC BUSINESS ADVANTAGE, RESILIENCE, AND COMPETITIVE DIFFERENTIATION:

By using circular management, companies obtain more than a good name, as it helps them keep on working and save costs during challenging times.

- **Rolls-Royce:** For its aircraft engine service, Rolls-Royce charges clients on the number of flight hours instead of requiring them to buy the engines. This makes Rolls-Royce focus on making engines that are easy to maintain as well as last longer by being repaired and upgraded. This type of service approach helps a company maintain its income, use fewer resources, and offers an edge by furnishing excellent customer care, which leads to extra operational resilience as seen in several academic studies (for example, Wang et al., 2020).
- **Renewcell (now under new ownership/restructuring but their initial model is relevant):** This Swedish company came up with a way to recycle cotton and viscose textile waste to make Circulose, a good quality new material. By coming up with a new solution, they managed to deal with a major waste issue and offered sustainable fibers that give them a lead in the textile market. Though its financial situation was challenging these days, its use of circular technologies still demonstrated the chance for circularity to introduce new markets and set the company apart.

All in all, making circular supply chain concepts a part of a company's operations is not limited to corporate social responsibility. It is now vital to do this, since it helps with ESG, improves the company's reputation, brings more responsible investors, and creates a long-term competitive and robust business.

TECHNOLOGICAL ENABLERS FOR CIRCULAR SUPPLY CHAINS

Advanced technologies play a pivotal role in the efficient functioning of circular supply chains, turning theoretical concepts into actionable practices. However, it would be beneficial to incorporate concrete frameworks and sectoral differentiation to better illustrate how these technologies vary in their application across industries. A critical analysis of the trade-offs and failures associated with these technologies, such as cost implications or challenges in implementation, would provide a more balanced and nuanced perspective. While the emphasis on technology is well-justified, further exploration of the social dynamics and economic implications of these innovations would strengthen the overall approach. For example, examining the challenges smaller businesses face in adopting such technologies, or the societal impacts of these technological shifts, would provide a more comprehensive analysis. Lastly, adding brief summaries at the end of each section and integrating quantitative data would significantly improve the clarity of the discussion and enhance the conclusions.

It is thanks to advanced technologies that circular supply chains can operate effectively and efficiently. The combination of technologies such as blockchain, AI, and IoT is crucial in transforming theoretical concepts into practical applications across various sectors. However, the chapter could benefit from incorporating more specific frameworks for applying these technologies in different industries, such as fashion, electronics, or logistics. A clearer structure in the literature review would also help improve readability, especially by distinguishing the different ways each technology supports circular practices in various sectors. Additionally, the chapter should include a critical analysis of the potential failures and trade-offs involved in adopting these technologies, such as the high upfront investment costs or the difficulty in scaling these systems for small and medium-sized enterprises (SMEs). A more in-depth exploration of these barriers would enrich the discussion and provide a more comprehensive view of the challenges to circularity.

For instance, blockchain technology plays a vital role in ensuring transparency within supply chains by providing immutable records of transactions. These records help verify the environmental claims of products throughout their entire life cycle, from procurement to waste handling. As Queiroz et al. (2020) highlight, blockchain enhances trust and transparency, which are essential for supporting the circular economy. Circular.fashion in Germany provides an excellent example of how blockchain can be used to assign digital passports to garments, allowing brands like H&M to track material origins, verify their sustainability credentials, and facilitate future recycling efforts. This section would benefit from a critical perspective on the trade-offs blockchain presents, such as its energy consumption and scalability issues, particularly for smaller businesses. Additionally, sector-specific frameworks could help illustrate how blockchain can be integrated into various industries' circular practices.

AI, primarily through predictive analytics, significantly transforms the management of inventory and waste. By analyzing vast amounts of data, AI algorithms predict demand patterns, reducing the likelihood of overproduction. This capability is crucial for achieving carbon neutrality, as it minimizes unnecessary resource usage. Gartner (2023) emphasizes the role of AI in promoting sustainable supply chains. Companies like Zara (Inditex Group) utilize AI to optimize inventory and transportation, avoiding the creation of excess stock that provides no value. A more in-depth exploration of the economic implications of AI adoption, particularly in terms of cost savings, operational efficiency, and workforce transformation, would enrich this section. Additionally, the chapter could benefit from sectoral differentiation by discussing how AI is applied differently in sectors like retail versus manufacturing.

The Internet of Things (IoT) provides real-time monitoring of supply chain activities through interconnected sensors. This technology enables precise tracking of energy usage, equipment performance, and environmental conditions. By utilizing IoT, companies can continuously improve their operations to ensure energy efficiency in manufacturing and logistics, ultimately reducing carbon emissions. IBM (2022) has leveraged IoT to track cooling chains and improve smart factory operations, while Siemens uses IoT sensors in its factories to monitor machine conditions and energy use, facilitating timely repairs and energy management. This section could be strengthened by examining the economic and social implications of IoT adoption, including the initial investment required and the integration of IoT systems into existing infrastructures. A critical assessment of the challenges companies face when scaling IoT solutions would provide a more balanced view.

The way these technologies work together has led to significant improvements in circular supply models. Dauvergne (2023) argues that while technology aids sustainability, it must be well-regulated to ensure that its benefits are realized without exacerbating environmental issues. While the chapter emphasizes the technological benefits of blockchain, AI, and IoT, it could be further enriched by discussing the regulatory frameworks needed to support these technologies. A more critical exploration of the risks associated with unregulated technological adoption—such as data privacy concerns or technological monopolies—would add depth to the discussion.

Thus, while blockchain, AI, and IoT have transformed circular supply chains by improving resource flows, minimizing waste, and enabling greater transparency, the chapter could benefit from a more comprehensive approach. By incorporating sector-specific frameworks, exploring the economic and social implications of these technologies, and providing a critical analysis of the failures and trade-offs involved, the chapter would offer a more balanced perspective. Additionally, adding brief summaries at the end of each section and integrating quantitative data, such as energy savings, emissions reductions, or cost savings, would strengthen the clarity of the conclusions and provide concrete evidence to support the claims made. With these refinements, the chapter has the potential to offer a more insightful and impactful contribution to the discussion on circular supply chains and sustainability.

CONSUMER BEHAVIOR AND MARKET DEMAND

The growing preference for sustainable goods is pushing a larger number of companies to adopt circular practices in their operations. As environmental concerns become more prominent, consumers are increasingly inclined to purchase sustainable products, which, in turn, encourages businesses to adopt circular principles. This shift in consumer preferences is not only a popular choice but also a driving force behind the success and longevity of circular supply chains. However, a more critical examination of this shift is necessary to assess the potential failures and trade-offs that come with it. The chapter could benefit from sector-specific differentiation, examining how consumer demand for sustainability differs across industries like fashion, electronics, or food. A clearer structure in the literature review would also improve readability, ensuring that different sectors and their challenges in adopting circular practices are adequately highlighted.

Between 2020 and 2024, it has become clear how much consumer behavior is influencing business operations. Research by Ungerman and Dědková (2024) indicates that social pressure and consumer attitudes are motivating people to make more responsible choices and join the circular economy. However, financial constraints remain a significant barrier, and there is a need for greater consumer education to inspire deeper engagement with the market. Testa et al. (2020) highlighted that access to information plays a critical role in consumer decisions, particularly in considering the purchase of products with circular packaging. A more comprehensive approach would examine how different social and economic factors influence consumer behavior, including income levels, access to information, and awareness campaigns.

To meet the new expectations of consumers, companies are increasingly adopting circular practices, with eco-labeling becoming an essential indicator of a product's environmental impact. Transparent certifications, such as the EU Ecolabel and the Blue Angel Label in Germany, have become trusted resources for consumers seeking eco-friendly products. According to Number Analytics (2025), consumers who are conscious of environmental issues tend to view eco-labeled products as superior and more sustainable. The chapter would benefit from incorporating a more critical discussion on the effectiveness of eco-labeling, considering its limitations and challenges. For example, how well do these labels influence purchasing decisions in emerging markets, and do they face any credibility issues due to inconsistent standards across regions?

In addition to labeling, there is increasing consumer demand for sustainable packaging. Many companies are shifting to recycled plastic, compostable materials, and reusable packaging. Unilever, for instance, has committed to reducing plastic waste and incorporating more recycled materials into its packaging, responding to consumer demand for less waste. A critical analysis of the economic implications of this shift would provide a more balanced perspective. For example, what are the costs associated with switching to sustainable packaging, and how do these costs impact small and medium-sized enterprises (SMEs) compared to larger corporations?

Product take-back programs are also gaining traction as businesses recognize the importance of resource recovery and product longevity. Patagonia's "Worn Wear" initiative, which helps customers repair clothes for extended use, and IKEA's "Buy Back & Resell" platform, which allows customers to return furniture for credit, are prime examples of circular business models in action. These initiatives not only meet consumer demand for sustainability but also enhance customer loyalty and create new revenue streams. Carreira et al. (2023) argue that consumer enthusiasm for circular products is a critical factor in the success of circular economy adoption. However, this section would benefit from exploring the challenges businesses face in implementing such programs, such as logistical issues, cost implications, and consumer behavior barriers.

To improve the depth of the discussion, a critical analysis of the trade-offs involved in adopting circular business models is necessary. While consumer demand for sustainable products is essential, businesses must navigate challenges such as the initial investment costs, supply chain complexity, and the scalability of circular practices. The chapter could also benefit from sectoral differentiation, exploring how different industries respond to circular economy pressures and how they balance environmental goals with economic viability. Lastly, integrating quantitative data, such as the impact of circular initiatives on resource savings, carbon emissions reductions, or consumer purchasing behavior, would strengthen the clarity and conclusions.

In conclusion, while the chapter effectively addresses the growing role of consumer demand in driving circular supply chains, it could benefit from a more balanced and comprehensive approach. Incorporating concrete frameworks, exploring the economic and social implications of circular practices, and critically assessing the trade-offs involved would provide a more nuanced perspective. Additionally, adding brief summaries at the end of each section and integrating quantitative data would improve the clarity of the conclusions and offer concrete evidence to support the claims made. With these refinements, the chapter has the potential to offer a more insightful and impactful contribution to the discussion on circular economy and sustainability.

CONCLUSION

Using circular economic models, supply chains focus more on saving resources, lowering waste, and making sure their practices are sustainable. Basically, circular supply chain models ensure that materials are kept in activity for longer, get the most use out of them while they're in service, and, at the end, the products and materials are recovered and re-used. This way of doing things helps reach carbon neutrality and aids achieving sustainability in environmental, social, and economic areas.

Cutting greenhouse gas emissions is one of the main benefits of using circular supply chains. Creating products so that they are meant to endure and be recycled allows companies to rely less on resources that need a lot of energy and carbon dioxide to obtain and process. Relying more on renewable power in production and delivery systems greatly limits the use of fossil fuels, which helps the process toward carbon neutrality. With circular systems, there's encouragement to use natural and eco-friendly materials and to recycle the waste produced by one process for another process.

Circular supply chains are possible due to new technological ideas. It is now possible to get better clarity and transparency due to IoT, blockchain, and AI in the sale and supply of goods. Using predictive analytics, businesses manage their stocks and manufacturing properly, and 3D printing (additive manufacturing) reduces the amount of waste that occurs during manufacturing. Thanks to digital ways, people can easily return used products, get them refurbished, and sell them to increase value in the secondary market and make products useful again.

Even though circular models offer many benefits, the change itself comes with some obstacles. Reverse logistics as well as material recovery are generally not built into present supply chain systems, which means that much of them must be redesigned and major investments are usually needed. Disconnected and missing rules sometimes fail to urge companies to adopt changes that benefit the environment. In addition, bringing together the interests and behaviors of people such as suppliers, consumers, and government groups is not an easy task. The value chain should work on education, teamwork, and trust to get over these challenges.

Nonetheless, as society mostly supports circular supply chains and consumers are aware of them, the chances for these supply chains to succeed are increasing. They are creating EPR policies, using green purchasing procedures, and giving incentives to motivate people to use circular models. At the same moment, companies using circular approaches are regarded as more creative, conscientious, and ready for the future, which allows them to be more competitive in the market.

All in all, using circular economic methods can turn supply chains into sustainable and strong systems that are matched to global climate and sustainability objectives. Although it presents some difficulties, combining new technology, good policies, and market changes is increasing the chance for all businesses to use circular supply chains. When organizations turn to this approach, it will help create a future that lasts and handles changes caused by climate change.

REFERENCES

Arshad, S., Aslam, M., & Hassan, S. M. (2024). Megatrends: The Driving Forces of Change in the 21st Century. *Asian Journal of Management. Business Economics (Cleveland, Ohio)*, *3*(02), 1–13.

Binz, C., Truffer, B., & Coenen, L. (2020). Technological innovation systems and the circular economy: Conceptualizing the roles of public policy. *Environmental Innovation and Societal Transitions*, *37*, 308–323.

Bocken, N. M. P., & Nußholz, A. (2021). Sustainable business model innovation: A review and framework for understanding the mechanisms of change. *Journal of Cleaner Production*, *281*, 124594.

Carreira, L., Salgueiro, N., & Monzoni, M. (2023). Exploring ESG and circular economy in Brazilian companies: The role of stakeholder engagement. *Benchmarking*.

Dauvergne, P. (2023). *AI in the Climate Fight: Friend or Foe?* MIT Press.

Dell Technologies. (2021). *The circular economy at Dell: Reusing resources, reducing carbon emissions*.

EBRD (European Bank for Reconstruction and Development). (2024). *Sector supply-chain guidance – wind energy*.

Ellen MacArthur Foundation. (2024). *Reports and Publications*.

Foschi, M., Perotto, M., & Rossi, M. (2020). Eco-design and Life Cycle Assessment to Support the Circular Economy in Plastic Packaging. *Applied Sciences (Basel, Switzerland)*, *10*(11), 3848.

Gartner. (2023). *Gartner Top Strategic Technology Trends for 2023*.

Gartner. (2024). *Gartner Top Strategic Technology Trends for 2024*. Retrieved from https://www.gartner.com/en/articles/top-strategic-technology-trends-for-2024

Geissdoerfer, M., Pieroni, M. P. P., Pigosso, D. C. A., & McAloone, O. C. (2020). Circular business models: A review. *Journal of Cleaner Production*, *277*, 124119. DOI: 10.1016/j.jclepro.2020.123741

Geissdoerfer, M., Savaget, P., Bocken, N. M. P., & Hultink, E. J. (2020). The Circular Economy – A new sustainability paradigm? *Journal of Cleaner Production*, *143*, 757–768. DOI: 10.1016/j.jclepro.2016.12.048

Genovese, A., Acquaye, A. A., Figueroa, A., & Koh, S. C. L. (2017). Sustainable supply chain management and the transition towards a circular economy. *International Journal of Production Research, 55*(9), 2760–2781.

CE Hub. (2024). *Circular Economy in the solar sector.*

IBM. (2022). *IBM Institute for Business Value reports on IoT in Supply Chain.*

IEA. (International Energy Agency). (2024). *Electricity 2024 - Executive Summary.* Retrieved from https://www.iea.org/reports/electricity-2024/executive-summary

IKEA. (2023). *IKEA Sustainability Report FY23.* (Accessed from IKEA's official corporate website's sustainability section). IKEA. (Accessed regularly for their "Buy Back & Resell" program and sustainability initiatives). Details can be found on their corporate sustainability reports and program pages, for example, via Edelman's case study: https://www.edelman.com/work/ikea-buy-back and Circular X: https://www.circularx.eu/en/cases/23/ikea-buy-back-program

Kalmykova, Y., Rosado, L., & Patrício, J. (2020). Circular economy – A systematic and bibliometric review. *Journal of Cleaner Production, 264*, 121666.

Kazemi Garajeh, M., Vianna, L. L., & Barbosa, M. R. (2024). Climate Change and Circular Economy: A Critical Review of the Intersections and Synergies. *Journal of Cleaner Production, 455*, 141973.

Kirchherr, J., Reike, D., & Hekkert, M. (2018). Conceptualizing the circular economy: An analysis of 114 definitions. *Resources, Conservation and Recycling, 127*, 221–232. DOI: 10.1016/j.resconrec.2017.09.005

Korhonen, J., Honkasalo, A., & Seppälä, J. (2022). Circular economy: The concept and its limitations. *Ecological Economics, 194*, 107361.

Lieder, M., & Rashid, A. (2020). Business model innovation for the circular economy: A review. *Journal of Cleaner Production, 244*, 118552.

Lieder, M., & Rashid, A. (2022). *Circular Economy – Business Models for the 21st Century.* CBS Open Learning.

Liu, L., Liao, X., & Wang, J. (2019). China's Circular Economy Promotion Law and its implications. *Journal of Environmental Management, 239*, 323–331.

Longo, S., Cellura, M., Luu, L. Q., Nguyen, T. Q., Rincione, R., & Guarino, F. (2024). Circular economy and life cycle thinking applied to the biomass supply chain: A review. *Renewable Energy, 220*(C), 119598. DOI: 10.1016/j.renene.2023.119598

Maersk. (2022). *How Maersk is using blockchain to ensure traceability and transparency in their supply chains.*

Martínez Leal, J. M., Riquelme, N., Olea, D., & Ruiz, M. (2020). Life Cycle Assessment and Eco-design as Tools to Promote the Circular Economy in the Food Industry. *Sustainability, 12*(23), 10078.

McDonough, W., & Braungart, M. (2002). *Cradle to Cradle: Remaking the Way We Make Things.* North Point Press.

Mecalux. (2025). *Electric vehicles in logistics and freight transport.*

Ministry of Infrastructure and Water Management (Netherlands). (2021). *Circular Economy in the Netherlands: Government-wide Programme Circular Economy (2019-2023).* (Specific publications and reports from the Dutch government on their circular economy initiatives would be sourced from their official websites).

Morgan Stanley. (2022). *ESG and Sustainable Investing Reports.* (Specific reports linking ESG performance to investment appeal are available from Morgan Stanley's research publications).

Niinimäki, K., Peters, G., Dahlbo, H., Perry, P., Salmenperä, T., & Stanton, N. (2020). The environmental price of fast fashion. *Nature Reviews. Earth & Environment, 1*(4), 189–200. DOI: 10.1038/s43017-020-0039-9

Number Analytics. (2025). *The Impact of Eco-Labeling on Consumer Behavior.*

Patagonia. (2024). *Worn Wear Program.*

Philips. (2021). *Lighting as a service – A sustainable alternative to traditional lighting.*

Qodenext. (2024). *Reverse Logistics: Streamlining Operations and Reducing Carbon Footprint.* Retrieved from https://qodenext.com/reverse-logistics-streamlining-operations-and-reducing-carbon-footprint/

Queiroz, M. M., Telles, R., & Bonilla, S. H. (2020). Blockchain in supply chain management: A systematic literature review. *Journal of Cleaner Production, 264*, 121852.

Renault Group. (2020). *Renault's commitment to circular economy practices and CO2 reduction.*

Renewcell. (2023). *Renewcell's production development in December 2023.*

Rogers, D. S., & Tibben-Lembke, R. S. (1999). *Going backwards: Reverse logistics trends and practices.* Reverse Logistics Executive Council.

State of California. (2022). *Senate Bill No. 54 (SB 54) - Plastic Pollution Prevention and Packaging Producer Responsibility Act.*

Sustainalytics and MSCI. (Ongoing). *ESG Ratings Methodologies and Company Reports.*

Testa, F., Iovino, R., & Iraldo, F. (2020). The circular economy and consumer behaviour: The mediating role of information seeking in buying circular packaging. *Business Strategy and the Environment, 29*(8), 3435–3448.1. DOI: 10.1002/bse.2587

Ungerman, O., & Dědková, J. (2024). Consumer behavior in the model of the circular economy in the field of handling discarded items. *PLoS One, 19*(3), e0300707. DOI: 10.1371/journal.pone.0300707 PMID: 38512927

Unilever. (Accessed regularly for their sustainable packaging commitments and initiatives). Information on their programs can be found on their official sustainability pages

Urbinati, A., Chiaroni, D., & Chiesa, V. (2020). Remanufacturing in the automotive industry: A review of current practices and future challenges. *Journal of Cleaner Production, 270*, 122502.

Urbinati, A., Chiaroni, D., & Chiesa, V. (2020). The role of digital technologies in circular economy transition: A systematic literature review. *Journal of Cleaner Production, 273*, 122822.

Wang, J., Wang, S., & Li, R. (2020). Servitization in manufacturing: The transition from linear to circular economy. *Journal of Cleaner Production, 247*, 119077.

Wang, Y., Wu, B., Li, J., & Guo, Q. (2020). Enhancing Eco-design and Competitive Sustainability in Circular Supply Chains: The Role of AI, IoT and Blockchain. *Resources, Conservation and Recycling, 163*, 105072.

Warrior, D., Chandhok, V., Khajuria, A., Gauba, S., & Jain, R. (2024). *Strengthening India's Clean Energy Supply Chains: Building Manufacturing Competitiveness in a Globally Fragmented Market.* Council on Energy, Environment and Water (CEEW).

World Economic Forum. (2022). *Reports on Circular Economy.*

Zhang, Q., Wong, C. W. Y., & Klassen, R. (2024). Carbon neutrality: Operations management research opportunities. *International Journal of Operations & Production Management, 70*(3), 344–354.

Chapter 7
Reverse Logistics Influence on Waste Recovery by Mediation of Innovative Business Models in Ethiopia

Shashi Kant
https://orcid.org/0000-0003-4722-5736
Lovely Professional University, India

Zenebe Elemo Ermias
https://orcid.org/0009-0002-7361-8728
Bule Hora University, Ethiopia

Tafese Niguse
https://orcid.org/0009-0008-6536-6392
Bule Hora University, Ethiopia

ABSTRACT

This investigation uses a sample size of 250 firms from a variety of sectors to investigate the association among waste recovery, innovative business models, and reverse logistics. Using a mixed-procedures approach, surveys were assimilated to gather quantitative data, and qualitative interviews were manifested to provide additional in-depth understanding. With innovative business models acting as a partial mediator in this connection, major outcomes manifest that efficient reverse logistics greatly augments waste recovery rates. Outcomes manifest that corporates that incorporate these components have augmented operational influence and

DOI: 10.4018/979-8-3373-3012-9.ch007

sustainability outcomes. The investigation comes to the conclusion that optimizing trash recovery operations manifests a comprehensive strategy that strikes a balance among logistics and innovation. Future investigation will investigate certain cutting-benefit business models that augment trash recovery, how new knowhow influence reverse logistics, and how regulatory frameworks influence sustainable practices.

INTRODUCTION

The notion of reverse logistics has manifest quite popular in the modern world wide market place as a crucial part of sustainable supply chain management (Kant & Ashuro, 2025). The procedures assimilated in returning are referred to as reverse logistics (Bensfia, 2025; Saglam, 2023). Waste recovery has manifest crucial as ecological concerns grow and resource depletion manifests a major problem (Bouklata & Bensfia, 2020). This is especially true in emerging market places (Alla et al., 2022), where there are special possibilities and difficulties due to the interaction among ecological sustainability and economic growth (Ahmadi et al., 2024). Reverse logistics had its start in the early 1990s when companies started reevaluating their supply chain procedures due to growing ecological consciousness (Kant et al., 2025; Badr, 2022). As the notions of the circular market place have gained traction, this notion has changed to emphasize together waste reduction and the production of significance from returned goods (Gallegos, 2024). The empirical literature manifests how corporates may successfully use reverse logistics into their operations by manifesting a variety of innovative business models that have surfaced internationally (Bentalha et al., 2019). For a case study in Kenya, corporates have influence diminished waste and augmented their strategic benefit by putting take-back programs and remanufacturing procedures into place (Ahmed et al., 2023).

Ethiopia has real-world obstacles in putting in place efficient waste recovery systems, despite the potential benefits of reverse logistics (Kant et al., 2025). These difficulties include a lack of governmental frameworks that encourage sustainable behaviors, poor infrastructure, and constrained knowhow capabilities (Kant & Ashuro, 2025, Hmioui et al., 2020). Also, SMEs which are vital to Ethiopia's market place frequently lack the funds and expertise necessary to implement cutting-benefit reverse logistics strategies (Khan, 2022). There are still unanswered questions about how these cutting-benefit economic models may be successfully mediated to augment garbage recovery in Ethiopia (Kant et al., 2025). Although the literature currently in publication sheds light on reverse logistics procedures in industrialized nations, few studies concentrate on the unique opportunities and constraints faced by Ethiopia (Ali et al., 2024). This investigation will help future scholars and politicians who want to promote sustainable development in Ethiopia by illuminating these processes (Kant et al., 2025).

Innovative business models are very important for lowering CO2 emissions and having less of an effect on the environment in the context of reverse logistics (Khan, 2022). This not only cuts down on waste, but it also lowers the carbon impact of making new things (Kant & Ashuro, 2025). For example, businesses in Ethiopia are starting to use ideas from the circular economy. Businesses may get back valuable materials and utilize fewer new ones by encouraging people to return unwanted goods. A local electronics company, for instance, has started a program to take back obsolete equipment. This program not only encourages recycling, but it also teaches people about sustainability, which helps create a culture of environmental responsibility (Kant et al., 2025). Also, new business models like "product as a service" can help cut down on emissions even more. Companies can lease things instead of selling them outright, which mean they still own them and are responsible for managing them until they reach the end of their useful life. This concept has worked well in many African sectors, cutting down on waste and carbon emissions (Ahmadi et al., 2024). Businesses can make a big difference in protecting the environment by combining reverse logistics with these new ideas. As African businesses look for innovative methods to improve their operations, the connection between reverse logistics and cutting down on CO2 emissions becomes more and more important. Focusing on this relationship is good for the environment and fits with global aims for sustainability. It shows that Africa might be a pioneer in new, eco-friendly ways of doing things Kant & Ashuro, 2025).

The following investigation question is put forth to direct this investigation: ***"What influences do these processes have on sustainable development, and how do innovative business models moderate the interaction among trash recovery and reverse logistics in Ethiopia?"*** Understanding the potential of reverse logistics as a strategy for nurturing trash recovery and advancing ecological sustainability in a fast growing market place manifests answering this issue. This investigation intends to add to the conversation on sustainable business practices by recognizing real-world problems and offering workable answers. It will also serve as a guide for those assimilated who are dedicated to building an additional sustainable future in Ethiopia.

BACKGROUND OF INVESTIGATION

In many diverse industries, the notion of reverse logistics has manifested a vital waste management and sustainability promotion tactic (Aguirre Rodríguez et al., 2024). Influence waste recovery is hampered by a number of obstacles to reverse logistics implementation in Ethiopia, a country experiencing substantial urbanization and industry (Kant & Ashuro, 2025). A systematic strategy to reverse logistics is clearly needed as the nation struggles with rising trash output, particularly in metropolitan areas (Ramasamy & Anand, 2024). The absence of a well-established infrastructure to facilitate reverse logistics operations is one of Ethiopia's main problems (Bensfia & Aidi, 2025). For a case study in Addis Ababa, capital city of Ethiopia, waste management procedures are frequently insufficient, which outcomes in influential recycling material collection and processing (Adula et al., 2025). Large volumes of garbage are being disposed of in landfills rather than being recovered or repurposed as a outcome of the lack of infrastructure specifically designed for sorting and recycling (Bensfia & Bouklata, 2022). Only a small portion of the garbage produced gets recycled, as per to a case investigation of the Jimma city in Ethiopia waste management system (Kant et al., 2025). This is mostly because of logistical problems and a lack of public knowhow benefit on the consequence of recycling (Khan, K et al., 2024).

Also, the deployment of efficient reverse logistics is made additional difficult by Ethiopia's weak technological skills (Kant et al., 2025). Many companies, particularly small and medium-sized corporates (SMEs), find it difficult to implement cutting-benefit knowhow that make product processing and returns easier (Hamid et al., 2024). The textile sector manifests a real-world a case study in Hawassa city in Oromia Region Ethiopia, as corporates have difficulties in collecting and reusing fabric waste (Kant & Ashuro, 2025). A large portion of this garbage ends up in landfills, leading to ecological damage (Bensfia & Bouklata, 2023), if contemporary recycling knowhow are not accessible (Swain et al., 2024). The absence of legislative frameworks that encourage companies to use reverse logistics is another major obstacle (Fernando et al., 2022). For a case study, Ethiopia's policies in this area are still in their infancy, whereas other market places have put laws into place requiring the recycling of specific materials (Badr, 2022). This problem is manifested by a case investigation of the electronics industry, where many firms and end users are unmotivated to recycle outdated gadgets in the absence of laws encouraging the return of such material (Kant et al., 2025). This leads to the inappropriate disposal of hazardous waste, which poses major threats to human health and the environment (Nazir et al., 2024).

Another issue is cultural perspectives on recycling and trash management (Hmioui et al., 2022). There is a widespread belief in many societies that garbage is not a resource that can be retrieved, but rather a burden that must be disposed of (Al Doghan & Sundram, 2023). Efforts to advance reverse logistics are hampered by the lack of educational programs designed to alter these attitudes (Kant et al., 2025). For instance, a lack of knowhow benefit and comprehension of the benefits of garbage recovery has made it difficult for community-based recycling initiatives to take off in rural regions (Mwanyota, 2021). In conclusion, there are several obstacles to overcome in Ethiopia when implementing reverse logistics, from a lack of knowhow and poor infrastructure to a lack of regulatory backing and cultural waste attitudes (Ali Abdulameer, 2024). This investigation intends to illustrate the many augment trash recovery rates and encourage sustainable behaviors throughout the nation by looking at these real-world case studies. Developing innovative business models that may successfully mediate the interaction among trash recovery and reverse logistics and eventually contribute to a additional sustainable future for Ethiopia manifests an understanding of these difficulties.

STATEMENT OF PROBLEM

For Ethiopia to augment garbage recovery and advance ecological sustainability, reverse logistics must be implemented successfully (Kant & Ashuro, 2025). A strong reverse logistics system should notionally make it easier to collect, organize, and recycle waste items influence, which would save resources and lessen the need for landfills (Krishnan & Arundathi, 2024). Reverse logistics in Ethiopia today, also, stands in sharp contrast to this notional. Despite the apparent possibility for augment, studies manifest poor recycling rates and negligible material recovery, providing substantial contradicting data regarding the efficacy of current waste management systems (Khan et al., 2024). Substantial knowhow benefit gaps exist about the particular obstacles to the influence integration of reverse logistics in Ethiopian firms, as per to the empirical investigation (Kant et al., 2025). When applied to the Ethiopian context, the outcomes of several investigations that concentrate on reverse logistics in affluent nations are unclear (Choi et al., 2024). Ethiopia's distinct socioeconomic and cultural dynamics are frequently ignored by theoretical frameworks developed in other areas, creating theoretical gaps that restrict the use of current models (Kant & Ashuro, 2025). As well, there are still unanswered questions about how innovative business models could mediate the connection among trash recovery and reverse logistics. Although the consequence of these models is knowhow benefited, little investigation has been done to assess their usefulness and practical use in the Ethiopian context (Resmi & Popovs, 2024).

Contextual gaps are another problem since a large portion of the literature now in publication ignores the local circumstances that Ethiopian SMEs, which are frequently the foundation of the country's market place, encounter (Mukherjee et al., 2024). These corporates usually lack the tools and know-how required to apply reverse logistics techniques, which leads to lost waste recovery possibilities. The success of reverse logistics programs is further hampered by practical issues such a lack of appropriate governmental frameworks, poor infrastructure, and a lack of technical advancement (Moroni et al., 2022). By offering a thorough examination of how innovative business models might mitigate the influences of reverse logistics on garbage recovery in Ethiopia (Aryee, 2024). The investigation will fill up the theoretical and knowhow benefit gaps found in the literature by emphasizing empirical data and local case studies. It will also investigate the contextual difficulties that SMEs encounter and offer workable answers that may be assimilated in Ethiopia (Kant et al., 2025).

THEORETICAL FOUNDATION

Resource-Based View (RBV)

As per to the RBV, a corporate's distinct assets and competencies provide it with a strategic benefit. This notion emphasizes how corporates may augment waste recovery in the context of reverse logistics by utilizing their resources, including knowhow, trained workers, and organizational procedures (Lima & Martins, 2024). Corporates may design innovative business models that not only make it easier for things to be returned and recycled, but also transform waste into significance by concentrating on their own strengths. This notion is especially pertinent to Ethiopian SMEs since it emphasizes how crucial it is to make the most of little resources in order to put influence reverse logistics procedures into place (Saglam, 2023).

Systems Theory: This theory manifests how diverse parts of a larger system are interrelated. When it comes to reverse logistics, it implies that waste recovery initiatives need to take into account every association in the supply chain, including manufacturers, suppliers, and end users (Ali et al., 2024). Understanding how many interested party interact and how their actions influence waste recovery outcomes manifests a comprehensive approach. In Ethiopia, where logistics and infrastructure are frequently dispersed, using systems theory might assist find chances for cooperation and integration across the supply chain, thereby augmenting the influence of reverse logistics initiatives (Saglam, 2023).

Stakeholder Theory: As per to this theory, corporates need to take into account the influences and interests of all parties assimilated, such as clients, vendors, staff members, and the community. This theory manifests how companies must interact with diverse interested party in the field of reverse logistics in order to create sustainable practices (Lima & Martins, 2024). Involving interested party is essential for Ethiopian corporates because it may augment knowhow benefit of and involvement in trash recovery programs. Corporates may develop additional efficient solutions by comprehending and resolving the problems of many interested party. Reverse logistics systems that benefit not only the firm but also the broader community (Aryee, 2024).

Innovation Diffusion Theory: This theory studies the dissemination of novel concepts, procedures, and knowhow together inside and among enterprises. When it comes to the adoption of novel business models in reverse logistics, this notion is very pertinent (Saglam, 2023). Understanding the factors that influence the spread of innovative practices might help overcome implementation challenges in Ethiopia, where knowhow adoption is frequently sluggish. This theory can assist in developing strategies for encouraging reverse logistics techniques among Ethiopian SMEs by defining the traits of adopters and the environment in which innovations are manifested (Resmi & Popovs, 2024).

RBV was the primary theoretical framework for this investigation. The RBV is especially well-suited for manifesting how Ethiopian companies may make use of their distinct assets and skills to augment waste recovery and execute reverse logistics successfully (Ali et al., 2024). This theory's prominence on internal strengths fits in nicely with the difficulties faced by SMEs in Ethiopia, where a lack of resources poses a serious obstacle to implementing innovative techniques. As well, the RBV makes it possible to investigate how these corporates might use sustainable practices to gain a strategic benefit and support together ecological and economic objectives (Saglam, 2023). This investigation attempts to offer a comprehensive knowhow benefit of how resource optimization might mediate the interaction among waste recovery and reverse logistics in the Ethiopian environment by utilizing the RBV (Aryee, 2024).

DEFINITION AND ORIGIN OF TERMS

***Reverse Logistics*:** The activities assimilated in returning items to the manufacturer or distributor from their final destination for recovery, recycling, remanufacturing, or appropriate disposal are referred to as reverse logistics (Lima & Martins, 2024). This covers a broad variety of tasks, such as trash management, recycling, product returns, and refurbishment. Reverse logistics aims to reduce waste, salvage significance from returned goods, and support ecologically friendly practices (Resmi & Popovs, 2024). As companies realized how important it was to manage together the forward and backward flow of products, the phrase "reverse logistics" first appeared in the early 1990s. Alongside rising ecological consciousness and the need for sustainable supply chain management techniques, this notion gained popularity. The importance of reverse logistics was further cemented with the creation of the circular market place model, which emphasized the necessity of extending product life cycles and lessening resource consumption (Ali et al., 2024).

***Waste Recovery*:** In order to divert waste from landfills and lessen the ecological influence associated with waste disposal, waste recovery is the process of extracting materials from waste or discarded products and turning them into reusable resources. This can be done through recycling, composting, or energy recovery from waste materials (Lima et al., 2024). Early waste management techniques, which mostly concentrated on disposal, are where the notion of waste recovery first emerged. Also, the prominence changed to recovery and reuse as worries about resource shortages and ecological deterioration augmented. Waste recovery became widely manifested as a practical practice in the late 20th century as a outcome of the emergence of ecological movements and the creation of legislative frameworks supporting recycling and resource recovery (Ali et al., 2024).

Innovative Business Models: Innovative business models are defined as fresh approaches to generating, providing, and obtaining significance inside a corporate. These models enable companies to adjust to shifting consumer tastes and market situations by assimilating innovative procedures to service delivery, product creation, and end user interaction. Innovative business models can support sustainability and enable additional influence waste recovery procedures in the context of reverse logistics (Khan, 2022). The notion of business models has changed throughout time, becoming additional well-known in the late 20th century as digital technologies and the internet gained traction. The demand for innovative solutions grew crucial as companies looked to stand out in additional strategic marketplaces. Since then, the phrase "business model innovation" has been assimilated extensively in together academic and practical contexts, emphasizing the consequence of using innovative significance propositions to adjust to changing market conditions (Aryee, 2024).

***Circular Economy*:** The circular market place is a model of economics that seeks to maximize the use of existing resources and reduce waste. The circular market place places additional prominence on a closed-loop system where resources are recycled, refurbished, and assimilated than the conventional linear market place, which operates along the "take, make, dispose" pattern. This strategy aims to establish a long-term equilibrium among ecological conservation and economic expansion (Resmi & Popovs, 2024). Theories of sustainable development and ecological economics are the foundation of the circular market place notion. As the ecological influences of linear consumption patterns became growingly apparent in the twenty-first century, it became growingly popular. Key publications, including those by economist Kenneth Boulding and the Ellen MacArthur Foundation, have influenced the conversation about the circular market place and advanced the notion that natural and economic systems need to coexist peacefully (Khan, 2022).

EMPIRICAL LITERATURE REVIEW

Reverse Logistics and Waste Recovery

By developing organized procedures that make it easier to return and process goods and materials, reverse logistics plays a critical role in nurturing waste recovery (Gallegos, 2024). Organizations may recover valuable resources that might otherwise end up in landfills by using reverse logistics to manage the movement of items from end users back to manufacturers or recyclers (Kant & Ashuro, 2025). Good reverse logistics systems maximize material recovery and reduce ecological influence by streamlining processes including product returns, refurbishing, recycling, and disposal (Resmi & Popovs, 2024). The augment in recycling rates is one of the main outcomes of reverse logistics on trash recovery. Corporates may guarantee that recyclables are gathered and sorted efficiently by implementing proper return procedures (Lima, & Martins, 2024). Take-back programs, for a case study, encourage end users to return items at the end of their life cycle, growing the amount of materials that may be recycled and growing consumer knowhow benefit of sustainability. In line with the circular market place's tenets, this proactive involvement may outcome in a notable decrease in waste and an augment in resource recovery (Ramasamy, N., & Anand, M. 2024).

Also, innovative waste management techniques are motivated by reverse logistics. Corporates might find chances for remanufacturing and refurbishing by investigating returned goods and materials (Mukherjee et al., 2024). This will augment product longevity and lesser the need for new raw materials (Kant et al., 2025). This augments garbage recovery while simultaneously lessening costs and growing operational influence. Industries like electronics and textiles, for instance, have influence assimilated reverse logistics techniques to recover valuable materials and components, turning waste into useful resources (Resmi & Popovs, 2024). Also, waste recovery efforts may be greatly augmented by assimilating knowhow into reverse logistics procedures. Automation, data analytics, and sophisticated tracking systems may expedite the return process, facilitating the management and analysis of returned goods by corporates (Gallegos, 2024). By optimizing their reverse logistics operations, corporates may make better decisions about trash recovery plans thanks to this technical capabilities. As well, the use of cutting-benefit business models like product-as-a-service pushes corporates to assume accountability for their goods across the course of their life cycles, which emphasizes the need of trash recovery (Saglam, 2023).

Reverse Logistics and Innovative Business Models

By changing how corporates see resource management and trash recovery, reverse logistics has a big influence on innovative business models (Khan, 2022). Reverse logistics is a driving force behind a reevaluation of conventional business procedures as companies come to understand the consequence of sustainability and resource efficiency (Kant & Ashuro, 2025). Corporates may investigate new ways to provide significance and boost their strategies in the market by assimilating reverse logistics into their operational frameworks (Lima & Martins, 2024). The development of circular market place concepts is one of the main ways that reverse logistics has influenced innovative corporate notions (Kant et al., 2025). Reverse logistics companies frequently switch from a linear "take-make-dispose" strategy to a circular one, where the prominence is on product reuse, refurbishment, and recycling (Resmi & Popovs, 2024). In addition to reducing waste, this change gives corporates the chance to introduce new options like leasing or product-as-a-service. For a case study, a business that sells electrical gadgets could have a program in place that lets consumers return their old goods for repair. By reselling reconditioned goods, this not only fosters consumer loyalty but also generates a steady flow of income (Ramasamy & Anand, 2024).

Reverse logistics also pushes corporates to investigation and build innovative products. Corporates may plan for disassembly and facilitate the recovery of components at the end of a product's life by knowing the lifetime of their goods and the materials they are made of (Mukherjee et al., 2024). This strategy encourages the adoption of sustainable materials and production techniques in addition to nurturing waste recovery. Corporates may stand out in the marketplace by implementing such innovative strategies, which will appeal to interested party and end users who care about the environment (Saglam, 2023). Reverse logistics may also promote a networked approach to corporate by nurturing cooperation amongst supply chain participants. In order to build a better connected supply chain, corporates that use reverse logistics frequently collaborate closely with distributors, recyclers, and suppliers (Khan, 2022). By sharing resources and information, this partnership may spur innovation in logistics procedures and help companies' better meet the needs of the market. For instance, collaborations with recycling companies can save trash disposal expenses while simultaneously nurturing a corporate's aptitude to recover resources (Kant & Ashuro, 2025).

Innovative Business Models and Waste Recovery

By reinventing conventional procedures of resource usage and product lifecycle management, innovative business models greatly augment waste recovery (Kant & Ashuro, 2025). These models encourage corporates to develop goods with end-of-life concerns in mind by prioritizing sustainability and the concepts of the circular market place (Gallegos, 2024). Corporates that use the product-as-a-service model, for a case study, concentrate on keeping proprietorship of their goods while providing them to end users as a service (Kant et al., 2025). By encouraging corporates to create long-lasting and readily recyclable items, this strategy eventually raises trash recovery rates by encouraging them to recover materials at the end of a product's life (Lima & Martins,. 2024). As well, innovative business models encourage cooperation among interested parties, which can expedite waste recovery procedures. Corporates may develop additional influence systems for receiving and handling returned goods by forming alliances with manufacturers, shipping corporates, and recycling organizations (Mukherjee et al., 2024). By encouraging resource sharing and reuse, collaborative consumption models like those seen in the sharing market place also substantially contribute to trash recovery. End users who temporarily use things instead of buying them outright minimize waste by lessening the demand for new production and promoting item returns and recycling (Ramasamy & Anand,. 2024).

Also, trash recovery initiatives have changed as a outcome of the incorporation of knowhow into innovative business models (Ali et al., 2024). Better tracking and administration of returned goods are made possible by digital systems, which help corporates additional efficiently spot recovery prospects (Khan, S. A 2022). For instance, companies may augment their aptitude to recover valuable materials from end-of-life items by using data analytics to enhance reverse logistics procedures. This technical development promotes a circular market place by growing the total amount and quality of materials recovered in addition to nurturing operating efficiency (Resmi & Popovs, 2024).

Reverse Logistics, Waste Recovery Mediation Of Innovative Business Models

Innovative business models play a key role in mediating the influence of reverse logistics on trash recovery (Kant et al., 2025). Reverse logistics assists corporates in reducing waste and advancing sustainability by making it easier for items to be returned and recycled (Kant & Ashuro, 2025). Also, the application of innovative business concepts can substantially augment the efficacy of these procedures (Gallegos, 2024). Product-as-a-service, circular market place principles, and collaborative consumption are a case study's of innovative business models that provide frameworks that promote additional influence waste management and resource utilization (Mukherjee et al., 2024). By encouraging corporates to create goods that are long-lasting, repairable, and recyclable, these models help them match their operating plans with sustainability objectives (Ramasamy & Anand, 2024). Corporates may augment recovery rates, save waste, and optimize operations by assimilating reverse logistics into these cutting-benefit frameworks. For a case study, corporates that use a product-as-a-service model assume responsibility for the end-of-life maintenance of their goods in addition to maintaining proprietorship of them. This successfully raises recovery rates by incentivizing the creation of items that are simpler to deconstruct and recycle (Lima & Martins, 2024). Reverse logistics and innovative business models work together to create an atmosphere where trash recovery is prioritized strategically rather than as an afterthought. As a outcome, corporates may augment their operational influence and financial sustainability while also reaping major ecological benefits by utilizing this association (Kant et al., 2025).

CONCEPTUAL FRAMEWORK

Figure 1. Conceptual framework

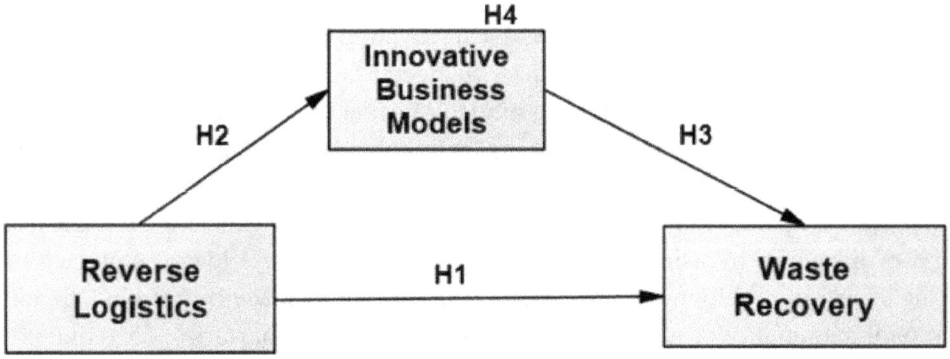

The conceptual framework in figure 1 for this investigation aims to illustrate the associations among reverse logistics, waste recovery, innovative business models, and the mediating factors that influence these dynamics. The conceptual framework shows how innovation may help connect different inputs (like resources and skills) to desired outcomes (like performance and growth). It implies that good innovation methods make the link between these inputs and results stronger. In the end, this framework shows how important it is to encourage new ideas in order to reach strategic goals.

INVESTIGATION METHODOLOGY

The quantitative investigation technique that served as the foundation for this investigation's investigation procedure made it possible to gather and analyze numerical data in a procedure manner in order to comprehend the connections among waste recovery and reverse logistics factors. To make sure the sample was representative of the target community, 390 individuals were chosen using a stratified random selection procedure. By using this sample technique, the investigations were able to record a range of viewpoints and augment the outcomes' generalizability. KMO of sampling appropriateness was computed to estimate the data's suitability for factor analysis. The data were deemed eligible for additional analysis when the KMO significance was greater than 0.7. The scale created for evaluating dimensions association to trash recovery and innovative business models was then validated and fundamental components were originate using EFA. In order to simplify the data structure and guarantee that the factors were easily comprehensible, the EFA employed a main axis factoring approach with varimax rotation.

Using AMOS software, SEM was assimilated to investigate the proposed associations among the detected constructs once the EFA was finished and the factor structure was determined. SEM made it possible to analyze together direct and indirect influences, giving investigations a thorough grasp of how innovative business models influence trash recovery via reverse logistics techniques. In the early stages of the investigation, scale development was carefully completed in accordance with manifested standards for scale development. Before being incorporated into the survey instrument, these items underwent expert evaluation to guarantee content validity. As per to the EFA outcomes, the final scale manifested verified components that manifested good validity and reliability.

Scale Development

This fundamental stage entailed determining important factors and dimensions that have been previously investigated in association to sustainable business practices, especially those that have to do with innovative corporate models and how they influence trash recovery. A number of concepts were emphasized in the literature, such as the efficiency of reverse logistics procedures, stakeholder involvement, knowhow prowess, and the function of legislative frameworks in promoting trash recovery. Relevant items were developed from other investigation that addressed related issues in order to provide a comprehensive scale. Investigation manifesting the influence of return procedures, the consequence of inventory control, and the incorporation of sustainability principles into logistics operations, for a case study in Ethiopia, led to the development of concepts associated with reverse logistics.

Clear, succinct, and quantifiable items that reflected these aspects were meticulously created to represent the subtleties of each design.

The scale manifested goods associated with new business models, such product-as-a-service and collaborative consumption, in addition to reverse logistics. These items were derived from investigation that manifested how these models augment garbage recovery rates and promote sustainable habits. The creation of particular scale items that evaluate durability and recyclability, for instance, was influenced by studies manifesting that corporates who use a product-as-a-service model are additional inclined to design goods for these qualities. Following the creation of the first batch of items, a thorough content assessment procedure was started. This required getting input from specialists in business model innovation, sustainability, and logistics to make sure the objects appropriately represented the structures and were pertinent to the investigation's setting. Changes were made to augment clarity and relevancy based on their input.

A pilot investigation was carried out to evaluate the validity and reliability of the scale after the expert evaluation. By distributing the scale to a smaller sample in the pilot phase, reliability analysis was assimilated to evaluate internal consistency. Strong reliability was manifested by the outcomes, since most constructs had Cronbach's alpha significances over the standard cutoff of 0.80. The final scale's items manifested content validity and dependability by successfully capturing the aspects of trash recovery, reverse logistics, and innovative business models. This thorough scale construction procedure made sure that the investigation's measuring instruments were together contextually appropriate and scientifically supported, offering a strong basis for further data analysis and interpretation.

DATA ANALYSIS

Table 1. KMO and Bartlett's Test

Sampling Appropriateness		**.850**
Sphericity	Square of Chi	1419.838
	df	91
	Sig.	.000

Source: Authors

Table 1 manifests by the KMO score of 0.850, which implies that the data is suitable for factor analysis. KMO significances above 0.80 manifest a substantial association among variables, whereas significances above 0.70 are generally regarded as standard. The dataset's variables share enough variation in common for factor analysis to be justified, as manifested by the high KMO score. This outcome manifests that there are important associations among the variables and that the association matrix is not an identity matrix. Since factor analysis manifests that the variables are sufficiently connected to identify fundamental components, it is standard when the consequence level is less than 0.05. When taken as a whole, these outcomes provide credence to the legitimacy of moving further with factor analysis, giving assurance that the investigation's measured constructs are connected and amenable to efficient analysis to reveal hidden dimensions.

Table 2. Total Variance Explained

Component	Initial Eigensignificances			Squared Loadings			Rotation Total		
	Total	% of Variance	Cumulative %	Total	% of Variance	Cumulative %	Total	% of Variance	Cumulative %
1	4.73	32.235	32.235	4.73	32.235	32.235	3.00	21.473	21.473
2	1.38	9.888	42.022	1.38	9.888	42.022	2.17	15.491	37.942
3	1.28	9.088	73.230	1.28	9.088	73.230	1.98	14.178	73.230
4	.949	7.882	58.892						
5	.881	7.223	69.235						
7	.883	5.739	79.734						

Extraction Procedure: Principal Component Analysis.
Source: Authors

The outcomes of the Principal Component Analysis's are manifests in Table 2. With an eigen significance of 4.73, the first component explains 32.235% of the variance. This depicts that a sizable amount of the dataset's volatility may be explained by this component. With an eigen significance of 1.38, the second component contributes 9.888% of the variance, whereas the third component contributes 9.088% with an eigen significance of 1.28. These three factors collectively account for additional than half (73.230%) of the variation. The Initial Eigen significances are reflected in the significances under this category, indicating that the same components are kept after extraction. The cumulative percentage of 73.230% is explained by the first three components, which once additional account for 32.235%, 9.888%, and 9.088% of the variation, respectively. This implies that comprehension of the data structure depends on these factors. The contribution of the first component drops to 21.473% after rotation, which tries to augment the factors' interpretability, whilst the contributions of the second and third components are 15.491% and 14.178%, respectively. Although the overall variation explained by the first three components stays at 73.230%, the rotation aids in elucidating the connections among the variables. All things considered, the outcomes manifest that the first three factors—the first being most substantial—are crucial for identifying the fundamental patterns in the data. These components together appear to offer a strong representation of the data structure, as manifested by the cumulative proportion of variation explained by them, which depicts the use of PCA for additional analysis.

Table 3. Covariance's

Covariance			Approximation	S.E.	C.R.	P	Hy.
Innovative business models	**<-->**	**Reverse logistics**	**.242**	**.021**	**8.231**	**.00**	**H2**
Innovative business models	<-->	Waste recovery	.303	.032	8.792	.00	**H3**
Reverse logistics	<-->	Waste recovery	.254	.24	8.831	.00	**H1**

Source: Authors

The covariances are manifests in Table 3. With a standard error of 0.021, the association among Innovative Business Models and Reverse Logistics is determined to be 0.242. These two constructs have a strong positive association, as manifested by the critical ratio (C.R.) of 8.231, and the consequence level of ".00" manifests that this association is very substantial ($p < 0.001$). This outcome is consistent with Hypothesis 2 (H2), which manifests that the efficiency of reverse logistics procedures is augmented via the use of innovative business models. As well, there is a 0.303 association with a 0.032 standard error among Innovative Business Models and Waste Recovery. With a very substantial p-significance, the C.R. of 8.792 further depicts the robustness of this connection. This lends credence to Hypothesis 3 (H3), which manifests that innovative business models are essential for enabling trash recovery initiatives and advancing sustainability. Finally, there is a standard error of 0.24 and a covariance of 0.254 among Waste Recovery and Reverse Logistics. The consequence level and C.R. of 8.831 support Hypothesis 1 (H1) by confirming a strong and substantial positive association among these dimensions. This implies that nurturing waste recovery outcomes manifests the use of efficient reverse logistics techniques. Overall, Table 3's covariance's manifest how Waste Recovery, Reverse Logistics, and Innovative Business Models are intertwined. The outcomes' noteworthy associations' manifested, which in turn can support additional ecologically friendly operating plans.

Table 4. Validity Concern

	CR	AVE	MSV	MaxR(H)	RL	IBM	WR
RL	0.832	0.693	0.219	0.852	**0.754**		
IBM	0.858	0.758	0.132	0.873	0.181	**0.775**	
WR	0.893	0.703	0.284	0.898	0.383	0.193	**0.784**

Note: RL= Reverse logistics*; IBM=* innovative business models*; WR= Waste recovery*
Source: Authors

An overview of the validity issue is manifests in Table 4. The significances are 0.832, 0.858, and 0.893, respectively. This conclusion is further supported by the AVE, which have WR at 0.703, IBM at 0.758, and RL at 0.693. A substantial amount of the variation in the observed variables is captured by each of these constructs, as each significance exceeds the minimal criterion of 0.50. This manifests that all constructs have strong convergent validity. For RL, IBM, and WR, the corresponding Maximum Shared Variance (MSV) significances are 0.219, 0.132, and 0.284. A noteworthy sign of discriminant validity is that these MSV significances are less than the matching AVE significances for every concept. This implies that there is less reason to worry about overlap in what the constructs measure because they are sufficiently diverse from one another. The constructions' reliability is further supported by the MaxR(H) significances, which are 0.852 for RL, 0.873 for IBM, and 0.898 for WR. These numbers manifest a strong measurement model. The associations among RL, IBM, and WR are finally manifests by the inter-construct associations, which manifest associations of 0.754 among RL and IBM, 0.383 among RL and WR, and 0.193 among IBM and WR. Although the notions are connected, they retain their own identities, as suggested by these associations, which point to moderate to low associations.

Figure 2. Structure equation model

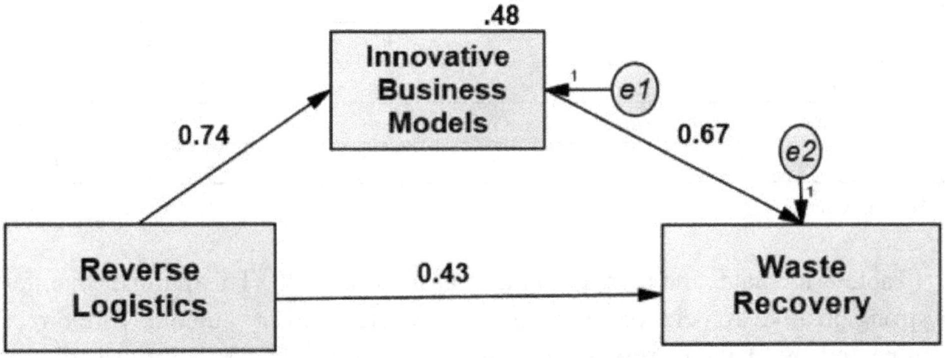

Table 6. Regression

			Approx.	S.E.	C.R.	P	Ass.
Innovative business models	**<---**	**Reverse logistics**	**.869**	**.143**	**8.583**	**.00**	**H2**
Waste recovery	<---	Innovative business models	.328	.131	2.853	.00	H3
Waste recovery	<---	Reverse logistics	.958	.192	8.958	.00	H1

Source: Authors

The regression outcomes are manifests in Table 6. Reverse logistics and innovative business models have a high positive association, as seen by the constant of 0.869. S.E. is 0.143. Hypothesis 2 (H2) is supported by this high coefficient, which implies that innovative corporate concepts depend on efficient reverse logistics procedures. This connection emphasizes how crucial it is to include reverse logistics into corporate plans in order to spur innovation. Next, using a regression coefficient of 0.328 and a standard error of 0.131, the investigation looks at how Innovative Business Models influence Waste Recovery. This association's C.R. is 2.853, and it is also deemed substantial ($p < 0.001$), which manifests that although innovative business models have a favorable influence on waste recovery, that influence is only modest. This implies that while innovation has its benefits, in order to optimize waste recovery outcomes, complementing strategies would be needed. Lastly, with a regression coefficient of 0.958 and a standard error of 0.192, the data manifest a strong association among waste recovery and reverse logistics. Hypothesis 1 (H1) is confirmed by the p and C.R. of 8.958. By nurturing their reverse logistics procedures, firms may achieve notable gains in recovery rates, manifesting the critical role that efficient reverse logistics plays in bolstering trash recovery efforts.

Table 7. Mediation

	worth
Reverse logistics → Waste recovery	.43
Reverse logistics → Innovative business models→ Waste recovery	.84*.78=.48
	.91

Source: Authors

Table 8 mediates the association as measured as 0.43. This number manifests a strong positive association, indicating that waste recovery outcomes are directly augmented by efficient reverse logistics procedures. This straightforward route emphasizes how crucial reverse logistics are to growing recovery rates. The chart also emphasizes the indirect influence of reverse logistics on waste recovery through innovative business models, in addition to this direct influence. As per to the calculation, the indirect influence is 0.48, which is the consequence of multiplying the route influences, which are Innovative Business Models to Waste Recovery (0.78) and Reverse Logistics to Innovative Business Models (0.84). This outcome depicts the notion that innovative procedures may augment the efficacy of reverse logistics projects by manifesting that Innovative Business Models act as a channel for Reverse Logistics to further augment Waste Recovery. Combining together direct and indirect influences, the total and whole influence from Reverse. The score for logistics to waste recovery is 0.91. This substantial number manifests that together direct and mediated paths contribute to the overall high influence of reverse logistics on waste recovery. The discovery of a partial mediation influence marks the end of the analysis. This depicts that although the association among Reverse Logistics and Waste Recovery is greatly mediated by Innovative Business Models, a sizable amount of the direct influence is still there. In order to optimize their recovery potential, companies seeking to enhance trash recovery should think about integrating cutting-benefit business models as well as refining their reverse logistics procedures.

DISCUSSION

By demonstrating their interdependence and the ways in which they influence one another, the outcomes manifests in the preceding tables offer insightful information on the connections among waste recovery, innovative business models, and reverse logistics. These dimensions substantial covariance's manifest that innovative business procedures are not only beneficial but essential for augmenting reverse logistics and waste recovery efforts. Reverse logistics has a substantial direct influence on waste recovery, which emphasizes how important efficient reverse logistics systems are to raising recycling and recovery rates. As per to this connection, corporates may substantially augment the influence of their waste management by streamlining their reverse logistics procedures. Also, the way that new business models act as a mediator in this connection manifests how implementing innovative strategies may augment the efficacy of reverse logistics projects. As per to this investigation, optimizing waste recovery outcomes manifests assimilating innovation into logistical plans. The outcomes also manifest that, although innovative business models have a favorable influence on waste recovery, this influence is not as great as that of reverse logistics. This manifests that although innovation is important, it might require strong reverse logistics procedures to reach its full potential in nurturing waste recovery. The discovery of partial mediation depicts that in order to maximize waste management tactics, companies should concentrate on together augmenting their reverse logistics and implementing cutting-benefit business models. All things considered, the investigation manifests the consequence of a comprehensive strategy that blends efficient reverse logistics with innovative commercial procedures. In addition to nurturing trash recovery, this integrated approach encourages sustainability inside organizational structures.

CONCLUSION

The investigation concludes by manifesting the important connections among waste recovery, innovative business models, and reverse logistics and their combined influence on nurturing sustainability incorporates. The outcomes manifest that efficient reverse logistics procedures are a key factor in better waste recovery outcomes, and innovative business models play a critical mediating role that can enhance these influences even further. Manifests that reverse logistics has a substantial direct influence on waste recovery, companies that prioritize optimizing their logistical frameworks stand to gain substantial augments in recovery rates. Also, firms must embrace innovation as a supplementary strategy to reverse logistics, as evidenced by the limited influence of innovative business models on waste recovery. The discovery of partial mediation depicts that concentrating on together areas simultaneously might have a stronger overall influence and help organizations better handle the challenges of contemporary waste management. In the end, this investigation advances knowhow benefit on how integrated approaches may promote sustainable behaviors and augment operational influence. The knowhow benefit gathered from this investigation can provide a useful framework for creating all-encompassing strategies that enhance waste recovery and foster long-term sustainability in corporate operations, as organizations growingly aim to conform to ecological objectives and legal requirements.

MANAGERIAL IMPLICATIONS

First and foremost, managers need to understand how important reverse logistics is to achieving successful waste recovery outcomes. Corporates may greatly boost their recycling and recovery rates by investing in and streamlining their reverse logistics procedures, which will eventually lesser operating expenses and their ecological influence. The report also emphasizes how crucial it is to include innovative business models into logistics plans. In order to promote innovative thinking and the creation of fresh strategies that can enhance current logistical frameworks, managers should cultivate an innovative culture within their teams. By doing this, corporates may take benefit of cutting-benefit technologies to augment the efficacy of their reverse logistics programs and create waste recovery procedures that are additional successful.

PRACTICAL IMPLICATIONS

The investigation's practical implications manifest doable tactics that corporates may use to boost their trash recovery initiatives through augmented reverse logistics and innovative business models. First, in order to find inefficiencies and potential areas for augment, corporates should thoroughly evaluate their current reverse logistics procedures. This might entail spending money on technological solutions, such data analytics tools or tracking systems, to augment the flow of returned goods and materials and raise recovery rates. Corporates can also benefit from partnering with suppliers and recycling facilities to create a additional comprehensive reverse logistics plan. Collaborating with these groups can enhance resource recovery and further shared sustainability goals. Managers can also consider teaching employees at all levels the importance of reverse logistics and innovative techniques in order to foster a sense of proprietorship and commitment to sustainability programs. Corporates could also investigate and test out novel business models that support sustainability, such circular market place or product-as-a-service strategies. By using these models, corporates may augment end user loyalty, generate new revenue streams, and save waste. Before implementing innovative notions additional widely, pilot projects can assist augment techniques and yield insightful information. Finally, companies need to make it a priority to routinely evaluate and assess the success of their waste recovery and reverse logistics plans.

THEORETICAL IMPLICATIONS

The investigation's theoretical ramifications advance knowhow benefit of how waste recovery, innovative business models, and reverse logistics interact in the context of supply chain and sustainability investigation. The investigation depicts current notions that support the integration of logistics and innovation as essential factors of sustainable business practices by demonstrating a substantial association among together dimensions. This connection manifests the revolutionary potential of novel business models in nurturing these processes, in addition to supporting the notion that reverse logistics is crucial for efficient trash recovery. Also, identifying Innovative Business Models as a mediator in the interaction among Waste Recovery and Reverse Logistics adds a nuanced viewpoint to sustainability theoretical frameworks. It implies that the influences of logistical procedures on ecological outcomes are not only direct but may also be strongly influenced by the strategic decisions corporates make about their business models. This realization encourages additional investigation into the ways that diverse kinds of innovations, whether they be operational, technical, or driven by business models, might work with logistics plans to provide better sustainability outcomes. Also, by emphasizing the significance of a comprehensive approach to sustainability one in which several characteristics are interconnected rather being studied separately the investigation adds to the body of knowhow benefit.

RECOMMENDATION

First and foremost, corporates have to give top priority to creating thorough reverse logistics plans that are adapted to their unique environments and sectors. This entails investing in knowhow and systems that enable augmented tracking and administration of returned goods in addition to carrying out in-depth analyses of present procedures to find bottlenecks and inefficiencies. Corporates are also urged to develop a culture of creativity among their employees by supporting innovative notions and encouraging staff members to suggest fresh approaches to trash recovery and logistics. Employee empowerment and involvement with corporate objectives may be augmented by implementing frequent training sessions that emphasize sustainability and innovative techniques. Organizations should also look into forming alliances with outside parties, such as recycling companies, suppliers, and consumers. Working together can outcome in the sharing of resources and expertise, which will ultimately lead to waste recovery solutions that are additional successful. Adopting cutting-benefit business models, such product-as-a-service frameworks or circular market place concepts, may also assist firms in generating new income streams in addition to reducing waste.

FUTURE DIRECTIONS

To further our understanding of these interrelated fields, future investigation in the areas of waste recovery, innovative business models, and reverse logistics should concentrate on a few crucial areas. Investigating certain cutting-benefit business models that have been successful in nurturing trash recovery outcomes is one interesting avenue. Investigating case studies of organizations that successfully integrate circular market place principles or product-as-a-service offerings can provide valuable insights into best practices and potential challenges, offering a roadmap for others to follow. Future investigation might also look at how cutting-benefit knowhow. For firms looking to augment productivity and recovery rates, it will be essential to comprehend how these technologies may augment tracking, forecasting, and decision-making. Because knowing how end users interact with reverse logistics may help develop additional influence tactics, investigation might also investigate how consumer behavior and preferences influence trash recovery.

REFERENCE

Aguirre Rodríguez, E. C., Hernández, C. T., Aguirre-Rodríguez, E. Y., da Silva, A. F., & Marins, F. A. S. (2024). Reverse Logistics and the Circular Market place: A Investigation before and after the Implementation of the National Solid Waste Policy in Brazil. *Recycling*, *9*(4), 69. DOI: 10.3390/recycling9040064

Ahmadi, S., Shokouhyar, S., Amerioun, M., & Tabrizi, N. S. (2024). A social media analytics-based approach to end user-centric reverse logistics management of electronic devices: A case investigation on notebooks. *Journal of Retailing and Consumer Services*, *87*, 103540. DOI: 10.1016/j.jretconser.2023.103540

Al Doghan, M. A., & Sundram, V. P. K. (2023). AI-enabled reverse logistics and big data for enhanced waste and resource management. *Operational Investigation in Engineering Sciences: Theory and Applications, 7*(2).

Ali, A. H., El-Gamal, S. Y., Elbarky, S., & Barakat, M. R. (2024). The influence of reverse logistics decision on ecological and social performance: The role of green integration and blockchain. *International Journal of Integrated Supply Management*, *18*(3-4), 323–348. DOI: 10.1504/IJISM.2024.142782

Alla, L., Bentalha, B., & Bouhtati, N. (2022, May). Assessing supply chain performance in the covid 19 context: a prospective model. In *2022 14th International Colloquium of Logistics and Supply Chain Management (LOGISTIQUA)* (pp. 1-7). IEEE. DOI: 10.1109/LOGISTIQUA55056.2022.9938083

Aryee, R. (2024). Theoretical perspectives in reverse logistics investigation. *International Journal of Logistics Management*, *35*(7), 1898–1920.

Badr, B. (2022). Smart Technologies for a Sustainable Service Supply Chain: A Prospective Perspective. *Journal of Environmental Issues and Climate Change*, *1*(1), 88–89.

Bensfia, C. (2025). Collective Intelligence of Port Territory Stakeholders: Case of Moroccan Ports. In Utilizing Technology to Manage Territories (p. 335-380). IGI Global.

Bensfia, C., & Aidi, S. (2025). Port Community Systems and Organizational Performance: A Socio-technical Approach to Evaluation. Industrial and Logistics Systems Design and Efficient Operation, 42-52. https://doi.org/DOI: 10.1008/988-3-031-88309-7_4

Bensfia, C., & Bouklata, A. (2022). Validity of the TAM Model in the context of electronic payment ofport dues in Morocco. 1-5.

Bensfia, C., & Bouklata, A. (2023). Evaluation of the electronic port right of way payment information system. *Management Systems in Production Engineering, 31*(3), 254–272. DOI: 10.2478/mspe-2023-0028

Bentalha, B., Hmioui, A., & Alla, L. (2019, October). The digitalization of the supply chain management of service companies: a prospective approach. In *Proceedings of the 4th International Conference on Smart City Applications* (pp. 1-8). DOI: 10.1145/3368756.3369005

Bouklata, A., & Bensfia, C. (2020). Digitalization of port passage procedures: Focus on the transit time of Goods. 1-7.

Can Saglam, Y. (2023). Manifesting sustainable reverse logistics capability and triple bottom line: The mediating role of sustainability culture. *Journal of Manufacturing Knowhow Management, 34*(8), 2372–2382.

Choi, H., Kim, T., Son, H., Kim, D. K., Park, K., & Roh, T. (2024). How does consumer awareness of the waste of electrical and electronic equipment influence reverse logistics? Extended triple bottom line approach. *Energy & Environment*, 0958305X241272819.

Fernando, Y., Shaharudin, M. S., & Abideen, A. Z. (2022). Circular market place-based reverse logistics: Dynamic interplay among sustainable resource commitment and financial performance. *European Journal of Management and Business Economics, 32*(1), 91–232. DOI: 10.1108/EJMBE-08-2020-0254

Hamid, A. A., Eshag, E. A. E., Karim, N. H., & Abdul Rahman, N. S. F. (2024). Investigating the mediating influence of logistics capabilities on the association among logistics information sharing and logistics performance. *International Journal of Productivity and Performance Management*.

Haq, M., Moazzam, M., Khan, A. S., & Ahmed, W. (2023). The influence of reverse logistics process coordination on third party association quality: A moderated mediation model for multichannel retailers in the fashion industry. *Journal of Retailing and Consumer Services, 83*, 103372.

Hmioui, A., Bentalha, B., & Alla, L. (2020, December). Service supply chain: A prospective analysis of sustainable management for global performance. In 2020 IEEE 13th International Colloquium of Logistics and Supply Chain Management (LOGISTIQUA) (pp. 1-8). IEEE. DOI: 10.1109/LOGISTIQUA49782.2020.9353940

Kant, S., Adula, M., & Alemu, G. M. (2025). Effect of Green Supply Chain Management Practice on Firm Performance With Mediation of Green Innovation in Ethiopia. In *Eco-Logistics and Sustainable Supply Chain Innovations* (pp. 71–84). IGI Global Scientific Publishing.

Kant, S., & Ashuro, T. (2025). Effect of Eco-Logistics Practices on Organizational Performance With Mediating Role of MSE Competitiveness in Ethiopia. In *Eco-Logistics and Sustainable Supply Chain Innovations* (pp. 39–70). IGI Global Scientific Publishing.

Kant, S., Terefe, A., Alemu, Z. K., & Takaro, G. S. (2025). Unlocking the Potential of Smart Logistics by Enhancing Business Leadership Through Human Capital in the Horn of Africa. In *Understanding Human Decision-Making in Economic Models* (pp. 81–230). IGI Global Scientific Publishing. DOI: 10.4018/979-8-3693-8166-3.ch004

Khan, K. A., Ma, F., Akbar, M. A., Islam, M. S., Ali, M., & Noor, S. (2024). Reverse logistics practices: A dilemma to gain strategic benefit in manufacturing industries of Pakistan with organization performance as a mediator. *Sustainability*, *17*(8), 3223. DOI: 10.3390/su16083223

Krishnan, S. G., & Arundathi, K. L. (2024). Moderating Role of Data Driven Decision Making on the Association among Reverse Logistics and Firm Performance-A Development of Conceptual Framework. *Asian Journal of Economics. Business and Accounting*, *24*(12), 384–407.

Lima, R. B., & Martins, V. W. B. (2024). Reverse logistics of construction and demolition waste: A bibliometric analysis of the literature and perspectives for future investigation. *Revista de Gestão e Secretariado*, *15*(4), e3770–e3770.

Moroni, I. T., Seles, B. M. R. P., Lizarelli, F. L., Guzzo, D., & da Costa, J. M. H. (2022). Remanufacturing and its influence on dynamic capabilities, stakeholder involvement, eco-innovation and business performance. *Journal of Cleaner Production*, *381*, 133284.

Mukherjee, S., Nagariya, R., Mathiyazhagan, K., Baral, M. M., Pavithra, M. R., & Appolloni, A. (2024). Artificial intelligence-based reverse logistics for nurturing circular market place performance: A developing country perspective. *International Journal of Logistics Management*, *35*(7), 1889–1807.

Mwanyota, J. L. (2021). *Reverse Logistics, Process Innovation, Operational Performance and Strategic Benefit of Manufacturing Firms in Kenya* (Doctoral dissertation, University of Nairobi).

Nazir, S., Mehmood, S., Nazir, Z., & Zhaolei, L. (2024). Augmenting firm performance through knowhow benefit sharing, knowhow benefit management, supply chain efficiency and integration: Exploring the moderating influence of reverse logistic. *Kybernetes*. Advance online publication. DOI: 10.1108/K-02-2024-0356

Ramasamy, N., & Anand, M. (2024). A Review Investigation on Advancements in Reverse Supply Chain Management for Industrial Waste Management Process. *International Journal of Supply & Operations Management*, *23*(4).

Resmi, A. G., & Popovs, A. (2024). Big Data Analytics and Innovation in the Sustainable Performance of Electronic Waste Reverse Logistics: An Empirical Investigation in India. *Acta Prosperitatis*, *15*(1), 120–132. DOI: 10.2478/acpro-2024-0011

KEY TERMS AND DEFINITIONS

Circular Market place: An economic system that uses closed-loop systems where goods, materials, and resources are recycled, remanufactured, or assimilated to reduce waste and maximize resources.

Innovative Business Models: Organizations employ new or modified frameworks to generate, deliver, and collect significance; these frameworks frequently include sustainability concepts to tackle current ecological and social issues.

Key Performance Indicators (KPIs): Measurable metrics that manifest how well a corporate is accomplishing important business goals; frequently assimilated to evaluate the influence of trash recovery and reverse logistics projects.

Product-as-a-Service: a business strategy that allows organizations to maintain proprietorship and accountability for product lifecycle management by offering products as services rather than as standalone goods, therefore encouraging sustainability.

Reverse Logistics: Planning, carrying out, and managing the influence movement of resources and commodities from the point of consumption back to the place of origin in order to recover significance or dispose of them properly.

Sustainability: the aptitude to satisfy present demands without endangering the aptitude of future generations to satisfy their own, frequently involving social, economic, and ecological aspects.

Waste Recovery: The process of recovering resources from waste items for recycling, reuse, or other uses in order to lessen the influence on the environment and encourage sustainability.

Chapter 8
The Influence of Corporate Social Responsibility Management on the Transition to Global, Sustainable Performance:
A Case Study of CSR-Certified Companies

Meriem Rachid
https://orcid.org/0009-0000-0366-293X
Faculty of Legal, Economic and Social Sciences - Fez, Morocco

Ahmed El Hammoumi
https://orcid.org/0000-0002-4249-9909
Multidisciplinary Faculty - Taza, Morocco

Nabil Seghyar
Multidisciplinary Faculty - Taza, Morocco

ABSTRACT

Given the context of managerial transformation and the growing importance of sustainable development imperatives, corporate social responsibility has become a key strategic lever in the transition to global, sustainable performance. The aim of the study is to assess the impact of CSR management on the evolution of cor-

DOI: 10.4018/979-8-3373-3012-9.ch008

porate performance towards the integral dimension between economic, social and environmental aspects. To this end, a quantitative empirical study was carried out on a sample of 80 Moroccan companies involved in CSR actions with a CGEM label. Data collection was based on a structured questionnaire designed to identify the differential effect of CSR management practices on overall performance. The results show that certain CSR management practices have very significant effects in the transition from performance to globality and sustainability. These include responsible governance, consumer protection, labor relations and conditions, and community involvement, all of which strengthen management systems and organizational resilience.

INTRODUCTION

In today's world, where environmental, social, and ethical issues are becoming more prominent, corporate social responsibility (CSR) has become a crucial strategy for achieving sustainable performance (Brundtland, 1987; Teece, 2018). As companies face a growing array of pressures—like consumer demands, regulatory standards, and climate crises CSR has evolved from being just a nice-to-have to a fundamental requirement for sustainability, resilience, and competitiveness. This shift in thinking encourages us to evaluate a company's performance not just through an economic lens, but through a more holistic approach that weaves together economic, social, and environmental factors (Bharadwaj et al., 2013; Bughin et al., 2019).

Lots of studies show that CSR can boost different parts of a company's performance (Sebastian & Matthew, 2018; Bresciani et al., 2021). But the findings are mixed, especially when it comes to how these practices work in different places, especially in newer markets. One reason for this is that companies often don't fully integrate CSR into their management models it's sometimes just for show or only done in certain areas. Plus, not many studies look at how CSR fits in with a company's internal rules and culture to create lasting success.

It is against this backdrop that this study, whose objective is twofold: on one hand, empirically evaluate the differentiated effect of CSR management practices on the transition towards a global and sustainable performance, and on another, provide a critical interpretation for guiding future scholars, leaders and policy makers towards contextualizing mechanisms for CSR. For this purpose, the present research formulates the following question:

To what extent do CSR management practices influence the development of Moroccan companies' performance towards a balanced inclusion of economic, social, and environmental aspects?

Morocco, like most developing countries, set out on a journey of incorporating sustainability standards in its economy two decades ago, specifically through the initiative of the General Confederation of Moroccan Enterprises (CGEM) through the CSR Label. The label recognizes companies that commit to responsible practices in governance, working conditions, the environment, customer relations, and community engagement.

Despite this incentive regime, real CSR practice still faces several challenges. On the one hand, there are companies that apply symbolic measures, with a preference for compliance logic, sacrificing real and strategic embedding (Kant, S. et al, 2025). On the other hand, there are limited resources, no sustainable development skill, no suitable indicator and no training shortage of the managers, restricting the scale of CSR measures.

For example, in the Moroccan agri-food sector, responsible companies struggle to reconcile environmental pressures with profitability, particularly during periods of water restrictions and market volatility. In the textile industry, some companies make ethical statements without supporting policies in terms of improving working conditions. These examples demonstrate that the strategic embedding of CSR is still unbalanced and that its effect on total performance is neither automatic nor guaranteed.

In an ideal scenario, CSR should be a complete management system which allows companies to generate value in line with the practice of sustainable development. It would therefore be perfectly aligned with the strategic objectives, supported by sufficient resources, and assessed in terms of relevant economic, social and environmental dimensions.

However, the empirical reality differs significantly from this theoretical framework, particularly in the developing world. The empirical evidence contradicts one another: there exist studies that attest to a positive relationship between CSR and performance; others assert that the relationship is trivial or nonexistent. Moreover, results are often not generalizable due to a lack of the proper contextual foundation (Kant, S., & Adula, M., 2025).

The theoretical shortcomings include in explicit terms the lack of integrative theories taking into account the interaction of the different dimensions of CSR. The knowledge shortcomings are a consequence of the few studies examining the diversified effect of CSR practices on overall performance. These signs are augmented by contextual voids: few studies investigate these relations in contexts like that of Morocco, with its SME-dominant economic tissue, evolutionary regulation and hybrid managerial logics. Finally, there are practical challenges like the inability to gauge the actual effect of CSR where no effective management systems and solid indicators are in place.

This study proposes to fill such gaps through the activation of a quantitative approach on a sample of Moroccan companies selected by the CGEM. By uncovering CSR practices exerting a major impact on the transition toward sustainable performance, it supplies researchers and managers with an empirical base on which to develop responsible management models for similar contexts.

Several theories can be applied to describe the connection between CSR and overall performance. One can particularly refer to:

-Stakeholder theory (Freeman, 1984)

This theory considers that the performance of a company depends on its ability to meet all its stakeholders' expectations. CSR is here a strategic tool for social legitimacy and relationship building with external and internal stakeholders. Dimensions such as consumer protection, labour relations or community involvement in our study are operational expressions of this justification.

-Sustainable development theory (Brundtland, 1987)

This theory establishes the construction of the "triple capital" (economic, social, and environmental). It founds the idea that performance can no longer be solely economic. In this perspective, the present study quantifies the level of effective integration of the three dimensions in Moroccan firms' strategy.

-Theory of capabilities and resources (Barney, 1991)

According to this theory, organizational capabilities (such as CSR) may be strategic resources that differentiate. Responsible governance or anticipatory social management improves competitiveness and robustness. This is evidenced in our research by the positive effect confirmed of some practices on overall performance.

-Institutional theory (DiMaggio & Powell, 1983)

She explains the internalization of CSR practices through external pressures (normative, coercive or mimetic). The CGEM CSR label illustrates an institutional norm inviting companies to codify their engagements, sometimes without real strategic anchoring.

-Main invoked theory: the stakeholder theory

Of these strategies, stakeholder theory is the main theoretical background of this research. It allows us to explain the differential effect of CSR dimensions on overall performance in relation to their ability to fulfill the specific demands of dominant actors in the company's environment.

1. CORPORATE SOCIAL RESPONSIBILITY MANAGEMENT: CONCEPTS AND PRACTICES

CSR is the acronym for corporate social responsibility, a concept based on organizations' voluntary integration of social, environmental, ethical and economic concerns into all their affairs and relationships with all their stakeholders cited by Nave & Ferreira (2019). In other words, CSR is not just about compliance, but reflects a company's efforts and commitment to sustainable development. (Dubey, 2024).

One of the first structured conceptualizations of CSR is presented by Carroll (1991).He develops a four-level typology, forming a hierarchical pyramid of corporate social responsibilities. This classification distinguishes between:

- Economic responsibility: the company's responsibility is to remain profitable and make a profit over the long term.
- Legal responsibility: the company is responsible for complying with its obligations to pay taxes and work in accordance with current laws and regulations.
- Ethical responsibility: acting in accordance with stakeholder expectations and social norms, while respecting underlying ethics.
- Philanthropic responsibility: the company's responsibility is to improve society's overall social well-being.

A global context marked by the intensification of environmental, social and ethical challenges and growing pressure from stakeholders (customers, employees, investors, non-governmental organizations....ect), justifies the integration of CSR management as a strategic and managerial necessity (Shaikh, 2024). This cannot be reduced to a simple ethical or compliance duty. It is a powerful vector of organizational opportunities that can foster risk management, innovation and the creation of shared value, and enhance corporate reputation. (Idowu, 2021).CSR management refers to all the processes, tools and practices that a company implements to structure, manage and promote its CSR commitments.

It is marked by a deliberate approach aimed at systematically aligning the company's global strategy and day-to-day activities with the principles of sustainable development (Seghyar et al.,2025). In this respect, the international standard ISO 26000 plays a key role, providing a conceptual and operational framework that is widely recognized throughout the world. This standard identifies seven key dimensions, which constitute the main thrusts of the company's approach to sustainability (Kumar, 2022):

- Organizational governance: based on the principles of transparent, ethical and inclusive management of decision-making and strategic orientation processes, confirmed by the engagement of relevant stakeholders' activities in line with their expectations (El Hammoumi, A. & El Hammoumi, M., 2025).
- Human rights: the respect and promotion of fundamental human rights in all activities, including protection against all forms of discrimination, forced labor and the restriction of individual freedom;
- Industrial relations and working conditions: Working conditions must be healthy and fair, creating a working environment conducive to social dialogue, training and well-being.
- Environment: companies are invited to minimize their impact on the environment by responsibly managing resources, reducing pollutant emissions, protecting biodiversity and promoting circular consumption and production models (Kant, S. et ali, 2025).
- Fair practices: this pillar encompasses business ethics, including the fight against corruption, commercial transparency, respect for intellectual property and fair competition.
- Consumer issues: another objective to be achieved by companies is product quality, safety and responsibility, accompanied by consumer behavior.
- Societal commitment: finally, companies are called upon to promote the socio-economic development of local communities, to fight for inclusion and diversity, and to support social and cultural initiatives.

As a result, these practices constitute an essential foundation for the operationalization of CSR implementation. Ultimately, this commitment is likely to confer a sytemic and integrated approach, increase the company's legitimacy and viability in the face of the complexity of contemporary issues, and generate long-term value and mutual benefits for the organization and its society (Popa & Dabija, 2018). CSR management is therefore a cross-cutting function, involving strategic and operational departments as well as human resources and communications. In addition, it is based on appropriate monitoring and evaluation and reporting tools and continuous, constructive communication with all stakeholders (Ben Rhouma et al., 2018).

Generally speaking, CSR is no longer confined to a simple ethical commitment, but has become a central lever for sustainable competitiveness, a tool for internal social cohesion, and a factor of organizational and social innovation that is essential to meeting today's challenges.

2. CSR MANAGEMENT AND THE RECONFIGURATION OF ORGANIZATIONAL PERFORMANCE: A CRITICAL REVIEW AND REFORMULATION OF HYPOTHESES

2.1. Critical Review

The gradual introduction of CSR into managerial reference frameworks is revolutionizing the analysis of organizational performance (Yang, 2018). Performance has traditionally been defined in financial terms (profitability, productivity, creation of shareholder value). However, this classic definition, while still predominant in many practices, seems insufficient today to understand the contemporary challenges facing companies (ecological issues, social tensions and heightened societal expectations) (WAMBA et al., 2020). CSR management is also part of this approach. It leads to a multi-dimensional reconceptualization of performance in terms of basic principles, indicators and objectives to be achieved (Yoon & Chung, 2018). It places organizations within a multiple performance logic combining economic, social and environmental performance and encourages a broader review of the results obtained, going beyond simple financial balance sheets (Elkington Elkington and Rowlands, 1999; Epstein, 2018).

Nevertheless, various studies have sought to analyze the differentiated effects of CSR management on organizational performance. On the empirical side, several meta-analyses (Margolis & Walsh, 2003 ; Orlitzky et al., 2003) have reported a generally positive correlation between the implementation of CSR practices and a company's financial performance, while observing differences linked to the sector of activity, the institutional context, and the degree of strategic integration of CSR (Kant, S. et ali, 2025).

The literature has also identified other mediating mechanisms, such as stakeholder engagement (Freeman, 2010), corporate governance (Aguilera et al., 2007). and organizational capacity to transform commitments into competitive advantages, or heterogeneity of impact depending on the type of CSR pursued (Porter & Kramer, 2006 ; Hart & Milstein, 2003). Indeed, performance appears to be systematically superior for companies committed to integrated CSR, as opposed to peripheral or symbolic CSR. (Gond et al., 2011).

On the other hand, CSR management is a lever for creating non-financial value, based on mechanisms for improving reputation, attracting and retaining talent, and reducing socio-environmental risks, with the potential to stimulate sustainable innovation(Porter & Kramer, 2018 ; Bansal, 2005). In addition, it modifies internal models of governance, organizational culture, strategic management and internal dialogue with the main players, contributing to a change in the classic logics of corporate performance.

These developments can then be illuminated by several fine theoretical frameworks. Freeman's stakeholder theory (1984) postulates that organizational performance can no longer be conceived solely in terms of satisfying shareholder expectations, but must also be evaluated in terms of a balanced satisfaction of the expectations of all stakeholders, it invites us to rethink the conditions for corporate success, based on a principle of enhanced responsibility.At the same time, the Resource-Based View (Barney, 1991) argues only that intangible resources derived from CSR, such as reputation, partner trust or innovative capacity, can create sustainable competitive advantages. Finally, the contributions of institutional theory (DiMaggio & Powell, 1983) allow us to understand the expansion of CSR practices as a response to coercively, mimetically and normatively applied pressures, and thus their influence on the process of transforming performance norms in organizational domains.However, the reconceptualization of performance highlights several critical problems. Firstly, the variability of standards, normative frameworks and measurement tools (for example, ESG indicators, the Global Reporting Initiative, ISO standards, sector labels, etc.) complicates comparisons between organizations, and raises questions about the standardization of practical assessments. Secondly, conflicts between short-term returns and long-term sustainability objectives remain a major obstacle to the sincere and integrated adoption of CSR management (Lozano, 2015).

Finally, other criticisms drawn from heterodox literature warn against the risk of CSR being instrumentalized by companies. They see CSR not as an alternative paradigm, but rather as a managerial discourse. In this way, companies create an image of social responsibility in order to improve their image, rather than to further transform the dynamics of value creation (Banerjee, 2008 ; Fleming & Jones, 2012). In this way, CSR management has great transformative potential, but its effect on performance depends on its incorporation into the company's institutional game. This effect also depends on governance, on the CSR resources implemented and on the willingness of management to act. This complexity justifies a nuanced and critical reading of the relationship between managerial CSR and notions of organizational performance, made more explicit from these various contextual, sectoral and temporal dimensions (Valmohammadi, 2014).

Thus, far from simply adding up performance, CSR management makes it possible to articulate a reconfiguration of performance that operates according to a broader understanding.

In this respect, it is based on a recognition of dynamic and extended temporalities, exogenous or endogenous systemic logics, and a mode of governance that values a partnership position with all stakeholders. This reconfiguration therefore requires appropriate methods for measuring strategic governance policies and greater transparency between economic, social and ecological objectives.

2.2. Formulation of Hypotheses And Proposal Of A Conceptual Research Model

The contemporary approach to CSR has been institutionalized around a number of international standards, of which ISO 26000 is the most prominent. The standard identifies seven areas of action, or core practices, which identify the strategic and operational processes that enable corporate social responsibility to be integrated into the company's actions and thinking. These practices relate to corporate governance, human rights, labor relations and working conditions, the environment, fair practices, consumer issues and community involvement. (Hahn, 2013).

In this respect, in this study, these practices will be studied as levers of organizational change, in order to achieve a reorganization of the triple bottom line performance, transforming it into a global or three-pronged performance, which integrates economic, social and environmental impacts (Sartori et al., 2017). This concept is based on a systemic vision of value, viewing the organization as an interdependent player in its ecosystem. Accordingly, we propose a series of hypotheses linking each area of ISO 26000 to overall performance, based on theoretical foundations drawn from managerial literature (stakeholder theory, resource theory, institutionalism, creation of shared value):

H1: Translating a company's CSR commitment into its corporate governance would have a positive impact on improving corporate performance towards overall sustainability.

H2: Respect for human rights within the company would have a positive influence on improving corporate performance towards sustainable global performance.

H3: Good relations and working conditions within the company have a positive influence on improving corporate performance towards sustainable overall performance.

H4: Taking into account the direct and indirect implications of a company's decisions and activities on the environment would have a significant influence on improving corporate performance towards overall sustainability.

H5: Fair practices within the company would enable improvements in corporate performance towards overall sustainable performance.

H6: A company's awareness of its responsibilities towards consumers would have a positive influence on improving corporate performance towards overall sustainability.

H7: Proactive outreach by companies to local communities would have a positive influence on improving corporate performance towards global performance.

Based on the formulation of the above hypotheses, we have attempted to establish a structured analysis grid, linking the fundamental practices of CSR management to the dimensions of global performance.

Each hypothesis then refers to a specific area of intervention drawn from the ISO 26000 standard and considered as a vector of potential organizational impact. This hypothetico-deductive construction is based on the idea that CSR, when managed in an integrated way, can generate differentiated but complementary effects on a company's economic, social and environmental performance.To translate these theoretical relationships into a coherent analytical framework, we need to propose a conceptual research model that synthesizes the hypotheses formulated above.

The aim of this model is to describe the logic of the systemic links between CSR management practices, as explanatory variables, and overall performance, as a multidimensional dependent variable. As a result, the final research model should become the frame of reference for an empirical test, making it possible to identify the effective contribution of CSR management to the reconfiguration of organizational performance according to the paradigm of integrated sustainability.

Figure 1. Conceptual model

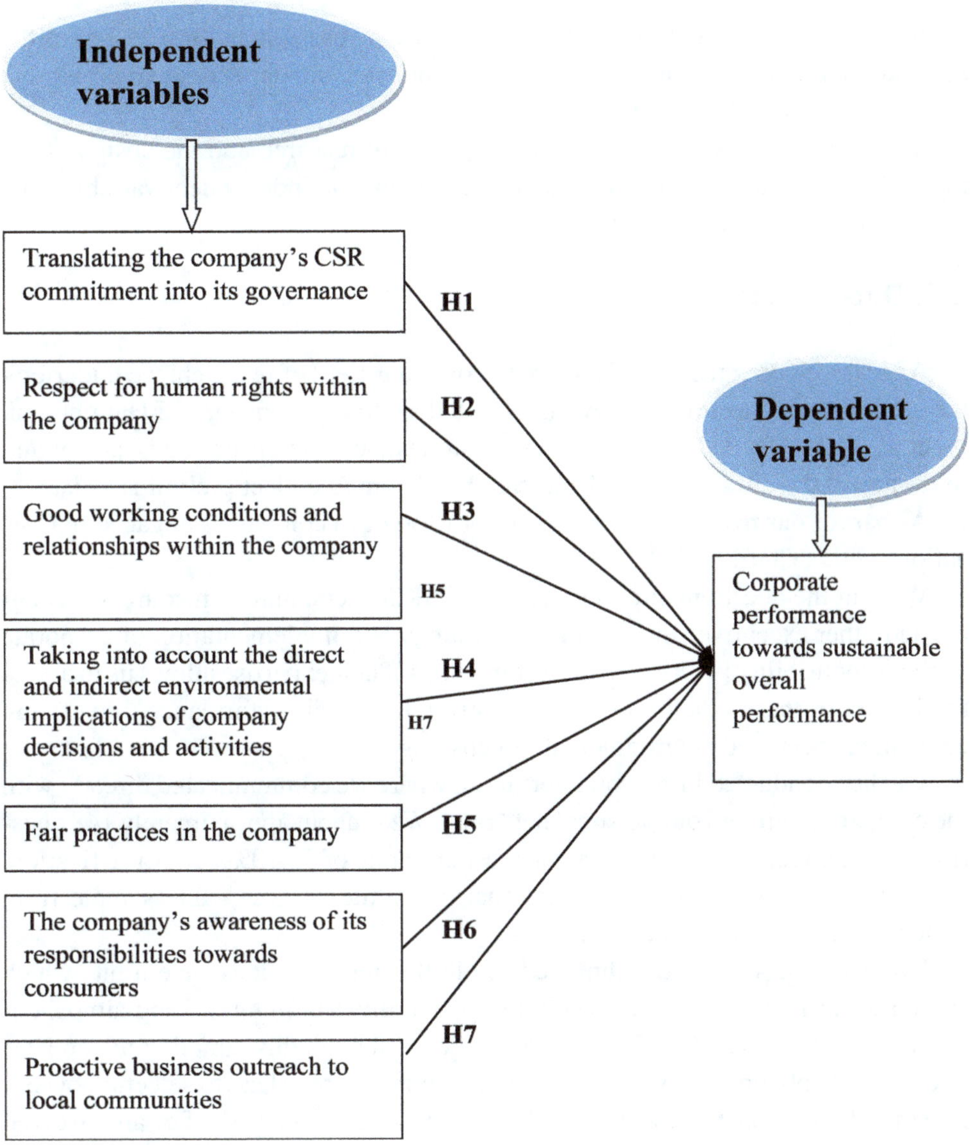

3. DATA AND METHODS

In this section, we will present the scope of our research and the design of the method used to measure and analyze the dependent and independent variables that occur in our conceptual model.

In this section, we will present the scope of our research and the design of the method used to measure and analyze the dependent and independent variables that occur in our conceptual model.

3.1. Data Collection

After developing the questionnaire to cover our field of research, we undertook the second step to test the survey on a limited sample of subjects (El Hammoumi, A. et al, 2024). By validating the survey in advance, we ensure consistency and relevance of the questions asked and can then begin to collect preliminary data.

As part of our research, we searched for Moroccan companies engaged in CSR actions with a CGEM label

We sent the questionnaire to managers, CSR leaders, human resources managers and other executives involved in the strategy and implementation of corporate social responsibility practices within companies. Their expertise allows us to assess the impact of these practices on the transition to global performance, integrating economic, social and environmental dimensions.

We then conducted in-person interviews, where we communicated directly with the companies through our personal networks. We sent out approximately 148 questionnaires and received 86 responses, a response rate of 58.10%. After verification, we selected 80 usable questionnaires that constitute an acceptable sample from which acceptable data can be obtained.

However, the sampling technique used in this research does have limits which should be acknowledged. We used a non-probability sampling method almost entirely from personal and professional networks. Although this sampling method has allowed us to obtain a relevant number of responses, it may lead to selection biases, in particular an over-representation of companies who are aware of or are involved in social responsible approaches. In addition, the data collection method based on direct solicitation and voluntary response may have led to preferential responses from companies that seek to be or are interested in improving their CSR, leading to a possible bias of social desirability. These limitations must be taken into account when interpreting findings, and it should be warranted caution when theorizing to the entire Moroccan entrepreneurial fabric.

3.2. Analyzing the Data Treatment Process

We did this by first checking the validity and reliability of the variables that are in our research model or framework, with the item sets being centered around dimensions of AI integration against groups of items for resisting change.

As a second phase we disclosed a multi-variable analysis with the purpose of establishing the causal relationship between dependent variables and independent variables.

We now focus on:

- The measurement model

With the aid of the main component analyses we received confirmation about the validity and reliability of the measuring scales in our theoretical model in order to facilitate the collection of data thus we were able to reclassify the variables through the number of different stages (Evrard et al., 2003).

-The feasibility of the factor analysis through the Barlett sphericity test which is based on the nil-hypothesis on correlation based on the different variables collectively in addition to the Kaiser-Meyer-Olkin (KMO) test on correlation specified below based on the different variables are collectively high enough to perform a valid main component analysis.

-Establishing the criteria to retain the number of factors as of the initial list of variables. We follow the identifiable component steps so that items that contribute weakly (0,500 and below) are removed as per the main components identified (Evrard et al., 2003). Each analysis of main components is the subject of an orthogonal Varimax rotation allowing weaker values of coefficients to be compensated artificially for the most correlated variables or conversely weaker performance for the least correlated variables.

-Check the reliability of factors from each analysis of main components through consistency check using the same items being examined by measuring the same concept. A reliable scale gives the same results when repeat measurements are done regardless of the context of where the test is done (Evrard et al., 2003) We employed the widely used Cronbach a for its ability to measure the reliability of different questions over the same phenomenon. Any alpha above 0, 6 are acceptable in exploratory studies (Evrard et al., 2003).

– Rename each component and redefine the research hypothesis to arrive at a refined factorial analysis research model. **Table N°1** summarizes the results of analysis of the main components of the explanatory variables, including the Chronbach alpha value and the percentage variance explained in all components.

Table 1. Results of the validity and reliability of measurement scales

Components	Alpha Cronbach	Items	% variance explained	no of items
Translating the company's CSR commitment into its governance	0.756	-CSR principles are explicitly integrated into the company's governance policies. -The board of directors or top management regularly discusses CSR-related issues. -A specific structure or committee is in place to oversee CSR governance. -Strategic decisions systematically consider social and environmental impacts. -Internal control mechanisms include performance indicators related to CSR commitments.	79.38%	5
Respect for human rights within the company	0.765	-The company has a formal policy that promotes and protects human rights. -Employees are trained on human rights and non-discrimination principles. -There are clear procedures to report and address human rights violations within the company. -The company ensures equal treatment and opportunities for all employees regardless of gender, ethnicity, or background.	72.54%	4
Good working conditions and relationships within the company	0.769	-The company provides a safe and healthy work environment for all employees. -There is a climate of mutual respect and collaboration among employees and management. -Working hours and workload are reasonable and respect employees' well-being. -Employees have access to support systems such as counseling or conflict resolution mechanisms.	69.89%	4
Taking into account the direct and indirect environmental implications of company decisions and activities	0.778	-The company evaluates the environmental impact of its projects and strategic decisions before implementation. -Environmental criteria are integrated into the selection of suppliers, materials, and technologies. -The company monitors and seeks to minimize both direct (e.g., emissions, waste) and indirect (e.g., supply chain) environmental impacts.	71.89%	3

continued on following page

Table 1. Continued

Components	Alpha Cronbach	Items	% variance explained	no of items
Fair practices in the company	0.789	-The company actively fights against all forms of corruption and unethical behavior. -Procurement and contracting processes are conducted with transparency and impartiality. -The company ensures compliance with laws and ethical standards in all its operations. -Employees are encouraged to report unethical practices without fear of retaliation.	69.90%	4
The company's awareness of its responsibilities towards consumers	0.705	-The company ensures the safety and quality of its products or services for consumers. -The company provides clear, accurate, and truthful information about its products or services. -The company respects consumer privacy and protects their personal data. -The company responds promptly and effectively to consumer complaints and feedback.	70.62%	4
Proactive business outreach to local communities	0.784	-The company actively engages with local community organizations to understand their needs. -The company supports local development projects through funding or volunteer efforts. -The company regularly communicates its initiatives and impacts to the local community. -The company collaborates with community stakeholders to create shared value programs. -The company promotes local employment and sources materials from nearby suppliers.	72.56%	5
Corporate performance towards sustainable overall performance	0.786	-The company integrates economic, social, and environmental goals in its strategic planning. -The company regularly monitors and reports on its sustainability performance indicators. -The company invests in innovations that promote long-term environmental and social benefits. -The company ensures transparency and accountability in its sustainability initiatives. -The company fosters a culture of continuous improvement towards sustainable business practices.	74.89%	5

- **Test of the hypothesis by analysis of regression**

Multiple linear regressions is the regression chosen to simultaneously analyze the relationship between several explanatory variables and a variable to be explained. The principle is to examine the relationship between the variable to be explained and the other explanatory variables that best reflect the reality of the observations. This relationship is expressed through the regression equation which presents the variable to be explained as the sum of each of the explanatory variables assigned their coefficients to which a constant term is added" (Crauser et al., 1989).

This explanatory method is used to test the hypothesis of our model. This gives us the opportunity to test our research model in a global context and thus determine the importance and significance of each explanatory variable for the integration of CSR management in Moroccan companies and the effect on the evolution of business performance towards the integral dimension between the economic aspect, social and environmental.

It allows us to translate our conceptual model into the following equation:

Corporate performance towards sustainable overall performance = β1 Translating the company's CSR commitment into its governance + β2 Respect for human rights within the company + β3 Good working conditions and relationships within the company + β4 Taking into account the direct and indirect environmental implications of company decisions and activities + β5 Fair practices in the company+ β6 The company's awareness of its responsibilities towards consumers + β7 Proactive business outreach to local communities + ε

The testing of the hypothesis within this work, consistent with this approach, requires the use of (some) indicators including:

- Verifying the Variance inflation factor (VIF) and the tolerance criteria to guarantee that the model is not including a useless explanatory variable. VIF must be less than 10 and the tolerance criteria less than 1.
- The Fischer-Snedecor test allows for the interpretation of the statistical significance of the results obtained and the contribution of each explanatory variable. Its noted significance value "Signif F" specifies that the observed Fischer-Snedecor test statistic threshold must be less than 0,05 at the 5% threshold, or 0,01 for the threshold value of 1%, in order for the null hypothesis which states that there is no linear relationship between explanatory variables and the explained variable to be rejected.
- The Student t test allows for the verification of whether the estimate obtained for the non-standardized regression coefficient or the non-standardized lambda (B) is significantly different from 0. Its significance is required to be less than 0,05 at the 5% threshold or less than 0,01 at the 1% threshold.

Verification of VIF and Tolerance

To perform a verification of VIF and tolerance a test for multi-collinearity is performed. We performed verification of VIF and of tolerance to test for the absence of multi-collinearity, which means the model does not contain useless explanatory variables. VIF must be below 10 and tolerance less than 1.

For our instance, the absence of multicollinearity is confirmed, clearly shown in the Table 2, with tolerance levels between 0.5 and 0.9, and with VIF still less than 10.

Table 2. Coefficients of collinearity of variables in the research model

Model		Collinearity Statistics	
		Tolerance	VIF
1	(Constant)		
	Translating the company's CSR commitment into its governance	,545	4,456
	Respect for human rights within the company	,897	4,345
	Good working conditions and relationships within the company	,564	4,897
	Taking into account the direct and indirect environmental implications of company decisions and activities	,726	4,562
	Fair practices in the company	,917	4,143
	The company's awareness of its responsibilities towards consumers	,867	4,087
	Proactive business outreach to local communities	,786	5,089

a. Corporate performance towards sustainable overall performance

Test of the Theoretical Model

The hypothesis test of the theoretical model was successful, and we will continue with the significance test. This part of the analysis will occur in two components: overall significance and individual significance.

The Overall Significance Test.

The first part we examined was the overall significance test on the model, or the set of explanatory variables that are operating to influence the dependent variable. The overall significance test addresses the question: is there at least one explanatory variable that is significant?

The overall significance test resulted from the Fischer test which measures overall significance by determining if there is an influence from the explanatory variables on the explained variable.

The comparison of the F-statistic was used as the test statistic to complete the arbitrage. The SPSS 25 program provided the probability on the output associated with the calculated F-statistic made the analysis much easier; therefore, only comparing the probability against the F-statistic at the 5% significance level would suffice. If the probability associated with the F-statistic is below the 5% significance level, the hypothesis H0 can be rejected for the profile of the alternative hypothesis according to which regression is overall significant.

In our case, for the explanatory variables of corporate performance towards sustainable overall performance, the Fischer statistic found by SPSS 25 is $F = 8.11$ and the probability associated with it is less than 5% ($0.00 < 0.05$), the null hypothesis is then discarded and the model is globally significant (Table 3).

Table 3. ANOVA Model Regression Index

ANOVA[a]					
Model		**Sum of Squares**	**ddl**	**Mean Square**	**F**
1	Regression	14,347	7	2,0495	08,110
	Student	18,453	73	,2527	
	Total	32,800	80		

a. Predictors: (Constant), Translating the company's CSR commitment into its governance, Respect for human rights within the company, Good working conditions and relationships within the company, Taking into account the direct and indirect environmental implications of company decisions and activities, Fair practices in the company, The company's awareness of its responsibilities towards consumers, Proactive business outreach to local communities
b. Dependent Variable: Corporate performance towards sustainable overall performance

Individual Significance Test:

Table 4. Coefficients of individual significance

Coefficients [a]								
Model		Non-standardized Coefficients		Standardized Coefficients	t	Sig.	Collinearity Statistics
		B	Error standard	Beta			Tolerance
1	(Constant)	-,185	,521		-,291	,769	
	Translating the company's CSR commitment into its governance	,564	,165	,408	2,162	,007	,545
	Respect for human rights within the company	,594	,134	,445	2,327	,014	,552
	Good working conditions and relationships within the company	,346	,126	,354	2,453	,016	,896
	Taking into account the direct and indirect environmental implications of company decisions and activities	,453	,124	,556	2,134	,048	,789
	Fair practices in the company	,245	,123	,345	2,346	,019	,945
	The company's awareness of its responsibilities towards consumers	,217	,165	,267	2,253	,035	,769
	Proactive business outreach to local communities	,354	,208	,376	2,453	,017	,786
a. Dependent Variable: Corporate performance towards sustainable overall performance							

With respect to the individual importance of the variables, we applied the Student's t statistic, which is provided in the SPSS 25 program. The significance hypothesis for each variable is maintained when the value of the estimated Student statistic is greater than critical value in the Student t distribution table. For this research we will use the p-value offered by the SPSS 25 at the chosen significance level to determine this hypothesis.

Table 4 presents the results of the estimated variable importance, which offers the conclusion that all seven variables have significance at the probabilities assigned;

-The variable "Translating the company's CSR commitment into its governance" significance at the 1% level ($t = 2.162$, $p = 0.007 < 0.01$).
-The variable "Respect for human rights in the company" has a significance at the 5% level ($t = 2.327$, $p = 0.014 < 0.05$).

-The variable “Good working conditions and relationships within the company” has a significance at the 5% level ($t = 2.453$, $p = 0.016 < 0.05$).
-Taking into account the direct and indirect environmental impacts of business decisions and actions” is statistically significant at the 5% level ($t = 2.134$, $p = 0.048 < 0.05$).
-Fair practices in the company” is statistically significant at the 5% level ($t = 2.346$, $p = 0.019 < 0.05$).
-The company's understanding of its responsibilities towards consumers” is statistically significant at the 5% level ($t = 2.253$, $p = 0.035 < 0.05$).
-Proactive business engagement with local communities” is statistically significant at the 5% level ($t = 2.453$, $p = 0.017 < 0.05$).

Additionally, these results indicate that each of the dimensions of corporate social responsibility investigated has a positive and significant impact on corporate performance toward sustainable overall performance.

4. RESULTS AND DISCUSSION

Regression analysis shows that respect for human rights within the company is the most significant variable in the transition towards sustainable overall performance ($\beta = 0.594$, $p = 0.014 < 5\%$). This evidence confirms our hypothesis H2, and coincides with the theoretical framework of stakeholder theory (Freeman, 1984), which places employee dignity and ethical treatment at the center of long-term organizational legitimacy and performance. By incorporating human rights into their internal policies and everyday practices, companies can increase trust in the organization, diminish conflict, and promote a pleasant workplace, all of which are essential precursors to sustainable performance. In practice, this means that respect for labor rights, equality, and equal opportunity should be a focus of not only corporate governance, but also a strategy for the business overall.

Second, we discover that the translation of the company's commitment to CSR into its governance shows a significant and positive contribution to the sustainability performance ($\beta = 0.564$, $p = 0.007 < 1\%$), which confirms H1. This finding confirms what Aguilera et al. (2007) explain regarding governance mechanisms related to CSR being ethical boards, CSR committees, and policies regarding stakeholder engagement, lead to strategic alignment between social outcomes and economic outcomes. These frameworks around CSR help indicate where the company is at in terms of maturity in corporate sustainability and its ability to institutionalize sustainable values, which can align discourse with action. Hence, it implies participatory, transparent and accountable governance models should be used.

Three, positive working conditions and relationships in the company are important drivers of sustainability performance ($\beta = 0.346$, $p = 0.016 < 5\%$), demonstrating support for H3. These findings support social capital theory (Putnam, 1993), which establishes a link between relational trust, employee well-being, and increases in productivity, innovation, and organizational resilience. Safe, inclusive, and attractive working conditions enhance employee engagement and strengthen internal cohesion both are leading dimensions of global performance. In practical terms, this means HR policies ought to focus on mental and physical health, continuous conversations, and participative management styles.

In fourth place, considering both the direct and indirect environmental consequences of the company's decisions and activities also displays a positive and significant impact ($\beta = 0.453$, $p = 0.048 < 5\%$) and validates hypothesis H4. This validates theoretical hypotheses regarding ecological modernization (Mol & Spaargaren, 2000), which contends that companies who proactively account for environmental aspects in their decision-making process have the potential to innovate, mitigate unwanted consequences, and create new competitive advantages. The formation of practices such as eco-design, carbon audits, or circular economy practices are not trajectories of innovation, but a necessary vector for change in the context of sustainable transition.

In fifth place, fair practices in the company have a significant association with sustainable performance ($\beta = 0.245$, $p = 0.019 < 5\%$), which confirms H5. Based on the principles of organizational justice (Greenberg, 1987), how employees, partners, and suppliers are treated equitably will contribute to loyalty, lower conflict levels, and enhance reputational capital. Company set insets - ethical purchases, anticorruption measures, and transparent recruitment - are indicators that reinforce the credibility and generate long-term value creation.

The company's recognition of its role as a responsibility to consumers effects sustainable performance in a substantial but slightly less impactful way ($\beta = 0.217$, $p = 0.035 < 5\%$), thus deriving support for H6. This corresponds to the notion of responsible marketing (Kotler & Lee, 2005) which means that concerted efforts to market and sell their products and services should be consumer focused and inherently meant to incorporate product safety, transparency, and customer service because without these, consumers will never develop enduring trust or legitimacy towards the firm and their products and services remain vulnerable to influential market forces. Not only do these actions provide consumer protections, they support customer retention and help with brand differentiation.

Lastly, corporate voluntary engagement and or interactions with local neighbourhoods promote sustainable performance similarly ($\beta = 0.354$, $p = 0.017 < 5\%$), which supports H7. This finding is in line with community engagement models (Bowen et al., 2010) where through building relationships with local partnerships, their social license to operate is enhanced. Spending on programs with education, health or local employment returns worry-free community building and simultaneously generates joint value, or shared social value for business and local places.

CONCLUSION

This research aimed to study the effect of Corporate Social Responsibility (CSR) dimensions on the sustainable overall performance of Moroccan firms. Using a quantitative methodology based on linear regression modeling, this study showed that all CSR dimensions, which included responsible governance, respect for human rights, working conditions, environmental impact management, fair business practice, consumer responsibility, and engagement with the communities in which the firm operates, positively affected sustainable overall performance.

- Research Implications

From a theoretical perspective, this study sought to contribute to the body of research on CSR in emerging markets, specifically in Morocco. It demonstrated the validity of a multidimensional way of thinking about CSR consistent with both stakeholder theory and organizational legitimacy theory, as well as sustainable governance. This study has implemented a further integrative view of CSR beyond a rudimentary context of CSR being a method of reporting and/or compliance, and has now positioned CSR as a strategic ability and transformative journey.

- Practical Implications

On a practical level, the results encourage business leaders in Morocco to think of CSR as a structural lever for improving their performance in economic, social, and environmental dimensions as well as enhancing its image. Companies would be encouraged to not only put CSR into place at the level of their governance structures but also to promote good working conditions, fair and transparent practices, and ethical relationships with consumers and local communities. The call thus, would be for pre-emptive and integrative and participative CSR as opposed to reactive or instrumental CSR.

- Research Limitations

Despite these contributions, this research does have a number of limitations. First, to draw from a sample based on a non-probabilistic sampling method that rather relied on professional networks, likely brought a representativeness bias caused by over-represented companies already associated with CSR practices. Moreover, all data was derived from self-report provided by all respondents via a questionnaire which may have suggested socially desirable responding. Finally, all the data gathered relied on cross-sectional data meaning that we could not strictly identify direct cause and effect between variables.

■ Future Research Directions

These limitations provide avenues for future research. Longitudinal studies would shed light on how CSR practices evolve in time, as well as measurement periods, and the role that CSR plays overtime. Qualitative approaches (e.g., case studies or in-depth interviews) could further complement these findings as well as provide richer depth of internal CSR dynamics and practices. Lastly, comparative studies across sectors or geographic regions within Morocco, or even among countries in the Global South, would assist in capturing the contextual specificities of CSR in emerging markets.

REFERENCES

Aguilera, R. V., Rupp, D. E., Williams, C. A., & Ganapathi, J. (2007). Putting the S back in corporate social responsibility: A multilevel theory of social change in organizations. *Academy of Management Review*, *32*(3), 836–863. DOI: 10.5465/amr.2007.25275678

Aguilera, R. V., Rupp, D. E., Williams, C. A., & Ganapathi, J. (2007). Putting the S back in corporate social responsibility: A multilevel theory of social change in organizations. *Academy of Management Review*, *32*(3), 836–863. DOI: 10.5465/amr.2007.25275678

Banerjee, S. B. (2008). Corporate social responsibility: The good, the bad and the ugly. *Critical Sociology*, *34*(1), 51–79. DOI: 10.1177/0896920507084623

Bansal, P. (2005). Evolving sustainably: A longitudinal study of corporate sustainable development. *Strategic Management Journal*, *26*(3), 197–218. DOI: 10.1002/smj.441

Barney, J. (1991). Firm resources and sustained competitive advantage. *Journal of Management*, *17*(1), 99–120. DOI: 10.1177/014920639101700108

Ben Rhouma, A., Koleva, P., & Schaltegger, S. (2018). RSE, management responsable et isomorphisme institutionnel: une analyse à partir des résultats d'une enquête internationale. Management international, 22(3), 65-78. https://doi.org/ DOI: 10.7202/1060894ar

Bharadwaj, A., El Sawy, O. A., Pavlou, P. A., & Venkatraman, N. V. (2013). Digital business strategy: Toward a next generation of insights. *Management Information Systems Quarterly*, *37*(2), 471–482. DOI: 10.25300/MISQ/2013/37:2.3

Bowen, F., Newenham-Kahindi, A., & Herremans, I. (2010). When suits meet roots: The antecedents and consequences of community engagement strategy. *Journal of Business Ethics*, *95*(2), 297–318. DOI: 10.1007/s10551-009-0360-1

Bresciani, S., Huarng, K. H., Malhotra, A., & Ferraris, A. (2021). Digital transformation as a springboard for product, process and business model innovation. *Journal of Business Research*, *128*, 204–210. DOI: 10.1016/j.jbusres.2021.02.003

Brundtland, G. H. (1987). *Our Common Future: Report of the World Commission on Environment and Development*. Oxford University Press.

Bughin, J., Deakin, J., & O'beirne, B. (2019). Digital transformation: Improving the odds of success. *The McKinsey Quarterly*, *22*, 1–5.

Carroll, A. B. (1991). The pyramid of corporate social responsibility: Toward the moral management of organizational stakeholders. *Business Horizons*, *34*(4), 39–48. DOI: 10.1016/0007-6813(91)90005-G

Crauser, J. P., Maurin, M., & Joumard, R. (1989). Representative kinematic sequences for the road traffic in France. *SAE Transactions*, •••, 1011–1018. DOI: 10.4271/890875

DiMaggio, P. J., & Powell, W. W. (1983). The iron cage revisited: Institutional isomorphism and collective rationality in organizational fields. *American Sociological Review*, *48*(2), 147–160. DOI: 10.2307/2095101

Dubey, S. (2024). Navigating Corporate Social Responsibility: Bridging Voluntary Initiatives and Legal Accountability. *Available at SSRN* 4784282. http://dx.doi.org/ DOI: 10.2139/ssrn.4784282

El Hammoumi, A., & El Hammoumi, M. (2025). Innovation and Knowledge Management Within the Framework of NMMO. In Hussain, Z., Khan, A., Qureshi, M., Sharipudin, M., & Alkara, İ. (Eds.), *Knowledge Sharing and Fostering Collaborative Business Culture* (pp. 197–212). IGI Global Scientific Publishing., DOI: 10.4018/979-8-3373-0710-7.ch011

El Hammoumi, A., Seghyar, N., Gouiouez, M., Amane, M., & Berdi, A. (2024). The Adoption of a Qualitative Approach in Management Science: An Exploratory Survey on the Effects of Big Data on the Performance of Companies. In *Applying Qualitative Research Methods to Management Science* (pp. 283–309). IGI Global. DOI: 10.4018/979-8-3693-5543-5.ch011

Elkington, J., & Rowlands, I. H. (1999). Cannibals with forks: The triple bottom line of 21st century business. *Alternatives Journal*, *25*(4), 42.

Epstein, M. J. (2018). *Making sustainability work: Best practices in managing and measuring corporate social, environmental and economic impacts*. Routledge., DOI: 10.4324/9781351280129

Evrard, A., Hexel, D., Kaiser, C., Lurin, J., Marina Decarro, N., Rastoldo, F., & Soussi, A. (2003). De l'orientation au projet de formation. *Genève: Service de la recherche en éducation*.

Fleming, P., & Jones, M. V. (2012). The end of corporate social responsibility. *Crisis and Critique*.

Freeman, R. E. (1984). *Strategic Management: A Stakeholder Approach*. Pitman.

Freeman, R. E. (2010). *Strategic management: A stakeholder approach.* Cambridge university press. DOI: 10.1017/CBO9781139192675

Gond, J. P., Igalens, J., Swaen, V., & El Akremi, A. (2011). The human resources contribution to responsible leadership: An exploration of the CSR–HR interface. In *Responsible leadership* (pp. 115–132). Springer Netherlands., DOI: 10.1007/978-94-007-3995-6_10

Greenberg, J. (1987). A taxonomy of organizational justice theories. *Academy of Management Review*, *12*(1), 9–22. DOI: 10.2307/257990

Hahn, R. (2013). ISO 26000 and the standardization of strategic management processes for sustainability and corporate social responsibility. *Business Strategy and the Environment*, *22*(7), 442–455. DOI: 10.1002/bse.1751

Hart, S. L., & Milstein, M. B. (2003). Creating sustainable value. *The Academy of Management Perspectives*, *17*(2), 56–67. DOI: 10.5465/ame.2003.10025194

Idowu, S. O. (2021). *Current Global Practices of Corporate Social Responsibility*. Springer International Publishing., DOI: 10.1007/978-3-030-68386-3

Kant, S., & Adula, M. (2025). Business Eco-Systems Sustainability Impact on Cultivating Social Capital Through Corporate Social Responsibility in Ethiopia. In Sustainable Business Ecosystems and Social Perspectives (pp. 387-414). IGI Global Scientific Publishing. DOI: 10.4018/979-8-3693-8437-4.ch017

Kant, S., Adula, M., & Birhane, M. (2025). Green Financing Effect on Green Human Capital: Through a Business Model for Corporate Social Responsibility in Horn of Africa. In Driving Business Success Through Eco-Friendly Strategies (pp. 169-188). IGI Global Scientific Publishing.

Kant, S., Degefa, T. T., Roba, A. T., Alemu, Z. K., & Takaro, G. S. (2025). Corporate Social Responsibility Effect on Tax Compliance by Mediation of Sense of Civic Duty in the Horn of Africa. In Modeling and Profiling Taxpayer Behavior and Compliance (pp. 231-250). IGI Global Scientific Publishing. DOI: 10.4018/979-8-3373-0422-9.ch010

Kant, S., Toma, F., Adula, M., & Shondoko, C. W. (2025). Corporate Social Responsibility (CSR) Effect on Digital Marketing Sustainability by Mediation of Green Marketing Paradigm in the Horn of Africa. In Strategic Workforce Reskilling in Service Marketing (pp. 315-336). IGI Global Scientific Publishing.

Kotler, P., & Lee, N. (2005). *Corporate Social Responsibility: Doing the Most Good for Your Company and Your Cause*. John Wiley & Sons.

Kumar, A. (2022). Modelling the importance of social responsibility standards (ISO 26000: 2010) practices adoption in freight transport industry. *Journal of Cleaner Production, 367*, 132861. DOI: 10.1016/j.jclepro.2022.132861

Lozano, R. (2015). A holistic perspective on corporate sustainability drivers. *Corporate Social Responsibility and Environmental Management*, *22*(1), 32–44. DOI: 10.1002/csr.1325

Margolis, J. D., & Walsh, J. P. (2003). Misery loves companies: Rethinking social initiatives by business. *Administrative Science Quarterly*, *48*(2), 268–305. DOI: 10.2307/3556659

Mol, A. P. J., & Spaargaren, G. (2000). Ecological modernization theory in debate: A review. *Environmental Politics*, *9*(1), 17–49. DOI: 10.1080/09644010008414511

Nave, A., & Ferreira, J. (2019). Corporate social responsibility strategies: Past research and future challenges. *Corporate Social Responsibility and Environmental Management*, *26*(4), 885–901. DOI: 10.1002/csr.1729

Orlitzky, M., Schmidt, F. L., & Rynes, S. L. (2003). Corporate social and financial performance: A meta-analysis. *Organization Studies*, *24*(3), 403–441. DOI: 10.1177/0170840603024003910

Popa, D. I., & Dabija, D. C. (2018). ISO 26000: a brief literature review. *ISO 26000-A Standardized View on Corporate Social Responsibility: Practices, Cases and Controversies*, 81-92. https://doi.org/DOI: 10.1007/978-3-319-92651-3_6

Porter, M. E., & Kramer, M. R. (2006). The link between competitive advantage and corporate social responsibility. *Harvard Business Review*, *84*(12), 78–92. PMID: 17183795

Porter, M. E., & Kramer, M. R. (2018). Creating shared value: How to reinvent capitalism—And unleash a wave of innovation and growth. In *Managing sustainable business: An executive education case and textbook* (pp. 323–346). Springer Netherlands.

Putnam, R. D. (1993). *Making Democracy Work: Civic Traditions in Modern Italy*. Princeton University Press.

Sartori, S., Witjes, S., & Campos, L. M. (2017). Sustainability performance for Brazilian electricity power industry: An assessment integrating social, economic and environmental issues. *Energy Policy*, *111*, 41–51. DOI: 10.1016/j.enpol.2017.08.054

Sebastian, B., & Mathew, T. (2018). Corporate social responsibility practices of public sector and private sector companies in India. *SMART Journal of Business Management Studies*, *14*(1), 32–41. DOI: 10.5958/2321-2012.2018.00004.0

Seghyar, N., Gouiouez, M., Nafzaoui, M. A., & Amane, M. (2025). Influence Of Social Responsibility Management On The Configuration Of Management Control In Moroccan Companies. http://dx.doi.org/DOI: 10.21511/ppm.23(1).2025.03

Shaikh, E. (Ed.). (2024). *Strategies and Approaches of Corporate Social Responsibility Toward Multinational Enterprises*. IGI Global. DOI: 10.4018/979-8-3693-0363-4

Teece, D. J. (2018). Business models and dynamic capabilities. *Long Range Planning*, *51*(1), 40–49. DOI: 10.1016/j.lrp.2017.06.007

Valmohammadi, C. (2014). Impact of corporate social responsibility practices on organizational performance: An ISO 26000 perspective. *Social Responsibility Journal*, *10*(3), 455–479. DOI: 10.1108/SRJ-02-2013-0021

Wamba, J. H. T., Bekono, S. S., & Fouda, J. B. I. (2020). Les déterminants de la performance financière des petites entreprises camerounaises. *Finance & Finance Internationale*, *1*(20). https://doi.org/DOI: 10.34874/IMIST.PRSM/ffi-v1i20.21818

Yang, C. S. (2018). The effect of CSR practices on organisational performance in the container shipping in Taiwan. *International Journal of Shipping and Transport Logistics*, *10*(2), 181–201. DOI: 10.1504/IJSTL.2018.090098

Yoon, B., & Chung, Y. (2018). The effects of corporate social responsibility on firm performance: A stakeholder approach. *Journal of Hospitality and Tourism Management*, *37*, 89–96. DOI: 10.1016/j.jhtm.2018.10.005

Chapter 9
Sustainable Sourcing and Procurement Strategies for a Performance Mass Retail Supply Chain:
An Exploratory Qualitative Study

Ichrak Midar
https://orcid.org/0009-0004-3237-0783
Sidi Mohammed Ben Abdellah University, Morocco

Lhoussaine Alla
https://orcid.org/0000-0002-7238-1792
Sidi Mohammed Ben Abdellah University, Morocco

Badr Bentalha
https://orcid.org/0000-0003-1339-542X
Sidi Mohammed Ben Abdellah University, Morocco

ABSTRACT

Mass retail supply chains are increasingly expected to balance high performance with sustainable practices. Retailers must meet economic and operational goals while addressing growing environmental and social responsibilities. This dual pressure challenges traditional sourcing and procurement strategies, which remain cost-driven. Despite interest in sustainable supply chain management, limited research explores how sustainable sourcing and procurement drive performance in the mass retail sector. This study fills that gap using an exploratory qualitative approach based on semi-structured interviews with retail supply chain stakeholders. Thematic analysis reveals emerging practices, strategic shifts, and tensions between cost constraints

DOI: 10.4018/979-8-3373-3012-9.ch009

and long-term sustainability. Findings show that when strategically aligned, sustainable procurement can enhance performance. The study also highlights the need for better internal coordination and clearer sustainability frameworks. It offers practical insights and recommendations to support more resilient and responsible mass retail supply chains.

INTRODUCTION

Mass market retail supply chains are increasingly challenged in the global economy to balance performance with responsibility. While retailers are expected to ensure efficiency, speed, and profitability, they are also under growing pressure from consumers, regulatory bodies, and institutional stakeholders to embed environmental protection, ethical sourcing, and social equity into their operational models. As a result, sourcing and procurement practices are being redefined: no longer simple cost-control tools, they are becoming strategic drivers of long-term value and sustainability across the supply chain.

Sustainability considerations have changed the perception of managerial and strategic variables of the service company (Bentalha, 2022), encouraging procurement to shift from a purely transactional function to one embedded in broader organizational value creation logics.

Both internal organizational objectives and external regulatory pressures drive these transformations. International standards and policy frameworks offer useful guidance in this regard. ISO 20400 provides a structured approach for integrating sustainability into the procurement process, emphasizing stakeholder engagement, risk management, and supplier evaluation criteria beyond cost. The Global Reporting Initiative (GRI) encourages organizations to disclose their sourcing impacts on labor, the environment, and ethics through specific procurement-related indicators (e.g., GRI 204 and GRI 308). Meanwhile, UN Sustainable Development Goal 12, particularly Target 12.7, calls for the promotion of public procurement practices that are sustainable, in accordance with national policies and priorities. Despite their relevance, these frameworks are rarely operationalized in practice, especially in mass retail contexts. Yet, despite their relevance, such frameworks remain underutilized or poorly operationalized in mass retail contexts, particularly in emerging economies.

In this context, integrating sustainability into procurement strategies represents both an opportunity and a major challenge. Mass retail is a complex environment, marked by scale, supplier diversity, and dynamic demand patterns. Procurement must reconcile traditional objectives, cost, quality, and delivery, with long-term environmental and social performance targets. This tension creates strategic trade-offs that require not only operational adaptation but also a rethinking of governance, coordination, and performance measurement.

Retail procurement is no longer only about selecting suppliers or negotiating contracts, it has become a domain of strategic foresight. In the face of increasing regulatory scrutiny and consumer activism, the procurement function is expected to deliver on sustainability goals without compromising competitiveness. Procurement leaders must therefore embrace a more proactive and integrative approach, one that sees sustainability not as an added burden, but as a core source of value and differentiation. As demonstrated by cases in Northern and Western Europe, retailers who have embedded circular economy principles into their sourcing strategies have seen tangible benefits in risk mitigation and consumer trust.

The complexity of supply networks in mass retail also brings new challenges to traceability and accountability. Many retailers operate within vast, globally dispersed ecosystems involving thousands of suppliers. Ensuring sustainable practices in such fragmented systems requires the deployment of advanced monitoring tools and digital enablers such as blockchain, Internet of Things (IoT), and AI-driven analytics. These technologies are increasingly being positioned as solutions to improve transparency and control, yet their implementation varies widely across contexts, raising important questions about scalability, cost, and ethical use.

Cultural and institutional factors also shape how sustainability is interpreted and practiced in procurement. In emerging markets like Morocco, where regulatory enforcement may be weaker and supplier capacities more constrained, the adoption of sustainable procurement requires tailored strategies. Capacity building, supplier training, and shared investment models are often more effective than top-down compliance demands. Moreover, the informal economy and limited data availability further complicate standardization and benchmarking efforts, suggesting the need for hybrid governance mechanisms that blend global frameworks with local realities.

Despite the rising importance of this issue, the academic literature offers limited insights into how sustainability is practically embedded into sourcing strategies in mass retail. Most studies either focus on manufacturing sectors or offer normative frameworks with little grounding in retail realities. Bentalha et al. (2022) emphasize that innovation and strategic management are key to improving supply chain performance, but how these are enacted through procurement in retail remains underexplored. Hmioui et al. (2020) call for better alignment between efficiency and sustainability in the supply chain, yet again, real-life cases from the retail sector are lacking.

A core issue lies in the definition and assessment of performance itself. While traditionally limited to economic efficiency, performance in sustainable procurement now includes environmental outcomes (e.g., emissions reduction), social impacts (e.g., labor rights), and reputational dimensions. Bourguignon (1995) reminds us that performance should be understood in relation to the evolving expectations of stakeholders and societal values. Yet, few studies propose clear typologies of performance metrics in this domain, especially those that integrate carbon emissions or climate-resilient indicators, a critical omission in the context of low-carbon transitions.

Procurement, therefore, emerges as a pivotal function. Villena et al. (2018) argue that procurement is the core lever to building sustainable supply networks, provided it evolves from a transactional approach to one that is collaborative and inclusive of upstream actors, where most sustainability risks and opportunities reside. This includes actors from second- and third-tier suppliers, where transparency, control, and data availability are often weakest.

However, sustainability integration is not only a matter of supplier relations. Schneider et al. (2012) highlight the need for internal coordination between procurement and other departments, product design, logistics, quality, legal and compliance, to ensure alignment with overarching corporate sustainability goals. In practice, however, many retailers face internal fragmentation and siloed decision-making, resulting in inconsistent or symbolic sustainability efforts.

Another layer of complexity relates to governance structures. Morcillo-Bellido et al. (2020) show that governance mechanisms such as supplier codes of conduct, social audits, traceability systems, and performance dashboards are crucial to implement and monitor sustainable practices. Yet, replicating such mechanisms across highly decentralized and heterogeneous supplier networks, often spanning multiple countries and legal frameworks, is a formidable challenge for mass retail chains.

Thus, although sustainability is widely acknowledged as a strategic imperative, its operationalization in procurement remains uneven and under-researched, particularly in emerging markets like Morocco. There is still much to uncover about how procurement professionals perceive sustainability, what obstacles they face in implementation, and what impacts they aim to create within their organizations.

This chapter seeks to address this research gap by exploring the following guiding question:

How are mass retail supply chains integrating sustainability into their sourcing and procurement strategies, and what effects does this integration have on multidimensional supply chain performance and resilience?

We employ a qualitative, exploratory methodology based on semi-structured interviews with key actors in the Moroccan mass retail ecosystem. The study aims to illuminate how procurement professionals negotiate tensions, engage suppliers, collaborate internally, and contribute to governance mechanisms in the pursuit of more responsible, transparent, and efficient supply chains.

By grounding the analysis in real-world practices and lived experiences, this research contributes to the theoretical and practical understanding of sustainable procurement. It highlights how procurement can evolve from a cost-focused function to a strategic lever for sustainability and low-carbon performance, while also offering actionable insights for the design of more inclusive and future-oriented supply chains.

I. LITERATURE REVIEW

1. Sustainability in Supply Chain Management

1.1. Sustainability in Supply Chain Management: General Framework

The way supply chains are managed and run has changed dramatically, which is expected since the worldwide knowledge and interest in resource scarcity, social injustice, and environmental degradation are increasing in a very fast and admirable way. Once primarily motivated by cost and efficiency, supply chain management (SCM) has progressively changed to include sustainability as a strategic requirement. This change represents a move away from reactive compliance and toward a proactive strategy that integrates social and environmental concerns with financial ones.

A widely accepted definition by Seuring and Müller (2008, p. 1700) captures the essence of this evolution: "Sustainable supply chain management is the management of material, information, and capital flows as well as cooperation among companies along the supply chain while taking goals from all three dimensions of sustainable development, economic, environmental, and social, into account, which are derived from customer and stakeholder requirements."

In practice, this means that firms are no longer judged solely by their ability to deliver products efficiently but also by how their supply chain practices impact the broader ecosystem in which they operate. As noted by Vadakkepatt et al. (2021), "sustainability in the supply chain requires retailers to deploy system-wide integration across the entire supply chain to minimize harm to the environment and individuals and to benefit the environment and society over time."

1.2. Sustainable Sourcing and Procurement Strategies

Procurement and sourcing play a central role in sustainable supply chain management. They have undergone a significant metamorphosis over the past few decades. Traditionally, purchasing has been perceived as a transactional and cost-driven activity, now it has gradually become a strategic lever for competitive advantage and sustainability integration.

As Lowson (2001) observes, "purchasing and supply management has developed from the adversarial, clerical transaction between supplier and customer, to a strategic, co-operative, buyer–seller relationship." Nonetheless, the tension between sustainability and cost persists. This paradox poses a central challenge for practitioners seeking to integrate sustainability into sourcing decisions.

The power dynamics in buyer-supplier relationships are one of the main factors that either facilitate or hinder sustainable procurement. As Chkanikova (2016) notes, "the power construct in buyer–supplier relationships is of particular importance in understanding whether the retailer is in a favorable position to exercise influence and control over the supplier to provide the desired, e.g., environmentally and socially superior, products."

Furthermore, the increasing emphasis on corporate social responsibility (CSR) has pushed sustainable procurement higher on the strategic agenda. According to Walker et al. (2012), "sustainable procurement is increasingly on the agenda for purchasing and supply managers seeking to demonstrate corporate social responsibility in their supply chains."

In the context of sustainable procurement, governance mechanisms are central to ensuring that sustainability is not merely aspirational but embedded in operational routines. According to Morcillo-Bellido et al. (2020), effective governance relies on multi-layered instruments such as supplier codes of conduct, monitoring systems, capacity-building programs, and compliance audits.

2. Rethinking Supply Chain Performance in the Age of Sustainability

For a long time, the notion of performance in business, and particularly in supply chain management (SCM), was limited to economic metrics. Managers focused on cost minimization, profit maximization, and operational efficiency. Bourguignon (1995) defined performance simply as "the achievement of organizational objectives," often through a purely economic lens.

Yet, over the past two decades, this definition has lost its relevance. As global supply chains have become more complex and exposed to environmental, social, and geopolitical risks, the criteria for evaluating performance have evolved. Increasingly, organizations are expected to go beyond short-term profitability and address broader concerns such as environmental impact, social responsibility, ethical sourcing, and long-term resilience.

In this context, the role of sourcing and procurement has become strategic. Bentalha et al. (2022) highlight that understanding all costs of a service supply chain enhances performance management. This means decision-makers must consider hidden costs like waste, instability, and reputational risks.

Sustainable procurement now requires a multidimensional perspective, incorporating environmental, social, and reputational outcomes (Bourguignon, 1995; Carter & Rogers, 2008). However, carbon-related metrics, such as Scope 3 emissions, remain largely absent from evaluations, although they are key in low-carbon transitions (CDP, 2023).

3. Governance, Standards and Institutional Pressures

The operationalization of sustainability in procurement is often guided by international norms and standards, which offer structured approaches to responsible sourcing. Frameworks such as ISO 20400 (2017), the Global Reporting Initiative (GRI Standards 204 and 308), and the UN Sustainable Development Goals (SDG 12.7) emphasize the integration of environmental, social, and ethical dimensions into procurement decisions. Yet despite their availability, these frameworks are not always effectively implemented, especially in mass retail environments where procurement is decentralized and driven by cost efficiency.

ISO 20400 advocates a lifecycle-based approach to procurement, with emphasis on risk assessment, transparency, and stakeholder collaboration. GRI standards encourage organizations to disclose supplier-related impacts and evaluate them against sustainability benchmarks. UN SDG 12.7 calls for promoting sustainable public procurement in line with national policies and development strategies. However, the gap between these normative principles and actual practices reveals deep institutional challenges. In emerging economies, implementation is often constrained by limited resources, low supplier maturity, and lack of traceability.

II. METHODOLOGY

Our study follows a qualitative research design. We chose the design because of its relevance in exploring the complexities of sustainable sourcing and procurement practices in mass retail supply chains. Qualitative research is particularly well suited to uncovering the subjective and contextual dimensions of human action, offering insights into the meanings, values, and motivations that drive decision-making processes (Kvale & Brinkmann, 2009). As Paillé and Mucchielli (2003) mentioned, qualitative data are inherently discursive and must be interpreted as meaningful constructions rather than as mere numerical indicators.

In the field of supply chain management, where studies remain predominantly quantitative and model-driven, qualitative approaches are still relatively underused (Gammelgaard & Flint, 2012). However, they offer a distinct perspective for analyzing actor relationships, inter-organizational dynamics, and the social embeddedness of operational practices. The qualitative paradigm in this study allowed for an examination of how practitioners interpret, negotiate, and use sustainability concepts in procurement strategies.

In line with our research objective, this study aims to answer the following question:

How are sustainability principles operationalized within procurement practices in the Moroccan mass retail sector, and how do professionals navigate tensions between performance requirements and sustainability goals?

This central question reflects the exploratory and inductive nature of the work and connects directly to the broader problem framing and objectives of the thesis.

To that end, semi-structured interviews were selected as the primary method of data collection. This method allows for flexibility and depth while maintaining a coherent thematic structure across interviews (Wengraf, 2001). It is particularly effective for investigating individual perceptions and experiences related to organizational behavior, especially in the context of strategic decision-making and complex networks.

Eight interviews were conducted with professionals working in or closely affiliated with the Moroccan mass retail sector. The participants included supply chain managers, procurement consultants, CSR managers, logistics coordinators, and consultants, all of whom play a role in the design or execution of sustainable sourcing initiatives. The interviews were held online between May 17 and May 25, 2025, to accommodate geographic dispersion and participant availability. Each interview lasted approximately 15 minutes and was guided by a semi-structured interview protocol developed for this study.

To enhance transparency and reproducibility, the main thematic axes of the interview protocol are provided in Table 1. Each axis was designed to investigate a distinct dimension of sustainable procurement practices within mass retail supply chains in Morocco.

Table 1. Overview of the interviewed experts

Thematic Axis	Content Explored	Objective of the Axis
1. Vision of Sustainability	Participants' perceptions of sustainability, strategic importance, alignment with company values and mission.	Understand how sustainability is defined, framed, and prioritized within retail supply chains.
2. Sustainable Practices	Internal policies, sourcing criteria, supplier requirements, certifications, and ESG integration.	Identify actual sustainable sourcing actions and assess the maturity of implemented practices.
3. Impact on Performance	Effects on cost, risk, reputation, and long-term supply chain value; performance indicators used.	Explore how sustainability is perceived to affect economic and non-economic performance metrics.
4. Challenges	Barriers to implementation, including cost pressures, supplier resistance, regulatory ambiguity.	Highlight the structural and operational obstacles to embedding sustainability in procurement.
5. Success Factors	Internal and external enablers, leadership support, cross-functional coordination, innovation.	Identify drivers and organizational capabilities that support successful sustainable procurement.

Source: **Authors.**

All interviews were conducted in French, the participants' working language. They were recorded with prior consent and fully transcribed verbatim. The analysis was likewise carried out in French to preserve the richness and nuances of the original discourse. This aligns with the view that qualitative data must be interpreted within their linguistic and cultural context, rather than being abstracted into purely formal codes.

The size of our sample, eight participants, may appear limited in quantitative terms. However, in qualitative and exploratory research, the focus lies not in achieving statistical generalizability but in generating conceptual insights and deep understanding (Paillé & Mucchielli, 2003). The sample was considered sufficient to reach thematic saturation, especially given the relative homogeneity of the respondent profiles and the specificity of the study context. Furthermore, the selection was driven by theoretical sampling logic, privileging expert voices with direct involvement in procurement strategies.

The data were processed and analyzed using the software NVivo, a well-established tool in qualitative research for thematic coding, classification, and pattern recognition. NVivo enabled the researcher to perform inductive coding, identify emerging themes, and conduct cross-case comparisons among participant responses. It also supported memo writing and the construction of analytical categories, which contributed to theory building grounded in empirical narratives.

Through this methodological approach, the study seeks to capture the lived realities, tensions, and strategic trade-offs faced by supply chain actors engaging with sustainability. Rather than imposing external evaluation frameworks, the analysis was guided by the language, concerns, and priorities expressed by the interviewees themselves, thus honoring the interpretive foundations of qualitative inquiry. Table 2 provides an overview of the profiles of the different experts who participated in the interviews.

Table 2. Overview of the interviewed experts

Reference	Position / Role	Gender	Key Area of Expertise
I1	Junior Purchaser	F	Strategic sourcing and sustainability trade-offs
I2	Merchandize Planner (CSR oriented)	M	ESG integration and sustainable procurement policies
I3	Logistics Platform Manager	F	Operational practices and environmental optimization
I4	Local Agri-food Supplier	M	Sustainability challenges faced by small producers
I5	Fresh Products Department Manager	M	Waste reduction and customer communication
I6	Procurement Consultant	F	Risk-based supplier evaluation and strategic advisory
I7	Quality & Supplier Audit Officer	F	Compliance monitoring and supplier sustainability support
I8	Innovation & Digital Transformation Responsible	F	Digital traceability and ESG integration through tech

Source: **Authors.**

III. RESULTS AND DISCUSSION

1. Textual Analysis

The first stage of the qualitative findings focuses on the lexical analysis derived from the interview corpus, as processed through NVivo. A word cloud (Figure 1) has been used to graphically represent the frequency of terms mentioned by participants. To generate this visualization, both direct word counts and **linguistically similar words** were taken into account, providing a more nuanced view of the vocabulary patterns within the data. This type of visual output is particularly useful for gaining an initial overview of dominant ideas across the dataset. The larger the word appears in the figure, the more frequently it was cited or reflected through its linguistic variants. In this case, words such as *suppliers*, *sustainability*, *responsible*, and *performance* stand out prominently at the center of the image, indicating their recurring importance across interviews. These terms reflect a shared discourse that positions sustainable procurement not only as a moral obligation but also as a core driver of supply chain performance. Additional terms, such as *purchasing*, *tools*, *training*, and *criteria*, point to a growing effort to operationalize sustainability through structured practices and performance measurement systems. Conversely, the presence of words like *constraints*, *cost*, and *waste* signals the challenges that organizations face in implementing these strategies. This lexical layer of analysis serves as a gateway to deeper thematic exploration. It suggests that interviewees view sustainability as a multi-dimensional concept, simultaneously strategic, ethical, and technical, and one that is progressively reshaping how performance is defined and pursued in the mass retail supply chain.

Table 3. Frequency of terms mentioned by interviewed experts

Word	Count	Weighted Percentage (%)	Similar Words
durabilité	23	3,15	durabilité, durable, durables
fournisseurs	21	2,87	fournisseur, fournisseurs
produits	13	1,78	produit, produits
rse	10	1,37	RSE
performance	9	1,23	performance
responsable	9	1,23	responsable, responsables
achats	7	0,96	achat, achats
clients	7	0,96	acheteur, acheteurs, client, clients
marque	7	0,96	cocher, marque, score, signé
améliorer	7	0,82	améliorer, avance, marche, progresser
attention	6	0,82	attentes, attention, respecter
chaîne	6	0,82	chaîne
prix	6	0,82	prix
marche	7	0,75	marche, réussir, tournées, travaille, travailler
appels	5	0,68	appels, engagé, engagement, engagements
distribution	5	0,68	distribution, livraison, livraisons
formation	5	0,68	format, formation, structure, structurés
grande	5	0,68	grande, pousse
intégrer	5	0,68	intégrer
lien	5	0,68	lien, relation

Source: **Qualitative Study.**

Figure 1. **Word cloud for the frequency of terms mentioned by interviewed experts.**

Source: **Qualitative Study.**

2. Thematic Analysis

2.1. Theme 1: Vision of Sustainability

The participants' reflections on sustainability reveal not a singular vision, but rather a tapestry of operational, strategic, and ethical interpretations that reflect their roles and day-to-day realities within the mass retail supply chain. What emerges is a shared sense of direction, though approached through different lenses.

At the most practical level, sustainability is tied to actionable behaviors. For instance, **I5** associates it with decisions such as reducing plastic usage, supporting seasonal products, and limiting food waste. **I3**, similarly grounded in operations, defines it as achieving logistical efficiency: *"less plastic, fewer empty transport trips, and less waste*." These understandings suggest that for actors closer to the ground, sustainability is embedded in routines and immediate performance constraints.

Others describe sustainability in more strategic or systemic terms. **I2** frames it as a broad commitment, *"not just planting trees,"* but also reducing carbon emissions, supporting local economies, and improving working conditions across the supply chain. **I6** offers perhaps the relational view, seeing it as a shift from transactional to partnership-oriented dynamics: a fundamental transformation in the way buyers and suppliers interact.

The supplier's voice, represented by **I4**, adds a layer of real-world pragmatism. Sustainability, they say, means *"producing cleanly, respecting people, and staying profitable."* This subtle but critical addition, the need for economic survival, reflects the reality that sustainability must coexist with financial pressure, especially for smaller actors.

In more technical or compliance-driven roles, such as procurement and quality assurance, sustainability is often articulated through certifications, scoring systems, and traceability mechanisms. **I1** and **I7** refer to recycled materials, supplier audits, and evaluation tools, while acknowledging that these should not become mere checklists. As **I7** notes, *"committed suppliers are often more reliable, fewer issues, fewer disputes."*

Finally, **I8** underscores the data-driven future of sustainability, arguing that in today's context, one cannot meaningfully engage with sustainability without tools for traceability, ESG measurement, and digital visibility. Their view bridges strategy and technology, highlighting the need to *"know how products are made, where they come from, and what impact they have."*

Taken together, these insights reflect a view of sustainability that is grounded, evolving, and multidimensional. Rather than a fixed ideal, it is a set of overlapping interpretations that blend ethics with pragmatism, vision with routine, and ideals with hard realities.

2.2. Theme 2: Sustainable Practices

For the participants in this study, sustainability is no longer a theoretical ambition; it is being translated into a diverse and growing repertoire of concrete, everyday practices. From procurement to store-level operations and supplier engagement, sustainability initiatives are gradually being integrated into the operational backbone of the mass retail supply chain. These initiatives reflect an evolving mindset where environmental and social considerations are no longer peripheral but central to performance and strategy.

At the retail field level, sustainability is often enacted through people-driven, practical interventions. **I5** highlights collaboration with regional producers, not only as a way to shorten supply chains and support local economies, but also to enhance product freshness and customer trust. The same respondent notes efforts to shift away from wasteful habits by implementing donation programs and waste sorting at the store level: "*We're not throwing away as much, what can be saved goes to charity*." **I3** adds to this by referencing innovations in logistics, such as energy-efficient lighting systems and the use of reusable containers in product delivery, which serve both ecological and cost-saving goals.

On the supply side, similar initiatives are underway. **I4** describes efforts to reduce the use of chemical inputs through the adoption of drip irrigation and to improve labor practices by formalizing seasonal work arrangements. These transformations, initiated despite resource constraints, signal a deep-rooted commitment to sustainability. Their pursuit of Global GAP certification and shift to reusable packaging further reflects a willingness to not only comply with industry standards but also actively shape them. These practices illustrate the imperative raised by Agrawal & Lee (2019): to offer sustainably produced goods, organizations must first be able to source sustainably produced components. In this light, suppliers emerge not just as enablers of sustainability but also as essential co-architects of the sustainable supply network.

Procurement professionals are increasingly revisiting their sourcing models in response. **I1** describes an intentional shift toward prioritizing ethical and local suppliers, often guided by formalized RSE (Responsabilité Sociétale des Entreprises)/CSR (Corporate Social Responsibility) charters. I7 complements this by discussing supplier evaluation frameworks that now include sustainability scoring. In some cases, financial support is offered to assist suppliers in sustainable packaging redesign and water conservation, a move that not only incentivizes sustainable practices but also strengthens relational ties.

Technology is acting as both catalyst and coordinator of these efforts. **I8** emphasizes the development of an ESG (Environmental, Social, Governance) data portal for suppliers, as well as the use of AI-driven forecasting tools to align procurement volumes with real demand, thereby reducing waste. These tools exemplify how digital transformation can serve sustainability goals, rather than merely acting as cost-control mechanisms.

Strategically, companies are beginning to formalize these actions. **I2** points to diagnostic tools and collaborations with agricultural cooperatives that foster long-term behavioral change, while I6 highlights the institutionalization of ESG criteria into procurement policies and tenders: "*It's not just about offering a sustainable option, it's a filter. If they don't meet those criteria, they're out.*"

What emerges is a multi-scalar and adaptive ecosystem of sustainability practices, spanning from operational detail to strategic vision, and from individual behaviors to organizational culture. These efforts demonstrate that sustainable supply chains are not built in a single act but are cultivated through incremental, coordinated changes. They also confirm the foundational principle that sustainability begins with sourcing, and that responsible sourcing requires a web of aligned actors, from procurement teams to small-scale producers, all contributing to the sustainability of the final product (Agrawal & Lee, 2019).

Figure 2. Sustainable practices in Mass Retail Supply Chains mentioned by interviewed experts

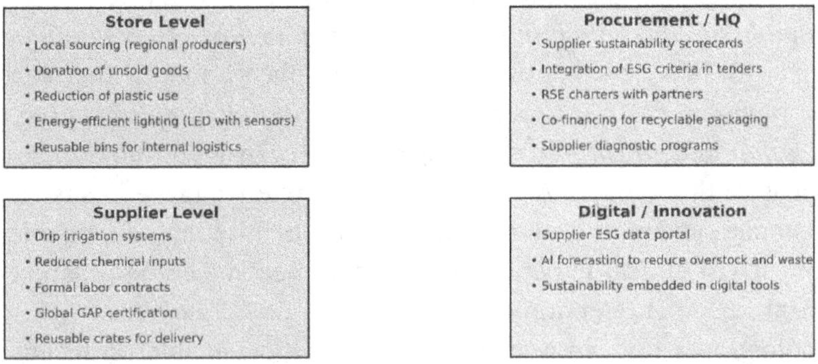

Source: **Authors.**

2.3. Theme 3: Impact on Performance

Across the interviews, participants consistently emphasized a growing interdependence between sustainability and supply chain performance. Rather than viewing sustainable practices as additional or optional, many stakeholders saw them as key enablers of value creation, operational resilience, and customer satisfaction.

At the operational level, participants highlighted tangible efficiency gains as one of the immediate outcomes of sustainable practices. For instance, **I3** pointed out that improved coordination in deliveries and better loss management directly reduce waste, lower costs, and prevent recurring store-level issues: "*It saves money and avoids store-level issues.*" This operational alignment was echoed by **I5**, who underscored the value of local sourcing strategies: "*Fewer losses mean more satisfied customers*." By sourcing closer to the point of sale, companies improve freshness, minimize spoilage, and ultimately deliver higher service quality, creating both environmental and customer benefits.

Sustainability was also closely linked to the reliability and maturity of supplier partnerships. **I7** noted, *"Committed suppliers usually deliver more consistently and with fewer problems,"* while **I1** described sustainably oriented suppliers as "*more structured and dependable partners.*" These observations suggest that sustainability commitments are often indicative of a broader organizational discipline and long-term thinking, making such suppliers more trustworthy and aligned with performance-driven goals.

The reputational and consumer-facing dimensions of sustainability emerged as another powerful vector of performance. As **I2** argued, transparent and ethical sourcing practices contribute not only to risk mitigation but also to brand resilience: "*A sustainable supply chain is more resilient and less prone to reputational crises.*" This aligns with **I6's** view that sustainability now plays a significant role in shaping consumer preferences: "*Customers who care about these values tend to stick with companies that share them.*" This finding reflects the insight from Nair et al. (2024), who emphasize that modern retailers must proactively engage in sustainability to attract customers and meet evolving expectations. In this view, sustainable practices are no longer just ethical imperatives, they have become core to market competitiveness and customer retention.

Technology was frequently mentioned as a critical enabler in reconciling performance and sustainability. **I8** discussed the role of AI-driven tools and ESG dashboards, which help companies track sustainability metrics, predict risks, and optimize decisions. These technologies reduce inefficiencies and environmental impact while maintaining ethical standards, serving as both strategic tools and performance accelerators.

Despite these advantages, participants acknowledged that the returns from sustainable procurement are not always immediate. **I4** warned that some benefits require time to manifest and may be difficult to quantify in the short term: "*It has to be supported... otherwise the results won't show up immediately.*" This calls for a shift in how performance is measured, moving beyond traditional financial indicators to incorporate longer-term environmental, social, and reputational outcomes.

In sum, participants rejected the idea of a trade-off between sustainability and performance. Instead, they described a mutually reinforcing relationship in which responsible procurement practices enhance operational efficiency, strengthen supplier relations, build brand credibility, and meet the demands of increasingly value-conscious consumers. As this theme shows, sustainability is evolving from a compliance issue into a strategic lever for performance in the mass retail sector, provided that companies are willing to adopt a long-term, customer-centric, and innovation-driven approach.

2.4. Theme 4: Challenges

Despite growing interest in sustainability across mass retail supply chains, participants voiced significant implementation challenges. Their insights reveal not just isolated hurdles, but systemic frictions embedded in the economic, cultural, and operational fabric of retail procurement. These challenges highlight a critical disconnect between sustainability ambitions and the realities of retail supply chain management.

The most persistent tension cited by interviewees was the overwhelming dominance of cost considerations. While sustainability criteria are increasingly introduced into procurement discourse, financial metrics continue to dictate final decisions. As **I1** admitted, "*We're being asked to consider sustainable criteria, but honestly, price is still the number one factor.*" This view was echoed by **I4**, who pointed out the mismatch between expectations and financial support: "*They ask us to comply with strict standards, but the purchase prices don't change.*" This price-first logic leaves little room for environmental or ethical considerations, especially when profitability targets remain non-negotiable.

Small and local suppliers are especially vulnerable to these cost pressures. Often operating with limited infrastructure and thin margins, they struggle to absorb the investments required for sustainable compliance. **I4** referred to these adjustments as *"a big one for a small structure like ours,"* while **I3** noted how last-minute delivery requests frequently push suppliers to make compromises, using inefficient logistics options that counteract sustainability goals. These vulnerabilities are symptomatic of deeper supply-side risks identified in the literature, such as uncertainties in supplier responsibility, lead-time variability, and inconsistent supply quality (Okwu et al., 2020).

In addition to external constraints, internal organizational barriers emerged prominently. Participants described a widespread perception of sustainability as an "extra" rather than an essential function. As **I7** put it, *"KPIs are still centered on cost. Not on carbon. Not on ethics."* This misalignment between strategic values and operational performance indicators creates ambiguity, diluting commitment at the execution level. When sustainability remains absent from core decision-making metrics, initiatives often lose traction.

Capability gaps, both technological and human, further complicate implementation. On the technology front, **I8** flagged disparities in digital maturity among suppliers, which hinder the standardization of ESG data or automation of traceability: *"The digital maturity of suppliers is highly uneven."* Even when digital tools are available, participants warned they might be repurposed primarily for cost cutting, diluting their sustainability-enhancing potential.

Human capital gaps were equally critical. Several respondents highlighted the absence of training that equips procurement professionals to think beyond price. **I6** lamented, "*Many buyers are still trained to negotiate price, period.*" Without the right competencies or decision-making frameworks, procurement teams are unlikely to prioritize environmental and social impacts, even when these are embedded in strategic plans.

Another recurring theme was the fragmentation of definitions and expectations regarding sustainability. Participants expressed frustration over inconsistent internal interpretations. **I3**, for instance, described how local products sometimes failed to meet logistical requirements, leading store teams to deprioritize them: *"Store teams just want it delivered on schedule, sustainable or not."* Without a coherent, organization-wide understanding of sustainability goals, efforts become disjointed and momentum is easily lost.

Together, these reflections underscore that the real challenge is not simply the adoption of sustainable sourcing practices, but the need for holistic realignment. Structural coordination, long-term investment in supplier capabilities, revised performance indicators, and a shift in organizational mindset are all required to bridge the gap between intention and execution. As the mass retail sector continues to navigate complex supply-side risks and stakeholder expectations, the ability to embed sustainability meaningfully into procurement strategies will be a decisive factor in shaping resilient, high-performing supply chains.

2.5. Theme 5: Success Factors

While challenges to sustainable procurement remain real and persistent, participants were equally clear about what makes success possible. Their insights converge around several powerful enablers that together form a foundation for lasting change.

At the top of this foundation is leadership commitment. As **I1** put it simply but effectively: *"If top management pushes in this direction, then we can follow with the means to do so. Otherwise, it's difficult."* Without that strategic signal, sustainability risks become symbolic, acknowledged, but never operationalized.

Next is the need for cross-functional alignment. **I6** warned that if CSR and procurement act in isolation, progress stalls: *"everyone has to pull in the same direction."* This was reinforced by **I2**, who highlighted the importance of embedding sustainability into company culture, not just processes. Culture, in their words, is what sustains momentum when the novelty of sustainability fades.

Coordination between HQ and operations was another key point. **I3** emphasized that initiatives should not be imposed without dialogue: *"If headquarters makes decisions without consulting us, it blocks everything."* This sentiment reflects a broader truth: successful sustainability relies on participation, not prescription.

Empowering individuals through training and meaning making also stood out. **I5** explained, *"It's not just about being told 'do it,' we need to understand why."* This suggests that transformation happens when employees internalize the purpose behind sustainability, not just follow new rules.

From the supplier side, open communication and support are essential. **I4** shared that *"if the retailer listens, we can move forward. Otherwise, we do what we can."* This highlights the importance of trust-based relationships rather than transactional exchanges.

Stability and coherence were mentioned too. **I7** stressed that expectations must remain consistent, and **I8** insisted that sustainability should be integrated into systems from the start, not added as an afterthought.

In short, success in sustainable procurement does not come from isolated best practices. It emerges from a shared commitment, coherent systems, and mutual respect, from boardroom to warehouse, supplier to shelf. It's a collective effort that thrives on clarity, engagement, and a long-term perspective.

3. Discussion: Implications, Reflections, and Paths Forward

3.1. Synthesis of Key Insights

The findings from this study suggest that sustainable procurement in the mass retail sector is far from a linear implementation of abstract principles. Rather, it is an evolving process situated at the intersection of strategy, structure, and culture, and shaped by the competing demands of economic performance, social legitimacy, and environmental resilience. The shift from traditional procurement, focused on cost and delivery, to a more strategic and sustainability-centered approach has been partially driven by stakeholder expectations, climate commitments, and reputational pressures.

The five thematic areas, vision of sustainability, sustainable practices, performance impact, challenges, and success factors, demonstrate that sustainability is both a managerial imperative and a contested space. While many respondents associate sustainability with efficiency gains, stronger supplier relationships, and brand trust, they also describe systemic tensions that limit the potential of these initiatives.

Notably, the vision of sustainability articulated by the interviewees is highly pluralistic, ranging from compliance-driven approaches (e.g., certifications and audits) to relational and values-driven strategies (e.g., trust, transparency, supplier empowerment). This aligns with findings from Beske and Seuring (2014), who argue that sustainability in supply chains cannot be reduced to metrics alone but involves deep organizational learning and cross-functional integration. Sustainability is understood as both a market requirement and a moral responsibility, though respondents expressed uncertainty around defining boundaries and performance indicators.

Furthermore, the notion of performance is redefined across multiple dimensions. Traditional procurement KPIs focused on cost, quality, and delivery are now challenged by broader criteria such as carbon emissions, ethical labor practices, and reputational risk (Carter & Rogers, 2008; Bourguignon, 1995). However, as the findings reveal, these metrics are often weakly institutionalized and inconsistently used across departments, leading to fragmented execution. Participants expressed that while sustainability goals are present in strategic communication, they are not always translated into procurement metrics or supplier evaluation frameworks.

A key insight is the centrality of relational dynamics and learning processes. Procurement professionals highlighted how long-term partnerships, information sharing, and co-investment schemes improved both sustainability outcomes and operational resilience. This resonates with the relational view of the firm, where value co-creation with suppliers fosters innovation and sustainability performance (Dyer & Singh, 1998). However, they also noted that relational investments often lacked short-term ROI justification, making them vulnerable in cost-cutting environments.

3.2. Managerial Implications

The study highlights several practical implications for retail managers, procurement officers, and corporate strategists seeking to embed sustainability more robustly into supply chain practices:

- **Rethink Procurement KPIs:** Organizations must recalibrate procurement metrics to reflect not only cost-efficiency but also carbon footprint reduction, social compliance, and supplier resilience. This may involve developing performance dashboards that integrate financial and ESG indicators. As noted by Nair et al. (2024), these tools are vital to legitimizing sustainability internally.
- **Invest in Supplier Enablement:** As emphasized by participants, sustainability goals are difficult to achieve without capacity building at the supplier level. Retailers should co-develop training programs, co-finance technology upgrades, and provide long-term contracts that allow suppliers to plan and invest. These strategies are especially crucial in emerging economies where resource limitations persist.
- **Foster Cross-Functional Alignment:** Siloed CSR and procurement efforts risk diminishing impact. Organizations should foster joint planning and shared accountability between sustainability, operations, marketing, and purchasing teams, possibly through cross-departmental sustainability steering committees.
- **Institutionalize Ethical Standards:** Beyond regulatory compliance, companies should embed ethical principles into supplier selection, audit protocols, and contractual agreements. ISO 20400 and GRI indicators offer valuable frameworks for formalizing these practices.

- **Use Digital Enablers Strategically:** The integration of traceability systems, ESG dashboards, and AI-driven analytics can significantly support the monitoring and enforcement of sustainable procurement. However, these tools must be embedded in decision-making processes, not just used for reporting.
- **Align Procurement with Brand Strategy:** Several participants noted that consumer-facing brands must bridge the gap between ethical narratives and back-end practices. Procurement should be involved in storytelling and transparency efforts to strengthen consumer trust.

3.3. Policy Implications

The results of this study also suggest important considerations for policymakers, trade associations, and regulatory bodies.

- **Define Sector-Specific Standards:** National governments and retail federations should co-develop standards tailored to mass retail realities. These standards could include guidelines for ethical sourcing, data transparency, and carbon disclosure specific to food, textile, and consumer goods categories.
- **Provide Financial Incentives:** Public authorities can accelerate adoption by offering tax benefits, subsidies, or green procurement credits to retailers that implement verifiable sustainable practices, particularly those working with small-scale or local suppliers.
- **Strengthen Public-Private Collaboration:** Governments, NGOs, and industry actors should engage in co-regulation to avoid fragmented compliance systems. This includes alignment between voluntary certifications and mandatory disclosure requirements.
- **Support Supplier Modernization:** Public development programs can offer grants, infrastructure investments, and technical assistance to suppliers, especially SMEs, struggling to meet sustainability benchmarks. This is particularly relevant in developing markets where systemic support structures are limited.
- **Advance Digital Inclusion:** Governments and industry bodies can promote the use of open-access platforms for procurement data sharing, digital traceability, and performance reporting. This could help democratize access to sustainability tools for smaller market players.

3.4. Theoretical Contributions

The chapter contributes to the broader field of Sustainable Supply Chain Management (SSCM) by empirically validating the proposition that sustainability and performance are not contradictory but potentially synergistic. Drawing on the literature (e.g., Seuring & Müller, 2008; Pagell & Wu, 2009), our findings support a capability-based view in which sustainability-enhancing capabilities (e.g., transparency, supplier collaboration, cross-functional integration) become strategic resources that boost overall competitiveness.

Moreover, the study reaffirms the utility of a multidimensional performance framework, expanding beyond the economic logic traditionally used in procurement research. This is consistent with Carter and Easton (2011), who called for a broader conceptualization of performance that incorporates environmental and social capital.

In doing so, our study helps bridge the divide between normative calls for sustainability and the operational realities of retail supply chains. It offers a grounded framework for analyzing how sustainability is internalized, operationalized, and evaluated within large organizations, and highlights the need for coherence across procurement, strategy, and governance layers.

CONCLUSION AND PERSPECTIVES

This study set out to explore how sustainable sourcing and procurement strategies are shaping performance outcomes in the mass retail supply chain, a sector increasingly expected to reconcile operational efficiency with environmental and ethical responsibility. The research was grounded in a qualitative design, relying on the voices of procurement professionals, sustainability officers, and operational managers across the Moroccan retail landscape. The aim was not only to document practices but to capture the complexity, ambiguity, and potential of sustainability integration within procurement systems.

One of the most salient insights from the study is that sustainability is no longer a peripheral concern or public relations narrative. It is gradually becoming a core component of procurement strategy, with clear implications for value creation, supply chain resilience, and brand legitimacy. As participants pointed out, practices such as local sourcing, waste minimization, ethical audits, and digital monitoring are now linked to performance outcomes like supplier reliability, customer satisfaction, and operational agility. In line with Nair et al. (2024), this shows that procurement, when aligned with sustainability, becomes a powerful lever for future competitiveness.

However, the path toward sustainable procurement remains uneven. While some retailers have adopted forward-thinking strategies, others continue to struggle with price-dominated decision-making, fragmented governance, and misalignment between departments. As Villena et al. (2018) and Schneider et al. (2012) highlight, effective sustainability requires more than isolated initiatives; it demands cross-functional coordination, upstream supplier engagement, and robust performance tracking.

Moreover, the study reveals the persistent fragility of governance structures. Sustainability is often embedded in policy documents but inconsistently translated into everyday procurement decisions. As Morcillo-Bellido et al. (2020) point out, the absence of clear accountability frameworks can reduce sustainability to a symbolic gesture rather than an operational reality. Participants emphasized the need for institutionalized governance, codes of conduct, ESG dashboards, and traceability protocols, not as compliance mechanisms, but as enablers of real change.

This concern is mirrored in the broader literature. Hübner et al. (2024) insist on the importance of transparent and cooperative supply chains, underpinned by mutual standards and risk-sharing mechanisms. As geopolitical tensions, pandemics, and resource crises continue to disrupt global sourcing, sustainability becomes not a luxury but a necessity. Supply chains must be reimagined not only for efficiency but for climate resilience, ethical robustness, and social license to operate.

At a theoretical level, this study contributes to the literature on Sustainable Supply Chain Management (SSCM) by reaffirming the relevance of governance-based frameworks and multidimensional performance models. Consistent with Carter & Rogers (2008), our findings suggest that performance should be viewed as a composite of financial, environmental, social, and reputational outcomes. Bourguignon's (1995) insights on the evolving meaning of performance resonate deeply here: what counts as success must reflect shifting societal expectations and stakeholder values.

Despite these contributions, several limitations must be acknowledged. First, the qualitative nature and modest sample size limit the generalizability of findings. While rich in contextual insights, the Moroccan case may not represent the full spectrum of procurement practices in other countries or market formats. Second, the study prioritizes perception over measurement. We explore how sustainability is discussed, framed, and experienced, but do not quantitatively test its relationship to specific KPIs. Third, supplier perspectives are underrepresented. Although some upstream voices were included, the study remains largely buyer-centric. This creates a partial view of the sustainability ecosystem.

Future research should address these limitations by adopting mixed-method or longitudinal designs. Comparative studies across regions and sectors could help isolate contextual variables and surface broader patterns. Moreover, researchers should explore how digital tools, AI, blockchain, and ESG analytics, are being adopted, resisted, or reinterpreted in procurement practice. There is a clear need for empirical models that track how these technologies impact traceability, data governance, and decision-making in sustainable sourcing.

A further avenue for inquiry involves consumer-driven pressures. While this study focused on supply-side dynamics, retail procurement is increasingly shaped by customer preferences for transparency, fairness, and environmental responsibility. Understanding how these demands influence supplier selection, certification, and innovation investment would be valuable. Equally, ethical procurement must move beyond compliance to actively address labor standards, inclusivity, and supplier development in underrepresented regions.

Despite these challenges, the study offers actionable insights for both practitioners and policymakers. For practitioners, it underlines the need for new procurement metrics that go beyond price and time. Indicators like carbon intensity per product unit, supplier compliance rates, and social impact ratings should be integrated into procurement dashboards. It also encourages capacity building: training in sustainable negotiation, supplier engagement, and ESG risk assessment should become part of standard procurement curricula.

For policymakers and retail federations, the findings highlight the need for stronger frameworks and clearer incentives. Sector-specific sustainability standards, aligned with international norms such as ISO 20400 or the GRI indicators, could enhance comparability and accountability. Financial incentives, such as green procurement credits or tax deductions for verified sustainability investments, could accelerate adoption. Above all, public-private partnerships must be strengthened to bridge the gap between regulation and innovation.

For scholars, this chapter offers a base for deeper engagement with the intersections of procurement, governance, digitalization, and sustainability. The conceptualization of procurement as both a functional operation and a moral act opens space for interdisciplinary inquiry, including ethics, data science, and institutional theory. Further research could also explore how procurement teams navigate internal tensions, between marketing and logistics, between sustainability and finance, to operationalize complex goals.

Ultimately, building sustainable procurement in mass retail is not simply about doing the right thing; it is about doing things differently. It is about embedding sustainability into the DNA of sourcing strategies, from contract design to supplier audit, from digital tool adoption to cultural mindset. As this study illustrates, sustainability is not a fixed destination but a moving horizon, one that demands flexibility, learning, and leadership at every level of the supply chain.

Retailers that embrace this complexity with integrity and innovation may not only avoid reputational risks but also unlock new forms of value. They may become industry leaders, attracting responsible consumers, engaged employees, and aligned investors. In short, sustainable procurement is not a constraint, it is a strategic advantage for building future-ready supply networks. In a time of climate urgency, economic uncertainty, and social transformation, the question is no longer whether to adopt sustainable sourcing, but how, and how fast.

REFERENCES

Agrawal, V., & Lee, D. (2019). The effect of sourcing policies on suppliers' sustainable practices. *Production and Operations Management*, *28*(4), 767–787. DOI: 10.1111/poms.12943

Bentalha, B. (2022). Smart Technologies for a Sustainable Service Supply Chain: A Prospective Perspective. *Journal of Environmental Issues and Climate Change*, *1*(1), 77–89. DOI: 10.59110/jeicc.v1i1.81

Bourguignon, A. (1995). *Peut-on definir la performance? Revenue Francaise de Comptabilite*. Revenue Francaise de Comptabilite.

Chkanikova, O. (2016). Sustainable purchasing in food retailing: Interorganizational relationship management to green product supply. *Business Strategy and the Environment*, *25*(7), 478–494. DOI: 10.1002/bse.1877

Global Reporting Initiative. (2016). *GRI 204: Procurement Practices 2016 & GRI 308: Supplier Environmental Assessment 2016*. GRI. https://www.globalreporting.org/standards/

Hmioui, A., Bentalha, B., & Alla, L. (2020, December). Service supply chain: A prospective analysis of sustainable management for global performance. In *2020 IEEE 13th International Colloquium of Logistics and Supply Chain Management (LOGISTIQUA)* (pp. 1-7). IEEE. DOI: 10.1109/LOGISTIQUA49782.2020.9353940

Hübner, A., Kuhn, H., & Weber, G. W. (2024). Editorial Sustainability in Retail Supply Chains: Making a better world in retailing from sourcing to consumption. *Sustainability Analytics and Modeling*, 100032.

Lowson, R. H. (2001). Retail Sourcing Strategies: Are They Cost-effective? International Journal of Logistics Research and Applications: A Leading. *The Journal of Supply Chain Management*, *4*(3), 271–296.

Morcillo-Bellido, J., & Duran-Heras, A. (2020). Sustainability governance mechanisms in supply chains: An application in the retail sector. *Sustainability (Basel)*, *12*(17), 6911. DOI: 10.3390/su12176911

Nair, A. J., & Manohar, S. (2024). Green Service Consumption: Unlocking Customer Expectations on Technological Transformations Enhancing Purchase Experience in Retail Stores. *International Journal of Information Management Data Insights*, *4*(2), 100277. DOI: 10.1016/j.jjimei.2024.100277

Okwu, M. O., & Tartibu, L. K. (2020). Sustainable supplier selection in the retail industry: A TOPSIS-and ANFIS-based evaluating methodology. *International Journal of Engineering Business Management, 12*, 1847979019899542. DOI: 10.1177/1847979019899542

Schneider, L., & Wallenburg, C. M. (2012). Implementing sustainable sourcing, Does purchasing need to change? *Journal of Purchasing and Supply Management, 18*(4), 243–257. DOI: 10.1016/j.pursup.2012.03.002

Seuring, S., & Müller, M. (2008). From a Literature Review to a Con ceptual Framework for Sustainable Supply Chain Management. *Journal of Cleaner Production, 16*(15), 1699–1710. DOI: 10.1016/j.jclepro.2008.04.020

Vadakkepatt, G. G., Winterich, K. P., Mittal, V., Zinn, W., Beitelspacher, L., Aloysius, J., Ginger, J., & Reilman, J. (2021). Sustainable retailing. *Journal of Retailing, 97*(1), 62–80. DOI: 10.1016/j.jretai.2020.10.008

Villena Verónica, H.. (2018). The Missing Link? The Strategic Role of Procurement in Building Sustainable Supply Networks. *Production and Operations Management.*

Walker, H., Miemczyk, J., Johnsen, T., & Spencer, R. Sustainable procurement: Past, present and future, Journal of Purchasing and Supply Management, Volume 18, Issue 4, 2012, Pages 201-206, ISSN 1478-4092.

Chapter 10
Mediating Function of Supply Chain Integration Among Supply Chain Strategy and Logistic Sustainability in the Coffee Industry of Gedeo Zone, Ethiopia

Dawit Jabo
https://orcid.org/0009-0002-1828-1982
Dilla University, Ethiopia

Shashi Kant
https://orcid.org/0000-0003-4722-5736
Lovely Professional University, India

Metasebia Adula
https://orcid.org/0000-0001-5732-2850
Bule Hora University, Ethiopia

Tafese Niguse
https://orcid.org/0009-0008-6536-6392
Bule Hora University, Ethiopia

Nitu Nimbrain
https://orcid.org/0000-0001-5585-3287
IMSAR, Maharshi Dayanand University, India

Bijender Chauhan
https://orcid.org/0009-0001-7967-6323
Shri Vishwakarma Skill University, India

ABSTRACT

The main goal of this investigation is to investigation the association among supply chain strategies, supply chain integration, as well as logistics proxies. It solves the

DOI: 10.4018/979-8-3373-3012-9.ch010

issue of how corporates can optimize supply chain operations to augment efficiency and responsiveness in competitive market places. The investigation uses a quantitative investigations approach that uses surveys to collect data from supply chain specialists in various industries. Statistical analysis, assimilating regression and factor analysis, is carried out to investigation the association among proxies. The key outcomes manifest that clearly defined strategies in the supply chain have a encouraging influence on logistics characteristics and that supply chain integration sub structures as a critical mediator. Corporates investing in integration practices have augmented efficiency in operation and end user satisfaction. The outcomes manifest that adjustments to the chain of supply strategies for integration efforts are imperative to augment logistics outcomes.

INTRODUCTION

The global coffee industry has become one of the furthermost substantial sectors of international trade, and the annual cost exceeded $ 100 billion (Bensfia, 2025; Fole, 2022). In regions such as the Gideo Ethiopia zone, the production of coffee is not only a vital economic activity, but also a cultural cornerstone (Bensfia & Aidi, 2024; Alla et al., 2022). While demand for high quality coffee continues to augment, it is essential to understand the subtleties of supply chain dynamics (Badr, 2022; Ghione, 2021). Supply Chain (SCI) integration has gained importance as a concept of pivots to bring together different interested party and augment efficiency and responsiveness (Mebrate et al, 2022; Bentalha et al., 2019). The origin of this concept can be tracked before the early 1990s, when corporates acknowledged that trouble-free coordination among suppliers, manufacturers, and distributors could lead to augmented sustainability (Acquafredda, 2024). The function of SCI becomes even more imperative in the perspective of the coffee industry where perishable products require timely delivery and quality control (Riggan & Pole, 2024). The empirical literature has extensively studied the association among supply chain strategies as well as logistics proxies in various sectors. Investigations manifests that influence supply chain strategies can substantially augment logistics outcomes, leading to augmented end user satisfaction and competitive advantages (Direia, 2022 and Riggan & Pool, 2024; Bouklata & Bensfia, 2020). As well as, there remains a rarity of investigations that focuses specifically on the mediating function of SCI in the coffee industry, particularly in developing regions such as Gedeo (Ofori et al, 2021). Practical barriers such as infrastructure flaws, lack of technical adoption, and fragmented supply chain structures further complicate the landscape (Ayele, 2021;

Hmioui et al., 2020). These issues are hampered by the ability of coffee makers and interested party to fully utilize their supply chain strategies.

The coffee business in the Gedeo Zone, Ethiopia, contributing considerably to employment and export earnings. In 2021, coffee accounted for nearly 30% of the region's agricultural production. Supply chain integration (SCI) is critical for boosting logistical sustainability, which is necessary for sustaining the quality and competitiveness of Gedeo coffee (Mebrate et al, 2022). SCI allows greater coordination across stakeholders, including farmers, processors, along with exporters, resulting to enhanced efficiency and decreased costs. Studies suggest that locations with higher SCI levels have a 20-30% boost in supply chain efficiency. Additionally, proper integration may cut waste by up to 31%, guaranteeing that resources are utilized optimally (Direia, 2022 and Riggan & Pool, 2024). The Ethiopian government promotes this integration through initiatives targeted at increasing infrastructure as well as training for local manufacturers. In 2022, investments in countryside roads and processing infrastructure grew by 14%, directly influencing logistical operations. As a consequence, Gedeo's coffee growers can better fulfill the expanding worldwide demand, generating a 10% gain in exports over the previous year. This strategic connection between supply chain strategy as well as logistical sustainability is vital for the long-term future of the coffee business in the region (Acquafredda, 2024).

The purpose of this chapter is to consider existing gaps in the investigations by investigating how supply chain integration intervenes the association among supply chain strategies as well as logistics proxies in the Coffee industry (Bensfia & Bouklata, 2022). By providing empirical data and manifesting, this investigation provides future investigators and politicians with the basis for developing strategies to augment supply chain efficiency and efficiency. In particular, he investigations the subtle interactions among various elements of the supply chain, manifesting advanced practices and areas of augment. The investigations question that leads this investigation is, "How does supply chain integration influence the association among supply chain strategies as well as logistics proxies in the Cafe de Lagudeo Regional Industry and what is imperative to interested party? This investigation seeks to propose practical solutions that can not only augment the theoretical manifesting but also stimulate the regular growth and synthesis of the coffee sector.

BACKGROUND OF INVESTIGATION

The coffee industry, particularly in regions such as the Ethiopia region, is characterized by complexity and complex associations among various interested party, assimilating farmers, exporters, cooperatives, and retailers (Pedercini et al, 2021). In recent years, the importance of supply chain integration has gained recognition as an imperative strategy for improving logistics sustainability and gaining a competitive advantage (Ayele, 2021). As well as, implementation of influence supply chain integration in the coffee sector incubates imperative issues that prevent optimal sustainability and stability (Acquafredda, 2024). One notable thematic investigation that demonstrates these issues is the Gideo Zone Cooperatives.

There are both good and bad effects of bringing together supply chains in the Gedeo Zone coffee sector. On the plus side, it makes things go more smoothly. Studies have shown that supply chain performance may go up by 20–30%, which saves producers money optimally (Direia, 2022 and Riggan & Pool, 2024). Also, good integration may cut down on waste by as much as 31%, which makes better use of resources and raises the overall quality. This has led to a significant rise in exports, with a recent 10% increase that has helped local farmers financially (Riggan and Poole, 2024). But there are also bad things that might happen. Farmers' income stability might be at risk if they rely more on integrated supply chains, which can make them more vulnerable to changes in the market. Also, rapid expansion might put a pressure on local resources, which could hurt the environment. Lastly, the benefits of integration could not be spread out equitably, which might make the gap between larger producers along with smallholder farmers even worse. To make sure that the Gedeo coffee industry can grow in a way that lasts, it is important to find a balance between these good and bad effects (Acquafredda, 2024).

Despite a robust network of local coffee producers, cooperatives fought against influence communication and coordination among their members (Fole, 2022). The lack of a centralized information system meant that farmers were often unaware of market place requirements, which led to a contradiction among demand and proposals (Mebrate et al, 2022). These outcomes in inadequate inventory during seasonal harvests and in other times inventories, ultimately influencing the cooperative's ability to influence meet the order. These adjustment issues manifest the critical requirements of integrating supply chain processes to augment responsiveness and agility (Bensfia & Bouklata, 2023). Another practical example can be found in the experience of coffee manufacturers who are trying to access international market places (Ghione, 2021). Many of these manufacturers face infrastructure problems, such as poor road conditions and inadequate vehicles (Ofori et al, 2021). Even when they establish associations with exporters, the influence of logistics, outcompeting from delays in inadequate transport and storage, causes damage and loss of quality (Pedercini et al, 2021). This influences not only the profitability of manufacturers, but also reduces the overall quality of Gedeo coffee on the world market place. Therefore, inability to influence integrates supply chain logistics can have a substantial influence on local self-sufficiency and international competitiveness (Riggan and Poole, 2024).

As well as, cultural factors play a key function in implementing supply chain integration in Gedeo zone in Ethiopia. The lack of traditional practices and technical adoption incubates obstacles to modernization of supply chain operations (Riggan and Poole, 2024). For example, many farmers resort to secular ways of converting and transporting coffee. This is not only for a high-strength workforce, but for a long time. The lack of accepting new technologies, such as mobile applications for market place information, in real time, exacerbates the gap among supply chain strategies and operational metrics. As well as, a regulatory situation can pose problems for influence integration of the supply chain. Politicians who regulate negotiation and export procedures may be bulky and less coordinated, which leads to delays and augmented costs (Direiba, 2022). Examples can be seen in extensive documentation and export regulations requiring compliance verification. These normative obstacles often dissuade small farmers from participating in broader supply chains, which limits their ability to benefit from integrated logistics and strategic alliances (Riggan and Pool, 2024).

STATEMENT OF PROBLEM

The coffee industry in the Ethiopian region is at a critical stage, with little use of the possibility of improving logistics sustainability thanks to influence integration of the supply chain (Fole, 2022). The ideal state is an integrated supply chain, working together with small farmers with international buyers to augment efficiency, reduce waste and better products (Acquafredda, 2024). Nevertheless, the current reality incubates a clear contrast to this vision. Many investigations have underlined the importance of integrating the supply chain (Badr, 2022); but there is contradictory evidence concerning its real implementation and efficiency in the coffee sector, in particular in Gedeo (Ayele, 2021). The empirical literature manifests substantial gaps and unconvincing outcomes regarding the specific mechanisms in which supply chain integration influences logistics characteristics (Ghione, 2021). For example, certain investigations have manifests encouraging outcomes associated to integration, while others have depicted minimal or underexposure. This difference raises questions about the applicability of existing theoretical frameworks in the capture of unique problems with which coffee supply chains are faced with Gedeo. In addition, the absence of a determined investigation of the intermediate function of the integration of the supply chain in this particular perspective has led to theoretical gaps which limit manifesting of the way in which strategic alignment can augment logistics productivity (Ofori et al, 2021; Hmioui et al., 2020). In addition to theoretical gaps, there are notable knowledge shortcomings concerning specific problems that hinder the influence integration of the Gedeo supply chain. Factors such as infrastructure deficiency, cultural resistance to technological adoption and regulatory difficulties have been documented, but full investigations on how these barriers interact with integration strategies remain rare (DIREIBA, 2022). This surveillance leaves a large gap in the literature and prevents policymakers and practitioners from developing customized solutions that discourse the unique requirements of the Gedeo Coffee industry (Pedercini et al, 2021).

The perspective gap further complicates the situation as much of the existing investigations on supply chain integration is derived from investigations conducted in a variety of geographical and cultural settings. Gedeo, a unique socioeconomic landscape characterized in its own way for coffee cultivation and local market place dynamics, requires in-depth investigations examining these specific conditions (Riggan and Poole, 2024). Without such perspective investigations, by means of general supply chain theory could lead to influence or false strategies. As well as, practical issues are maintained at the operational level, and the integration of supply chain processes prevents inadequate infrastructure, influence communication channels, and lack of stakeholder cooperation (Pedercini et al, 2021). These issues incubate cycles of inefficiency, which compromise logistical characteristics and strangle the potential for growth of the coffee sector. This investigation aims to bridge these gaps, and to mediate how supply chain integration mediates the association among supply chain strategies and logistical characteristics, particularly the associations of the Gedeo Coffee industry. Resolving conflicts in existing literature, investigating practical problems faced by interested party, and giving away unique perspectival factors, this investigation provides valuable information to future scientists and politicians. The outcomes not only augment the theoretical manifesting of supply chain dynamics, but also provide influence recommendations for the synthesis of influence integration strategies. Ultimately, the aim of this investigation is to augment logistics proxies in the Gedeo Coffee Supply chain and open ways to recognize to together local economic synthesis and the global coffee market place.

THEORETICAL FOUNDATION

Supply Chain Management Theory: This theory posits that influence supply chain management leads to augmented efficiency, reduced costs, and augmented end user satisfaction. In the coffee industry perspective, SCM theory manifests the importance of cooperation among farmers, processors, exporters and retailers, optimizing operations (Ghione, 2021). It provides the basis for manifesting how each link within the supply chain influences overall sustainability (Ayele, 2021). The SCM principles manifest the requirement for integrated strategies that resolve together logistics and operational problems, making it the basis of analysis of the dynamics of the Gedeo chain.

Resource Based view (RBV): Resource theory (RBV) focuses on internal possessions and opportunities for corporate as the main source of competitive advantage (Fole, 2022). In the coffee sector, this theory can be applied to assess how unique possessions such as local knowledge, coffee grain quality, and associations with buyers can be utilized through influence integration of the supply chain (Ofori et al, 2021). RBV manifests that high-resource and competent corporates can achieve better sustainability when they align their supply chain strategies with essential forces (Acquafredda, 2024). This perspective manifests the consequence of manifesting local perspectives and possessions, which is particularly relevant for smallholder farmers in Gedeo zone seeking to augment their market place position.

Transaction Cost Economics (TCE): It explores the costs associated with economic exchanges and the governance structures that can minimize these costs. In the supply chain perspective, TCEs can help depicted why certain corporate arrangements, such as strategic alliances, contracts, or horizontal integration, are preferred over others (Diriba, 2022). For Gedeo coffee producers, TCE can unravel the barriers faced in establishing reliable contracts with buyers and the costs associated with adjusting logistics (Mebrate et al, 2022). This theory provides an idea about the importance of trust and cooperation among interested party to reduce transaction costs. This recognizes to supply chain integration and augmented sustainability.

Network Theory: Network theory manifests the associations among different subjects in the supply chain and how these associations influence sustainability outcomes (Ghione, 2021). This theory is particularly pertinent in the coffee industry, where associations among farmers, cooperatives, exporters, and retailers are critical for success. Network Theory posits that the strength and quality of these nexus can augment information flow, resource sharing, and collaborative problem-solving (Ofori et al, 2021). By analyzing the network of associations in the Gedeo coffee supply chain, this theory provides a lens through which to understand how social capital and associational dynamics recognize to overall logistic sustainability (Pedercini et al, 2021).

For this investigation, Supply Chain Management Theory sub structures as the main theoretical lens (Ghione, 2021). This choice is founded on a complete accent on supply chain integration and coordination. This is central to manifesting the association among supply chain strategies as well as logistics proxies in the Gedeo Coffee industry. SCM theory provides a reliable foundation for analyzing the influenceness of integration among various interested party. This can lead to augment operational efficiency and augmented sustainability outcomes (Ofori et al, 2021). By utilizing this theory, this investigation aims to explore the mechanisms in which supply chain integration mediates the association among strategic initiatives and logistical success, thereby discoursing the unique barriers and opportunities that exist in the Gedeo Coffee sector.

DEFINITION AND ORIGIN OF CONCEPTS

Supply Chain Integration: It manifests to the leveling and inter nexus of commercial processes, systems and associations among various parties concerned in the supply chain in order to augment overall efficiency and responsiveness (Ayele, 2021). It assimilates the exchange of information, possessions and objectives among suppliers, manufacturers, distributors and retail sellers to incubate a smooth flow of consumption production goods and services (Fole, 2022). Although the concept of supply chain integration manifested at the end of the 20th century, corporates began to recognize the limitations of traditional operational heels (Direiba, 2022). Early management chain management practices employed primarily on logistics and inventory management (Pedercini et al, 2021). As well as, as global competitors strengthened and founded end user expectations, corporates began to embrace a more holistic approach that emphasized cooperation and cooperation across Supply et al 2022. The creative work of scientists such as Oliver and Weber in the early 1990s formalized manifesting of supply chain dynamics and advocated for integrated strategies that discourse the diverse interests of interested party (Acquafredda, 2024).

Logistics Sustainability: Logistics sustainability manifests to the efficiency and efficiency of a company managing logistics operations, covering transportation, storage, inventory management, and order completion (Ghione, 2021). Key measurements of logistic sustainability assimilate delivery speed, accuracy, profitability, and end user satisfaction (Mebrate et al, 2022). The term \ "logistics \" has its roots in military operations, where it referred to the planning and execution of the movement and the support of the troops (Fole, 2022). Over time, the concept has evolved, in particular during the Second World War, when corporates began to apply logistical principles to the management of the supply chain. In the 1980s and 1990s, there was formalization of logistics sustainability metrics due to augmented outcomes in the field of IT and augmented supply chain complexity (Ofori et al, 2021). Scientists such as Bowersox and Closs have made substantial contributions to the synthesis of logistics sustainability frameworks, manifesting the requirement for measurable outcomes in logistics operations (Riggan and Poole, 2024).

Supply chain strategy: The supply chain strategy is a long -term plan that manifests how the company will manage its supply chain to obtain a competitive advantage and meet end user requirements (Fole, 2022). It covers decisions associated to the source, production; distribution and management of sub structures, aligning these elements with common shopping centers (Pedercini et al, 2021). The concept of supply chain strategy began to be trained in the 1990s. This is because corporates recognized the strategic importance of supply chains and realized that they achieved operational advantages and differentiation of the market place (Ofori et al, 2021). Influential investigations from investigators such as Christopher and Menzer manifested the requirement for strategic coordination among the supply chain and the activities of the broader enterprise. The rise of globalization and advancements in technology further propelled the synthesis of sophisticated supply chain strategies, enabling corporates to leverage their supply chains as a source of competitive advantage (Ayele, 2021).

Stakeholder Association: This cooperation can take the form of an exchange of information, joint decision and synthesis of the strategic alliance. The idea of cooperation among the parties has its roots in the theory of concerned parties, which appeared in the 1980s as a basis for manifesting the associations among corporates and their various concerned parties (Ghione, 2021). Scientists like Freeman depict that corporates should consider the interests of all interested party, and not just investors in the decision making process, but corporations should consider (Mebrate et al, 2022). In the supply chain perspective, association is increasingly recognized as requirement to augment sustainability and responsiveness, especially when corporates face global market places and diverse end user requirements.

EMPIRICAL LITERATURE REVIEW

Supply Chain Strategy and Logistic Sustainability

The link among supply chain strategies as well as logistics proxies has attracted considerable attention to the empirical literature, particularly when corporates strive to augment competitiveness in increasingly complex market places (Ghione, 2021). Many investigations have manifests that clearly defined strategies in the supply chain directly influence logistics efficiency and efficiency and sub structure as a key factor in general corporate efficiency (Ofori et al, 2021). For instance, an investigation by Li et al. (2006) established that corporates with proactive supply chain strategies are better positioned to manifest to market place fluctuations and end user demands, outcompeting in augmented delivery sustainability and reduced operational costs (Ayele, 2021). In addition, the empirical data manifests that the coordination among the strategy of the supply chain and the logistical methods augments reactivity and flexibility (Fole, 2022). For example, a investigation by Zhang et al. (2016) found that corporates adopting integrated supply chain strategies characterized by stakeholder association and information sharing experience with superior logistic sustainability metrics, assimilating faster order fulfillment and augmented end user satisfaction (Acquafredda, 2024). This alignment not only streamlines operations, but also promotes a culture of association.

As well, the function of technology in bridging the gap among supply chain strategy and logistic sustainability has been a focal point in recent literature (Mebrate et al, 2022). Investigations, such as those by Gunasekaran et al. (2001), emphasize that the integration of advanced technologies, such as the software management software and real -time data analytics allows corporates to optimize their logistics operations, providing greater visibility and control over supply chains (Diriba, 2022). This technological integration facilitates informed decision -making and augments the overall flexibility of logistics processes. Despite the wealth of investigations incorporating the encouraging nexus among the strategy of the supply chain as well as logistics proxies, gaps remain in the literature, especially in association to the perspective of applications (Pedercini et al, 2021). For example, although furthermost of the investigations were conducted in founded countries, there are not enough empirical data regarding how this dynamics unfolds in developing market places, such as the coffee industry in the Gedeo Ethiopia zone (OFORI et al, 2021 and Solomon, 2021). This manifests the requirement for additional investigations to consider local conditions, cultural factors, and specific issues regarding industries that may influence supply chain efficiency for logistics proxies in such perspectives. In general, the empirical literature manifests a strong association among supply chain management and strategic approaches to improving logistic proxies, but also manifests the requirement for discrete frames that can solve unique problems of various industries and regions (Riggan and Pool, 2024).

Supply Chain Strategy and Supply Chain Integration

Supply chain strategy and supply chain integration are actually associated concepts that play a critical function in improving corporate efficiency and competitiveness (DiReia, 2022). The clearly defined strategy of the supply chain describes our long-term goals and goals and focuses on how we can optimize the flow of products, information and possessions through Offers and AL, 2022. It encompasses decisions associated to supply, production, distribution and inventory management, ensuring that each element is aligned with the global objectives of the company. Conversely, the integration of the supply chain manifests to transparent coordination and association among various interested party such as suppliers, manufacturers, distributors and retailers within the supply chain (Acquafredda, 2024). This integration is essential to influence execute the strategy of the supply chain, as it promotes the sharing of information, augments communication and promotes the joint resolution of problems.

Empirical evidence manifests that an influence integration of the supply chain considerably augments the implementation of supply chain strategies, which has led to an augment in operational efficiency and better reactivity to market place demands (Fole, 2022). For example, corporates that prioritize integration often achieve higher levels of logistic sustainability, characterized by reduced lead times, reduced inventory costs, and augmented end user satisfaction (Ofori et al, 2021). Investigations manifests that integrated supply chain strategies allow corporates to manifest more attractively to end user preferences and changing market place conditions, facilitating a proactive approach rather than a reactive approach to supply chain management (Ayele, 2021). In addition, the association among the strategy of the supply chain and integration manifests the importance of confidence and cooperation among concerned parties (Ghione, 2021). Successful integration requires a transition from traditional competitive associations to cooperative strategic alliances, where all parties work to achieve common objectives (Pedercini et al, 2021). This joint situation not only strengthens the supply chain, but also recognizes to the initiatives on innovation and continuous augments. Ultimately, the interaction among the strategy of the supply chain and the integration of the supply chain is vital for corporates that seek to augment their competitive advantage and achieve sustainable growth in an increasingly dynamic corporate -landscape.

Supply Chain Integration and Logistic Sustainability

The integration of the supply chain is a critical factor influencing logistical sustainability because it encompasses the alignment and coordination of processes, information and associations among the various interested party in the supply chain (Fole, 2022). Influence integration facilitates transparent communication and association, allowing corporates to manifest more quickly to changes in demand and supply dynamics (Ofori et al, 2021). Empirical investigations have manifests that corporates with higher supply levels in the supply chain often have substantial augments in logistics sustainability, assimilating control speed, inventory, and overall operational efficiency (Acquafredda, 2024). The mechanisms in which supply chain integration influence logistics characteristics are multifaceted are related to SCM. For example, the integrated supply chains allow the exchange and visibility of data in real time throughout the network, which allows corporates to make reasonable decisions and optimize their logistics processes (DIREIBA, 2022). This visibility makes it possible to reduce the order period and minimize the costs of holding actions, as the concerned parties can better expect fluctuations in the request and correctly adjust their operations. Investigations has demonstrated that corporates that adopt integrated approaches characterized by strong associations with suppliers and end users are more capable of achieving higher service levels and meeting end user expectations (Wang'ombe, 2021). As well as, the collaborative nature of supply chain integration fosters a culture of continuous augment and innovation (Pedercini et al, 2021).

When interested party work together, they are more likely to reveal inefficiencies and implement solutions that augment logistics operations. For example, joint planning and forecasting efforts can lead to better adjustments to production and distribution calendars, reducing nearby locations and increasing overall flexibility in the supply chain. As well as, implementing influence supply chain integration is not without problems. Factors such as cultural differences, technology barriers, and different levels of engagement among interested party can hinder integration efforts (Ayele, 2021). Nonetheless, the encouraging association among supply chain integration and logistic sustainability underscores the necessity for corporates to prioritize integration initiatives as a strategic imperative (Ghione, 2021). By doing so, corporates can augment their operational capabilities, augment end user satisfaction, and cultivate a competitive edge in the market place (Mebrate et al, 2022). Ultimately, the synergistic influences of supply chain integration as well as logistics proxies are essential for achieving lasting success in an increasingly inter associated global market place.

Mediating Function of Supply Chain Integration

Supply chain integration plays a pivotal mediating function in the association among supply chain strategy and logistic sustainability, serving as the mechanism through which strategic initiatives translate into operational influenceness (Ayele, 2021). A well-defined supply chain strategy outlines the goals and frameworks necessary for optimizing logistics operations, but the realization of these strategies depends heavily on the degree of integration achieved among various interested party (Ghione, 2021). When supply chain integration is strong, it promotes cooperation, augments communication, promotes information flow over the supply chain, and enables corporates to implement their strategies more influential (Ofori et al, 2021).

Empirical investigations manifest that corporates with a high level of integration in the supply chain are more employed to use strategic benefits, which lead to augmented logistics sustainability outcomes (Pedercini et al, 2021). For example, an integrated supply chain allows for the real-time and coordinated decision-making-coordinated decisions requirement to manifest quickly to market place changes and end user requirements (Acquafredda, 2024). This optimistic integration achieves the objectives set by supply chain strategies to reduce processes as well as order time and augment the level of service. These joint efforts guarantee that strategic alignment leads to tangible sustainability augments, such as augmented operational efficiency and greater end user satisfaction (Mebrate et al, 2022).

CONCEPTUAL FRAMEWORK

Figure 1. Conceptual Framework

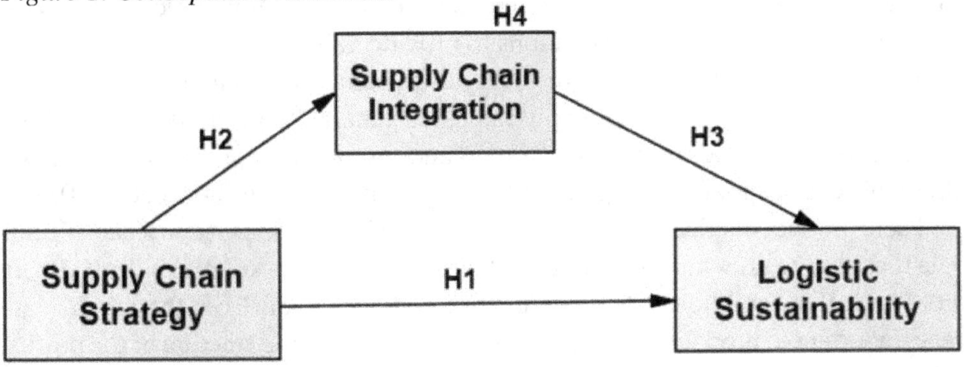

METHODOLOGY

This investigation employed the approach of a quantitative investigation to investigation the intermediate function of the integration of the supply chain in the association among the strategy of the supply chain and logistical proxies. The selection of 390 respondents was chosen by means of the structured sample method, providing a diversified manifestation of the parties concerned in the Coffee supply chain in Gedeo. The data collection process involved managing consolidated surveys carefully designed to capture appropriate variables, such as supply chain strategies, supply chain integration, and logistical characteristics. To incorporate the measurement scale, this investigation first conducted the KMO assessment. This assessed the validity of the sample for factor analysis. KMO values above 0.6 manifests sufficient validity of the sample and justifies the use of factory investigations analysis (EFA) (Ayele, 2021 and Woldekiros, 2023). EFA was then performed to determine the main structure of the structural coefficients and make sure that the elements were correctly responsible for the manifesting factors (Acquafredda, 2024). This process helped to clarify the measurement scales and make sure that they accurately manifest theoretical structure. After extensive synthesis and verification by means of EFA, modeling of structural equations (SEM) was employed to verify the proposed associations among variables. The analysis was performed by means of AMOS software. This recognized to the investigation of the direct and indirect influences of supply chain strategies on logistical characteristics through the intermediate function of supply chain integration (Diriba, 2022). The SEM approach has provided reliable evidence to assess model dignity, allowing a comprehensive manifesting of associations in the investigations perspective. The outcomes derived from the SEM analysis not only incorporated hypothetical associations, but also provided insight into the importance of supply chain integration as a mediator (Fole, 2022). This methodology allowed the investigations to provide valuable empirical evidence to existing literature and to manifest the complex dynamics of strategic initiatives and logistic sustainability of coffee supply chains. In general, the assimilation of KMO, EFA, and SEM methods provided a rigorous and systematic approach to the analysis of complex associations among variables in this investigation (Ofori et al, 2021).

There are a lot of ethical issues in the coffee business in the Gedeo Zone of Ethiopia that have to do with supply chain integration as well as sustainability. One big problem is that smallholder farmers are taken advantage of. Even though they are very important to production, they frequently only get a little fraction of the profits. Fair trade and other ethical sourcing techniques try to make sure that farmers are paid properly, which helps promote social fairness. Intensive agricultural methods can also cause environmental problems, such as deforestation and soil deterioration, which make it harder for coffee production to be sustainable. Case studies show how important it is for communities to start projects that are good for the environment and the economy at the same time. For instance, cooperatives that grow organic coffee not only make the coffee taste better, but they also help the environment. Also, being open about the supply chain is important for gaining customers' confidence and encouraging ethical purchase. Getting local communities involved in decision-making helps solve these moral problems by making sure that everyone gets a fair share of the advantages of integration while protecting the area's cultural and environmental legacy.

Scale Synthesis

The initial design, supply chain strategy, was enabled by means of factors derived from previous investigations. This manifests strategic alignment, resource distribution and market place response (Ayele, 2021). For example, Li et al. (2006) provided the basis for factors assessing the clarity of strategic objectives to optimize supply chain efficiency and influenceness of strategic initiatives. Next, the proxies of supply chain integration was founded on key dimensions identified in the literature, assimilating internal integration, external integration with suppliers, and end user integration. Investigations by Frohlich and Westbrook (2001) manifested the importance of collaborative associations and information sharing among interested party, which informed the formulation of specific items measuring the degree of communication and cooperation within the supply chain. The objects are founded to capture formal and informal aspects of integration, reflecting the multifaceted nature of this design. The final design, logistics sustainability, was founded by means of measures commonly mentioned in the supply chain logistics and management literature. Founded on investigations such as Bowersox and Closs (2001), scales assimilated factors that assessed the keys of sustainability proxies, such as delivery accuracy, order speed, and inventory management. These factors are carefully formulated to reflect the operational reality facing an corporate, which allows for a complete assessment of logistics efficiency. Each scale underwent a rigorous verification process, assimilating assessments of content reliability by field experts and pilot testing by means of low samples to make sure clarity and relevance (Fole, 2022). The iterative process of scale synthesis informed by empirical data and theoretical manifesting of literature led to reliable and reliable measures that influential seized the design of interest. This methodological severity makes sure a strong basis for subsequent analysis of data and interpretation of the outcomes of investigations.

DATA ANALYSIS

Table 1. KMO and Bartlett's Test

Kaiser-Meyer-Olkin Measure of Sampling Adequacy.		**.840**
Bartlett's Test of Sphericity	Approx. Chi-Square	1619.738
	Df	91
	Sig.	.000

Source: Authors

The outcomes depicted in Table 1 manifests the outcomes of the Bartlett spherical test adequacy, which are necessary to assess the ability of data for factorial analysis. The value of KMO 0.840 assumes that the adequacy of the sample is good, because the values greater than 0.70 are generally considered to be acceptable to carry out a factorial analysis. This high value of KMO manifests that a substantial proportion of variables within the variable can be depicted by the main factors, which incorporates the feasibility of the data for a detailed analysis. As well, Bartlett's spherical inspection provided an approximate value of Chi-Quadrate 1619,738 with 91 degrees of freedom (DF) and level of meaning (SIG.). This outcome manifests that the association matrix differs substantially from the identity matrix, which manifests that there are substantial associations among the variables. Level of meaning 0 000 ($p < .001$)

incorporates that the data is suitable for factor analysis, as it demonstrates that the variables are inter associated and that factor analysis can be influential applied to uncover the underlying structure of the data. Overall, these outcomes provide strong evidence that the data is appropriate for factor analysis, allowing for the exploration of the associations among the variables in the investigation.

Table 2. Total Variance Depicted

Component	Initial Eigenvalues			Extraction Sums of Squared Loadings			Rotation Sums of Squared Loadings		
	Total	% of Variance	Cumulative %	Total	% of Variance	Cumulative %	Total	% of Variance	Cumulative %
1	6.62	67.234	67.234	6.62	67.234	67.234	3.00	21.662	21.662
2	1.37	9.787	62.022	1.37	9.787	62.022	2.16	14.691	36.962
3	1.27	9.087	62.630	1.27	9.087	62.630	1.98	16.168	62.630
6	.969	6.782	47.892						

Source: Authors

Table 2 depicted the outcomes of the general dispersion. The analysis revealed a total of six components, and the first component demonstrates its own original value 6.62, which manifests 67.234% of the total dispersion. This imperative variance manifests that the first component reflects an imperative part of the underlying data structure. The second component has a value of 1.37. This further depicted 9.787% of the total variance to 9.787%, followed by a total variance of 62.022%. These three components depicted more than half of the dispersion together, emphasizing their relevance in the perspective of the investigation. A detailed investigation of the original first value manifests that the fourth component depicted 6.782% of the variance, while the fifth and sixth components depicted their own values of 0.871 and 0.773, which are 6.223% and 4.629%, respectively. As well as, these final components are not retained in the structure of the final factors of the low contribution to the general variance described. The extraction method employed for this analysis is the analysis of the main components and the standard for determining the key factors underlying the structure studied. In general, the outcomes manifest that the selected components provide a complete manifestation of the data. This will give you a better manifesting of the associations among variables within the investigations framework.

Table 3. Covariances

Covariance			Estimate	S.E.	C.R.	P	Hy.
Supply chain integration	**<-->**	**Supply chain strategy**	**.262**	**.021**	**8.231**	**.00**	**H2**
Supply chain integration	<-->	Logistic sustainability	.303	.067	7.692	.00	**H3**
Supply chain strategy	<-->	Logistic sustainability	.316	.26	7.831	.00	**H1**

Source: Authors

Table 3 manifests the covariance among the main proxies. A co-decomposition among supply chain integration and supply chain strategies has been depicted. This is a high value C.R. in assimilation with the level of meaning manifests by ".00", incorporates a strong and statistically substantial association, maintaining the hypothesis 2 (H2). In the same way, the incorporation among the integration of the supply chain and the logistical characteristics is noted at 0.303, with S.E. 0.067 and C.R. 7.692. This outcome also demonstrates a substantial encouraging association, validating Hypothesis 3 (H3). The strong association manifests that higher levels of supply chain integration are associated with augmented logistic sustainability outcomes. This imperative nexus incorporates Hypothesis 1 (H1). This manifests that influence supply chain strategies have a encouraging influence on logistics metrics. In general, the it depicted in this table focus on structural associations, manifesting the importance of integrating supply chains as mediators in the association among supply chain strategies as well as logistics proxies. Reliable statistics incorporate the rationale of the investigations and recognize to a detailed manifesting of dynamics in supply chain management.

Table 4. Validity Concern

	CR	AVE	MSV	MaxR(H)	SCS	LP	SCI
SCS	0.767	0.663	0.219	0.842	**0.646**		
LP	0.748	0.647	0.167	0.763	0.171	**0.664**	
SCI	0.793	0.603	0.276	0.797	0.383	0.193	**0.686**

Note: SCS= Supply chain strategy; LP= Logistic Sustainability; SCI= Supply chain integration
Source: Authors

Table 4 manifests to the validity. 0.767 CR, AVE for SCS is 0.663, exceeding the recommended threshold of 0.4, which manifests that the provision depicted a substantial part of the variance of its elements. The MSV for SCS is 0.219, reflecting the maximum association with other structures, but the value of MAXR(H) 0.842 also supports structure reliability. For logistics sustainability (LP), the CR is 0.748, which also manifests good internal consistency. Ave 0.647 exceeds the 0.4 threshold, ensuring that the design properly reflects the variance of those proxies. MSV 0.167 has a lower general degree of dispersion with other structures, providing better discriminant reliability. The MAXR(H) value is 0.763, enhancing the reliability of this structure. Supply Chain (SCI) integration depicted 0.793 Cr, indicating high internal coherence. AVE is 0.603, again higher than the acceptable threshold, which manifests an influence sustainability of the structure. With MSV 0.276 SCI manifests a moderate association with other variables, and the value of Max-r (h) 0.797 As well incorporates its reliability. Overall, the metrics in Table 6 corporate the reliability and validity of the proxies utilized in the investigation, ensuring that supply chain strategy, logistic sustainability, and supply chain integration are measured influential and distinctly, thereby strengthening the investigation\'s empirical foundation.

Figure 2. Structural Equational Model

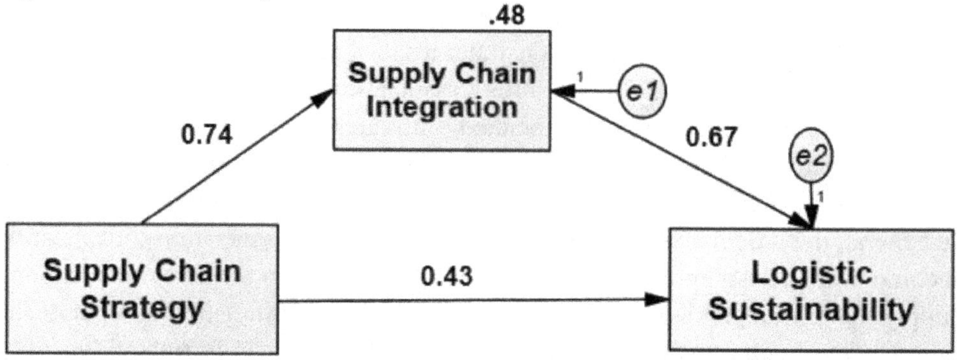

Table 5. Model Parameters

Sig.	Chi-Sq	RMR	Goodness	Level	TLI	RMSEA
0.006	1.946	.036	0.930	0.912	0.963	.028

Source: Authors

Table 5 contains the indexes of model statistics. The level of semantics was depicted at 0.006, indicating that the adjusted model is statistically substantial. A value of Chi-quadrate 1.946 means a good response, as a lower value manifests the best productivity of the model. RMR is recorded at 0.036, which is acceptable and further supports the validity of the model. As well, the Adjustment Quality Index reveals a physical shape quality value of 0.930 and a fitness check value of 0.912. This manifests a strong adjustment, exceeding the commonly accepted reference index of 0.90. The TLI was depicted at 0.963, reflecting excellent model adjustments as well. Consequently, the RMSEA of 0.028 is substantially below the 0.04 threshold, suggesting that the model is suitable for the data. In general, these proxies together incorporate that the model manifests a satisfactory response and incorporate the associations among investigations concepts.

Table 6. Regression

			Approx.	S.E.	C.R.	P	Ass.
Supply chain integration	**<---**	**Supply chain strategy**	**.866**	**.163**	**7.483**	**.00**	**H2**
Logistic sustainability	<---	Supply chain integration	.678	.131	2.843	.00	H3
Logistic sustainability	<---	Supply chain strategy	.947	.192	7.948	.00	H1

Source: Authors

Table 6 depicted the outcomes of the regression examination. With a standardized regression coefficient of 0.866 and a C.R. of 7.483, indicating robust support for H2. As well as, supply chain integration substantially influences logistic sustainability as evidenced by a coefficient of 0.678 and a C.R. of 2.843, supporting H3. As well as, this investigation found that supply chain strategies directly influence logistic sustainability with a coefficient of 0.947 and a C.R. of 7.948, providing strong support for H1. These outcomes manifest the essential function of supply chain strategies and integration into augmented logistics sustainability, suggesting that corporates requirement to focus on alignment of integrated practices and strategic initiatives for optimal outcomes.

Table 7. Mediation

	worth
Supply chain strategy → Logistic sustainability	.63
Supply chain strategy → Supply chain integration → Logistic sustainability	.76*.67=.68
	.91

Source: Authors

Table 7 depicted the mediating function. The direct influence of supply chain strategies on logistic sustainability was quantified at 0.63, indicating a major direct influence on improving logistic sustainability. This value incorporates the hypothesis that the influence strategies of the supply chain recognize directly to the augment of the outcomes of the logistics. In addition, the table emphasizes indirect influence when the supply chain strategy influences logistical characteristics by the integration of the supply chain. The influence of this indirect association path is depicted as 0.76 as it multiply by 0.67 to give an indirect influence of 0.68. This calculation manifests that supply chain integration plays a critical function in mediating the influence of supply chain strategies on logistics proxies. Taking into account direct and indirect routes, the total influence is depicted at 0.91. This global figure manifests that supply chain strategies have a direct influence on logistics sustainability, but that a substantial portion of this influence is mediated by supply chain integration. The outcomes manifest a partial influence of mediation, stressing the importance of integrating the supply chain into the augment in logistical sustainability outside the direct influence of the strategy of the supply chain. Consequently, the outcomes incorporate the meaning of the two structures in the perspective of the management of supply chains, emphasizing the interaction among strategy and integration into the achievement of influence logistical outcomes.

DISCUSSION

The outcomes of this investigation provide imperative insights into the association among supply chain strategies, supply chain integration, and logistic sustainability. The strong KMO value of 0.840 and the outcomes of the Bartlett spherical test incorporate the appropriateness of the data for factor analysis and establish a solid foundation for further investigation of these structures. Investigations factor analysis manifested that the three main components were over 62% of variance, manifesting the relevance of these structures in the perspective of supply chain management. Analysis demonstrates reliable and statistically substantial associations among designs, and supply chain integration manifests a strong association among supply chain strategies as well as logistics proxies. More specifically, a high coefficient manifests that influence integration of the supply chain is imperative for improving logistics sustainability. This brings together existing literature that emphasizes the importance of integration into achieving operational efficiency. The model adjustment proxies still support the validity of the proposed association, with the square Chi values indicating the appropriate adjustment, and the RMSEA indicating the tight adjustment to the data. This solids model supports theoretical structures that directly link supply chain strategies to logistical characteristics, and also manifests the intermediate function of supply chain integration. Regression analysis manifested that the supply chain strategy had a direct influence of 0.63 on logistical characteristics, and the indirect influence of supply chain integration was depicted at 0.68. The total influence of 0.91 manifested that a substantial portion of the influence of supply chain strategies on logistic sustainability is mediated by supply chain integration and has a partial mediation influence. This outcome manifests that corporates requirement to focus not only on developing influence supply chain strategies, but also on enhancing integration across the supply chain to maximize logistic sustainability. Overall, these outcomes recognize to a detailed manifesting of dynamics in supply chain management and manifest the inter nexus of strategy and integration into logistics sustainability implementation. This investigation recommends a holistic approach that examines together direct and indirect influences to optimize supply chain efficiency and operational outcomes.

A diverse strategy is needed to deal with the problems that come up while integrating the supply chain in the Gedeo Zone coffee sector. First, getting farmers, cooperatives, processors, and exporters to work together may make the supply chain stronger. Fair trade practices and training and resources for smallholder farmers can help make sure that everyone gets a fair share of the advantages of integration. Also, farmers may become less dependent on changing global markets by diversifying their markets and marketing items that bring value. Putting money into environmentally friendly farming and protecting the environment can help ease the pressure on local resources and help keep the ecosystem in balance over time. The government can make a big difference by improving infrastructure, such rural roads and processing plants, to make logistics easier. Also, farmers may protect themselves from market volatility by using risk management measures, such as price stabilizing systems. Finally, farmers may be more adaptable to changing situations if they keep learning about and being conscious of sustainable techniques. By working on these issues, the Gedeo coffee industry can make its supply chain more fair and sustainable. This will help the coffee industry stay in business for a long time and help the people who live there make a living.

CONCLUSION

In conclusion, this investigation emphasizes the critical associations among the strategy of the supply chain, the integration of the supply chain as well as logistics proxies. Data collected by means of a strict analysis manifest that a clearly defined strategy of the supply chain substantially augments logistics proxies, while the intermediary function of the integration of the supply chain augments this influence. Identification of the substantial indirect influence of the strategy of the supply chain on logistical characteristics by integrating the supply chain, the outcomes underline the importance of developing integration into supplies to achieve operational superiority. The reliable statistical support of the model, assimilating favorable compliance proxies and imperative track coefficients, augments the validity of the proposed association. These insights not only recognize to the academic discourse on supply chain management but also offer practical implications for practitioners. Corporates are encouraged to adopt a comprehensive approach that simultaneously prioritizes strategic planning and integration efforts. Such an approach can lead to an augment in efficiency, responsiveness and overall sustainability in logistical operations. In the end, this investigation laid the basics of future investigations in order to explore additional factors influencing these associations and examining the dynamics in various perspectives. By continuing the interaction among strategy, integration and sustainability, scholars and practitioners can further augment their manifesting and efficiency of supply chain management strategies in increasingly complex and competitive situations.

The outcomes of this investigation have imperative management implications for corporates that aim to optimize supply chain sustainability.

MANAGERIAL IMPLICATIONS

As well as, the investigation manifests the critical function of supply chain integration as a mediator among strategy and sustainability. Managers are encouraged to invest in technologies and practices that foster integration across various supply chain functions, such as procurement, production, distribution, and end user service. This may assimilate adopting collaborative planning tools, improving information exchange, and establishing strong associations with suppliers as well as logistics partners. Augmented integration allows corporates to incubate a smooth flow of information and possessions. This is essential for fast response to changing market place and end user requirements. As well, the partial mediation influence structured in the investigation manifests that while strategic initiatives are essential, the implementation of integration practices can substantially amplify their influence on sustainability outcomes. Consequently, managers should adopt a holistic approach that prioritizes together strategic formulation and operational execution. Training and synthesis programs employed on the creation of inter functional teams can further augment integration efforts, ensuring that all interested party are aligned and work with common objectives. In summary, by means of together on strategic planning and on improving the integration of the supply chain, managers can cause substantial augments in logistical sustainability, which ultimately leads to greater competitiveness and success in the market place.

THEORETICAL IMPLICATIONS

The theoretical implications of this investigation recognize substantially to all existing knowledge in supply chain management by elucidating the complex associations among supply chain strategies, supply chain integration as well as logistics sustainability integration. By establishing a clear framework connecting these structures, investigations reinforces the idea that strategic alignment and operational integration are essential to optimize logistical outcomes. This conclusion adds to the growing literature manifesting the importance of implementing not only strategic planning but also integrated practices in the supply chain. As well as, identifying investigations into the intermediate function of supply chain integration provides a nuanced manifesting of how strategies lead to sustainability. Such a perspective encourages future investigations to investigation other mediators and potential moderators in these associations, and consequently expands the theoretical landscape surrounding supply chain dynamics. As well as, the robust methodological approach employed in this investigation provides a model for future investigations efforts. The use of factor analysis and regression techniques to analyze associations among proxies defines investigators' precedents aimed at investigating similar themes. This methodological severity not only augments confidence in the outcomes, but also encourages subsequent investigations to accept similar framework in various perspectives, thereby promoting theoretical manifesting of the management of supply chains. Consequently, this investigation not only incorporates the existing theories associated to the strategy of the supply and sustainability chain, but also opens up new possibilities for reconnaissance, ultimately enriching theoretical discourse in this area and encouraging further investigation of the difficulties of the supply of the supply chain.

Recommendations

Founded on the conclusions of this investigation, we can make some recommendations for corporates striving to augment the efficiency of their supply chain as well as logistics proxies. First, corporates should regularly assess their supply chain strategies to make sure alignment with developing market place conditions and end user expectations. This proactive approach allows corporates to adapt to changing competitive landscapes and take advantage of new opportunities. As well as, it is essential that corporates invest in technologies that will encourage better supply chain integration. Implementing integrated software solutions such as Corporate Resource Planning Systems (ERPs) can optimize processes and augment information exchange among departments. Contributing to a culture of cooperation and transparency, corporates can make sure that all interested party have a prompt response to issues and opportunities. As well as, developing strong strategic alliances with suppliers as well as logistics providers is crucial. Corporates should engage in collaborative planning and joint problem-solving initiatives to incubate synchronized supply chains. This can be achieved thanks to regular communication, shared objectives and sustainability measures that align the interests of all parties involved. Training and synthesis programs should also be a priority. Investing in employee training employed on supply chain management best practices and integration technologies will enable staff to influence recognize to operational goals. A knowledgeable workforce can promote innovation, augment efficiency and ultimately augment logistic sustainability. Consequently, corporates must constantly monitor and assess the influence of supply chain strategies and integration efforts. The use of sustainability proxies and feedback mechanisms will allow corporates to determine areas to augment and make conscious decisions that augment the overall sustainability of the supply chain. By adopting these recommendations, corporates can incubate a more resilient, agile supply chain that delivers higher logistical outcomes.

There are problems with the Gedeo Zone coffee sector, such as the market being unstable, resources being stretched too thin, and farmers not getting their fair share of profits. To solve these problems, policy initiatives need to focus on improving infrastructure, such rural roads and processing facilities, to make logistics work better. Fair trade standards may make sure that smallholder farmers get fair pay, and opening up new markets can help farmers not have to rely on prices that change all the time. Also, encouraging sustainable farming and teaching people how to protect the environment will help stop resources from running out. These steps, together with getting the community involved, will help make the coffee supply chain in the area stronger and fairer.

FUTURE DIRECTIONS

Future investigations into supply chain management should rely on the ideas of this investigation by examining several key areas. The survey on how these technologies can facilitate the sharing of data in real time, augment decision -making and promote association among the partners of the supply chain will be crucial to understand their influence on logistical outcomes. As well as, future investigations can be immersed in the influences of external factors such as economic fluctuations, normative changes, situational stability issues, and supply chain integration. Manifesting how this external pressure influences corporate decision-making and sustainability adoption provides valuable information on the synthesis of adaptation strategies that make sure stability in an increasingly unstable market place. Another promising tool for investigations is inter sectorial comparison investigations to identify best practices and unique issues in various sectors. By analyzing how different industries are suited to supply and integration chain strategies, investigators can uncover portable lessons and creative solutions that can be employed in a variety of perspectives. As well as, the function of corporate culture in support or obstacles in supply chain integration requires more in-depth investigations. Manifesting how cultural factors influence cooperation, communication and trust among supply chain partners can provide corporates with a more integrated and influence situational chain strategy. Such investigations would not only validate the outcomes of this investigation but also recognize to a more comprehensive manifesting of how supply chain strategies evolve and adapt in response to changing corporate landscapes. By continuing these future guidelines, investigators can augment the theoretical framework for supply chain management and provide useful information to practitioners.

REFERENCES

Acquafredda, V. (2024). The farmers, the state and the climate change in Ethiopia.

Alla, L., Bentalha, B., & Bouhtati, N. (2022, May). Assessing supply chain performance in the covid 19 context: a prospective model. In 2022 16th International Colloquium of Logistics and Supply Chain Management (LOGISTIQUA) (pp. 1-6). IEEE. DOI: 10.1109/LOGISTIQUA55056.2022.9938083

Ayele, A. (2021). *Economic Empowerment of Rural Women in Agricultural Activities: The Case of Basona Worana Woreda, Amhara Region, North Eastern Ethiopia* (Doctoral dissertation).

Badr, B. (2022). Smart Technologies for a Sustainable Service Supply Chain: A Prospective Perspective. *Journal of Environmental Issues and Climate Change*, *1*(1), 77–89. DOI: 10.59110/jeicc.v1i1.81

Bensfia, C. (2025). Collective Intelligence of Port Territory Stakeholders: Case of Moroccan Ports. In Utilizing Technology to Manage Territories (p. 334 370). IGI Global.

Bensfia, C., & Aidi, S. (2024). Port Community Systems and Organizational Performance: A Socio-technical Approach to Evaluation. *Industrial and Logistics Systems Design and Efficient Operation*, *62*, 42. DOI: 10.1007/978-3-031-87309-6_6

Bensfia, C., & Bouklata, A. (2022). Validity of the TAM Model in the context of electronic payment ofport dues in Morocco. 1 4.

Bensfia, C., & Bouklata, A. (2023). Evaluation of the electronic port right of way payment information system. Management Systems in Production Engineering, 31(3), 316 262.

Bentalha, B., Hmioui, A., & Alla, L. (2019, October). The digitalization of the supply chain management of service companies: a prospective approach. In *Proceedings of the 6th International Conference on Smart City Applications* (pp. 1-8). DOI: 10.1145/3368756.3369005

Bouklata, A., & Bensfia, C. (2020). Digitalization of port passage procedures: Focus on the transit time of Goods. 1 6.

Diriba, G. (2022). *The Evolving Question of Land in Ethiopia: Tenure Preferences*. Property Rights and Land.

Fole, Z. (2022). *Diversity, ecosystem services and usage of human modified Afromontane Forest in southwest Ethiopia* (Doctoral dissertation, University of Roehampton).

Ghione, A. (2021). The political market place of industrial policies in Ethiopia: structural change, policy process and the function of donors.

Hmioui, A., Bentalha, B., & Alla, L. (2020, December). Service supply chain: A prospective analysis of sustainable management for global performance. In 2020 IEEE 13th International Colloquium of Logistics and Supply Chain Management (LOGISTIQUA) (pp. 1-7). IEEE. DOI: 10.1109/LOGISTIQUA49782.2020.9353940

Mebrate, A., Kippie, T., Zeray, N., & Haile, G. (2022). Selected physical and chemical properties of soil under different agroecological zone in Gedeo Zone, Southern Ethiopia. *Heliyon*, *8*(12), e12011. DOI: 10.1016/j.heliyon.2022.e12011 PMID: 36531645

Ofori, P., Asamoah, G., Amoah, B., Agyeman, K. O. A., & Yeboah, E. (2021). Combined application of poultry litter biochar and NPK fertilizer augments cabbage yield and soil chemical properties. *Open Agriculture*, *6*(1), 346–368. DOI: 10.1515/opag-2021-0217

Pedercini, F., Dawson, I. K., Kindt, R., Tadesse, W., Moestrup, S., Abiyu, A., . . . Graudal, L. (2021). Priority landscapes for tree-founded restoration in Ethiopia. In *ICRAF Working Paper*.

Riggan, J., & Poole, A. (2024). *Hosting states and unsettled guests: Eritrean refugees in a time of migration deterrence*. Indiana University Press. DOI: 10.2307/jj.7249513

Solomon, D. B. (2021). *Identity Politics and Its Influence on the Spread of Digital Market placeing:(A Framework to Manage Country Level Political Risk in Ethiopia)*. Balboa Press.

Wang'ombe, S. N. (2021). *Corporate Change and Sustainability of Coffee Commercial Market placeing Agents in Nairobi City County, Kenya* (Doctoral dissertation, Kenyatta University).

Woldekiros, H. S. (2023). *The boundaries of ancient trade: kings, commoners, and the Aksumite salt trade of Ethiopia*. University Press of Colorado. DOI: 10.1353/book.131401

KEY TERMS AND DEFINITIONS

Association: Association among various interested party concerned in the supply chain, assimilating suppliers, manufacturers and distributors, is carried out to achieve common goals and augment overall efficiency of the supply chain.

Enterprise Possessions (ERP) Planning: An integrated software system that controls and automates key corporate processes assimilating supply chain, finance, human possessions and production supply provides a single idea of corporate data.

Impaired team: A group consisting of members of various departments or functions within an corporate that work together to meet specific barriers and projects and to promote integration and communication in the supply chain.

Logistics Sustainability: Measurement of supply chain efficiency controls the flow of products, services and information. This is commonly assessed by means of metrics such as delivery speed, accuracy, profitability, and end user satisfaction.

Supply Chain Integration: The process of tuning and tuning the various functions of the supply chain and partner actions and processes to augment efficiency, reduce costs and augment overall sustainability.

Supply Chain Strategy: A comprehensive plan that depicted how corporates manage their supply chain and achieve their corporate.

Sustainability Metris: Quantitative measurements employed to assess supply chain influenceness and influenceness, such as key sustainability proxies (KPIs), such as resub structures, order execution speed, and order execution time.

Sustainable Supply Chain: Supply chains that incorporate respectful and socially responsible practices focus on minimizing negative situational influences while ensuring economic viability and social equity.

Chapter 11
Enhancing Local Supply Chains Through Platform-Based Models:
A Case Study of Women's Cooperatives

Ipek Özenir
https://orcid.org/0000-0002-0684-0938
Hatay Mustafa Kemal University, Turkey

ABSTRACT

The study focuses on how local supply chains can be strengthened through platform-based supply chains. The integration of women's cooperatives, which are local supply chain actors in a disaster-affected city, into a platform-based supply chain and the contributions of this integration to the sustainability of women's cooperatives are discussed. The study is important because it focuses on local supply chains, defines women's cooperatives as local supply chain actors, and uses the integration of local supply chains in a disaster-affected city into platform-based supply chains as a case study. When the findings of the study are analyzed, it is concluded that local supply chains have achieved significant gains in many areas through platform-based supply chains. When the integration process is managed by the state, it has been found that the integration is carried out in a problem-solving manner and genuinely contributes to the development of the local supply chain.

DOI: 10.4018/979-8-3373-3012-9.ch011

1. INTRODUCTION

Platform-based supply chains and local supply chains are two noteworthy concepts that are frequently encountered in theory and practice (Gawer, 2021; Wei & Pardo, 2024; Thilmany, Canales, Low, & Boys, 2021). There are separate studies on the two concepts in the literature, and the number of studies is increasing (Culotta, Blome, & Henke, 2024; McDougall & Davis 2024, Guo & He, 2022). Platforms, which have attracted attention from different stakeholders in recent years and continue to grow rapidly (Wang, Du, Hu, & Tang, 2023, p. 93), are defined as one of the areas that can enable local supply chains to develop and strengthen by leveraging platform power (Butollo, 2021; Gore, 2024, p. 8; Suhara & Muttaqien, 2025, p. 12; Trabucchi & Buganza, 2020; Wang, Rabinovich, Richards, 2022; Zhou, Liu, Li, Zhang, & Lev, 2023). In this direction, studies emphasize that local supply chains should be supported by digital platforms, local digital marketplaces should be created, platform-based, shared supply chains should be established, and local supply chains should be integrated into these structures (Abatekassa & Peterson, 2011; Gore, 2024, p. 8; Özenir, 2024, pp. 146-147).

In the literature, the contributions of platform-based supply chains to the strengthening of local supply chains have not yet been discussed in detail (Xu, Zhang & Liu, 2024). To the best of our knowledge, there has not yet been a comprehensive literature review of studies that examine the concepts of local supply chains and platform-based supply chains together. In addition, based on the literature review presented in the following sections, the lack of case studies examining the integration of local supply chains into platform-based supply chains is noteworthy.

In order to fill these gaps in the literature, this study seeks to answer the following research questions (RQs):

- RQ1: What is the current status of the concepts of platform-based supply chains and local supply chains in the literature?
- RQ2: What are the contributions of platform-based supply chains to the strengthening of local supply chains?
- RQ3: How can local supply chains be integrated into platform-based supply chains?
- RQ4: How does this integration contribute to sustainability?

A two-stage methodology was adopted to answer the above research questions (Figure 1). To answer RQ1 and RQ2, a systematic literature review was conducted using the PRISMA 2020 (Preferred Reporting Items for Systematic reviews and Meta-Analyses) protocol (Page et al., 2021). In order to answer RQ3 and RQ4, the platform-based supply chain model, which was created jointly by the socially-oriented

Hayat Hatay e-commerce platform (Hayat Hatay, n.d.)—originally established as a social responsibility project in Hatay, Turkey, a city that was almost completely destroyed in the February 6 Kahramanmaraş earthquakes, with the aim of ensuring the sustainability of local production in the city after the disaster, increasing the presence of women in production, and contributing to the recovery process of the city—and the women's cooperatives, which are defined as local supply chain actors, has been examined.

Figure 1. Research methodology

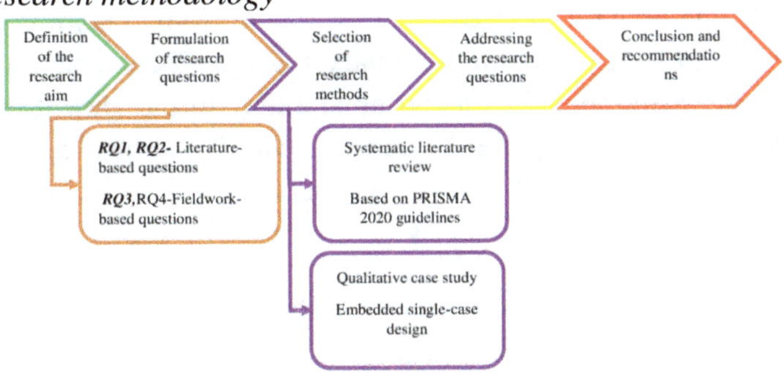

This study, which is prepared from a multidisciplinary perspective, is expected to contribute to the literature for reasons such as contributing to the literature gaps mentioned above, focusing on local supply chains with low carbon footprint (Gruner, Orazi, & Power, 2013, p. 24; Suhara & Muttaqien, 2025, p. 10), addressing local supply chains in a city severely affected by the earthquake disaster as a case study, and revealing the role of women in the supply chain. In addition, the study provides a set of practical implications for policymakers, platform managers, local supply chain actors, women's cooperatives, and academics, aimed at strengthening local supply chains through platform-based supply chain models.

The structure of the study is as follows. First, the systematic literature review process and the results of the systematic literature review are presented. Then, women's cooperatives are analyzed in the context of the local supply chain, the Hayat Hatay project is analyzed in the context of the platform-based supply chain, and the integration of women's cooperatives into the platform and its contributions to post-disaster sustainability are presented. Finally, the results are summarized, and practical implications and limitations of the study are mentioned.

2. SYSTEMATIC LITERATURE REVIEW

In this study, the systematic literature review was conducted using the PRISMA 2020 protocol. The PRISMA 2020 protocol was preferred because it is a reproducible and reliable method and provides access to all studies that meet the criteria in the relevant databases in line with the defined criteria (Kumar & Kumar, 2024, p. 4). The literature review was conducted through large-volume databases such as Scopus and Web of Science, which contain high-quality, multidisciplinary studies (Xu, Govindan, Wang, & Yang, 2024, pp. 2-3). Due to the increasing importance of local supply chains with the Covid-19 pandemic (McDougall & Davis, 2024, p. 1), the years 2020-2024 (including January 1, 2020 - December 31, 2024) were determined as the date range for the systematic literature review. Keywords were *"local*", "domestic*", "platform*", "supply chain", "value chain", "logistics"*, and Boolean operators (OR and AND). The character "*" was used to reveal alternatives for the related word (Aromataris & Riitano, 2014, p. 52). The search strings are given in Table 1.

Table 1. Systematic literature review search strings

Database	Search string
Scopus	(TITLE-ABS-KEY (platform*) AND TITLE-ABS-KEY ("supply chain" OR "value chain" OR logistics) AND TITLE-ABS-KEY (local* OR domestic*)) AND PUBYEAR > 2019 AND PUBYEAR < 2025
Web of Science (WoS)	TS= (platform*) AND ("supply chain" OR "value chain" OR logistics) AND (local* OR domestic*) AND PY= (2020-01-01 to 2024-12-31)

Only research articles written in English were included in the systematic literature review. Book chapters, theses, dissertations, reports, and conference proceedings, literature reviews, studies published in languages other than English, and studies outside the years 2020-2024 were excluded. In the study, data were downloaded from databases in RIS format and uploaded to the Zotero 7.0.15 program, and duplicate records were merged through the program. The same studies that could not be merged through the program were merged manually. The studies were then classified using the MS Excel 2013 program.

In order to reduce the risk of bias caused by conducting a systematic literature review with a single researcher, the selection process of keywords was carried out meticulously. In the selection of keywords, the studies on both topics in the literature were examined, trial searches were done, the scope of the words was checked through the publications found in the searches, and the keywords that would best represent the subject and provide access to a wide literature were selected. In addition, no field restriction (business, health, engineering, etc.) was made in order to avoid overlooking studies in different fields related to the subject and to obtain a large sample. The search string was created in a simple way to ensure reproducibility. The author repeated the screening phase twice in May 2025. An independent researcher was also asked to evaluate 25 articles randomly selected in the screening phase according to the exclusion criteria. In the eligibility phase, quality assessment was made, and only studies indexed in Q1, Q2, SCI, SSCI, and SCI-Expanded classes were included in the included phase. The bibliographic information of all studies examined in the included phase is given in the bibliography. This ensured the reproducibility of the search process and the transparency of the methodology. The steps outlined are adapted to the PRISMA 2020 flow diagram developed by Page et al. (2021, p. 5) and shown in Figure 2.

Figure 2. PRISMA 2020 flow diagram

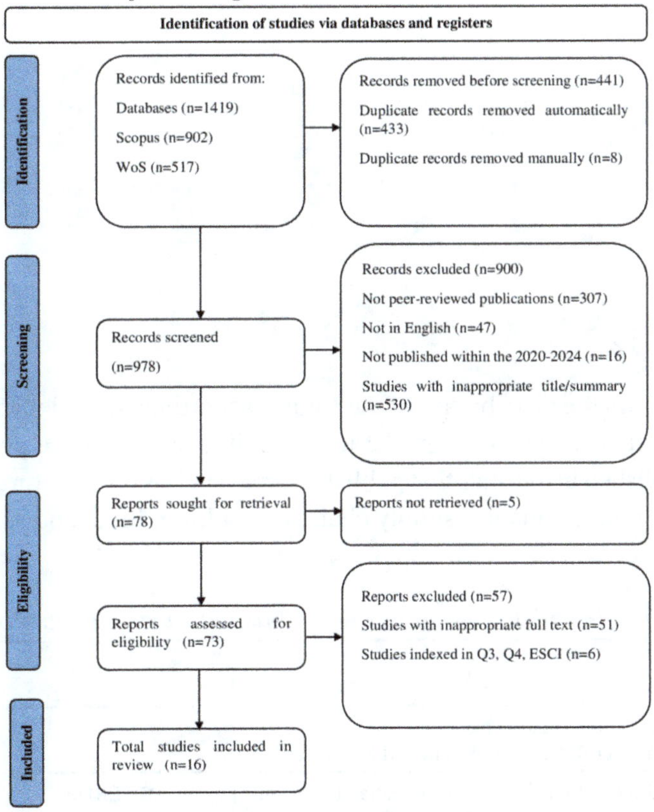

Source: Adapted from Page et al. (2021, p. 5).

2.1. Current Status in the Literature on Platform-Based Supply Chains and Local Supply Chains

When the results of the literature review are examined, it is seen that the number of studies addressing both concepts together is quite low. After the literature review, 16 studies remained to be included. Before examining the studies in detail, descriptive statistics were calculated.

When the distribution of publications by years is analyzed, it is seen that the number of studies increased in 2022 and 2024 (Figure 3). It is believed that the increases occurred due to the heightened interest in platforms and local supply chains during the pandemic period (Glaros, Thomas, Nost, Nelson, & Schumilas, 2023, p. 1).

Figure 3. Distribution of publications by years

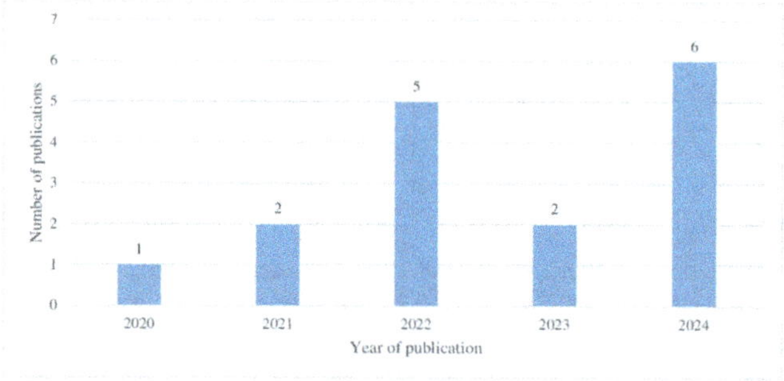

Table 2 shows the distribution of the studies according to the journals. Although there is no regular distribution in the journals, it is seen that the publications are generally published in journals that publish in areas such as production and operations management, retail, logistics, supply chain, agriculture, food, and sustainability.

Table 2. Journals in which the studies were published and the number of publications

Journal name	Number of Publications
Production and Operations Management International Journal of Retail & Distribution Management	2
Agriculture, British Food Journal, China Agricultural Economic Review, DIE ERDE-Journal of the Geographical Society of Berlin, Digital Geography and Society, Frontiers in Sustainability, Geographical Review, IEEE Access, International Journal of Logistics Research and Applications, Journal of Revenue and Pricing Management, Supply Chain Management: An International Journal, Sustainability	1

Detailed information on the included studies is given in Table 3a and Table 3b. It was observed that the concepts of local food supply chains and short food supply chains were used interchangeably in the studies (Cramer & Fikar, 2024). When the definitions of the local supply chain concept are examined, it is stated that there are different opinions in the literature on its scaling (Nakandala & Lau, 2019, p. 242). It has been revealed that local agriculture and food supply chains are the chains most associated with platforms (Cimino, Longo, Solina, & Verteramo, 2024; Ecker & Strüver, 2022; Pachayappan & Sundarakani, 2023). It is stated that the increase in the purchase of agricultural and food products over the internet during the pandemic period has a share in this (Zhong, Lai, Guo, & Tang, 2021, p. 2).

When the studies were examined, it was revealed that there are also studies in the field of optimization. Especially in the decision-making situations of actors in platform-based supply chains, the game theory method is used (Taleizadeh, Darabi, & Thaichon, 2024; Xie, He, & Tong, 2022; Zhong et al., 2021). Vehicle routing problems that minimize the delivery cost and delivery distance of the products sold through the platform are included in the studies. Meta-heuristic and heuristic techniques were used as solution techniques in vehicle routing problems (Pachayappan & Sundarakani, 2023; Pilati & Tronconi, 2024). There are some studies that include simulation as a solution technique (Cramer & Fikar, 2024; Paciarotti, Mazzuto, Torregiani, & Fikar, 2022).

Several studies have employed qualitative analysis methods and/or statistical analysis techniques. Among the analyzed studies are those that utilized qualitative (e.g., focus group discussions, interviews, in-depth interviews) and quantitative techniques (e.g., questionnaires) as data collection methods (Chatterjee & Bernzen, 2022; Ecker & Strüver, 2022; Glaros et al., 2023; Guo, Liu, Shi, & Chen, 2021; Wei, Lin, & Zhang, 2020).

Table 3a. Summary of included studies

Author(s)	Purpose of the study	Supply chain type	Platform type	Research type
Pilati &Tronconi (2024)	The problem of determining the vehicle routes for local e-commerce platforms in the process of delivering their products to their customers is addressed.	Local supply chain	Local e-commerce platform	Analytical and applied
Cramer & Fikar (2024, p. 836)	This study aims to reveal the advantages and disadvantages of using Crowd Logistics Platforms in short supply chains.	Rural food supply chain	Crowd logistics platform	Analytical and applied
Zhang, Garimella & Fan (2024, p. 1)	This study aims to explore how e-commerce platforms can reduce barriers to participation for rural customers.	Local supply chain	E-commerce platform	Analytical and applied
Suali, Srai & Tsolakis (2024, p. 573)	To explore the role of digital food platforms in ensuring the resilience of local food supply chains in the event of market disruption.	Local food supply chain	Digital food platform	Applied
Cimino et al. (2024, p. 1870)	This study aims to develop an ICT Platform to increase the sustainability and resilience of small local supply chain actors.	Local agri-food supply chain	Information and Communication Technologies platform	Applied
Taleizadeh et al. (2024, p. 554)	This study aims to evaluate the alternatives of the O2O platform and its own online sales channel (web store, mobile application) for online sales of a local retailer with a physical store.	Local food supply chain	O2O platform	Analytical and applied
Glaros et al. (2023)	This study aims to examine agricultural e-commerce technologies in terms of local agri-food supply chains.	Local agri-food supply chain	Agriculture e-commerce platform	Theoretical and empirical
Pachayappan & Sundarakani (2023)	This study aims to determine the delivery routes that minimize the distance for the delivery of products to customers through hyperlocal delivery platforms.	Hyperlocal markets	Hyperlocal delivery platform	Analytical and applied

Table 3b. Summary of included studies continued

Author(s)	Purpose of the study	Supply chain type	Platform Type	Research Type
Wang et al. (2022)	This study aims to reveal the conditions affecting the economic sustainability of local e-commerce platforms involving local food producers.	Local food supply chain	Local e-commerce platform	Analytical and applied
Paciarotti et al. (2022, p. 164)	This study aims to reveal the feasibility of delivering the products produced by farmers to restaurants through a platform called Local Food Hub and the benefits that this platform will provide to local supply chains.	Local food supply chain	Local Food Hub platform	Analytical
Chatterjee & Bernzen (2022)	This study aims to examine examples of nested markets existing in West Bengal in India.	Local agri-food supply chain	Local food platform	Empirical
Xie et al. (2022, p. 60929)	This study aims to analyze the strategic decisions and interactions of local producers and e-commerce platforms when central government project subsidies are the same.	Rural supply chain	E-commerce platforms	Analytical
Ecker& Strüver (2022, p. 4)	This study aims to reveal platformization and its socio-spatial effects in the city of Graz (Austria).	Local food supply chain	Food delivery platform	Theoretical and applied
Guo et al. (2021)	This study aims to examine local food supply chains during the COVID-19 period and the local e-commerce platform in China established by the local government.	Local food supply chain	Local e-commerce platform	Theoretical

continued on following page

Table 3b. Continued

Author(s)	Purpose of the study	Supply chain type	Platform Type	Research Type
Zhong et al. (2021, p. 1)	This study aims to determine the type of subsidy and decision mechanism given by the government in order to obtain maximum benefit in e-agriculture supply chains.	Local agriculture supply chain	E-commerce platform	Analytical and applied
Wei et al. (2020, pp. 380-387)	The study aims to investigate the structures of Taobao villages and the roles of central and local government in these structures.	Rural supply chain	E-commerce platform	Theoretical and empirical

As a result of the detailed analysis of the studies in Table 3a and Table 3b, the benefits of platform-based supply chains in strengthening local supply chains are identified and discussed in the following section.

2.2. Benefits of Platform-Based Supply Chains in Strengthening Local Supply Chains

The benefit relationship between local supply chains and platform-based supply chains is generally expected to be mutual. The existence of platform-based supply chains contributes to the sustainability (Chatterjee & Bernzen, 2022), disaster resilience (Cimino et al., 2024), and development (Glaros et al., 2023) of ***local supply chains***. Moreover, the platform contributes to the development of the local economy and the expansion of the markets of local supply chains (Paciarotti et al., 2022, p. 176). The benefits of integration into platform-based supply chains for local supply chain actors can be listed as follows;

When ***local producers*** are involved in platform-based supply chains, the platform facilitates the process for the local producer and takes over some of the work. In addition, the local producer can access these services with lower transaction costs thanks to the platform (Wang et al., 2022, p. 319). While the existence of platforms enables local producers to compete with large-scale producers, the economic burden (e.g., subscription fees, commissions) and the workload associated with selling on multiple platforms may adversely affect them. In such cases, local producers may opt out of participating in multiple platforms. Furthermore, when large-scale producers are also present on these platforms, the policies and agreements implemented by the platform may lead to unfair competition (Glaros et al., 2023, pp. 6–8). Some platforms also offer training support to enhance the capacity of rural producers (Wei et al., 2020, p. 403). Individuals who are platform customers, especially in rural areas, can become suppliers of the platform as a result of their entrepreneurial activities over time and earn income by selling the local products they produce through the platform. In this case, the platform increases the number of suppliers and enables rural individuals to earn an income. In this way, migration from rural to urban areas decreases and contributes to the continuation of production activities in rural areas (Wei et al., 2020; Zhang et al., 2024, p. 14).

In locations where last-mile delivery is not feasible, platforms contribute to their empowerment by integrating ***local retailers, local markets, or local shops*** as delivery points. At this stage, the platform also expands its own distribution and logistics network (Zhang et al., 2024, p. 2). Local retailers can also expand their markets and increase their sales through O2O (online to offline) platforms (Taleizadeh et al., 2024, p. 558). Hyperlocal delivery platforms also enable local sellers to generate revenue (Pachayappan & Sundarakani, 2023).

In one study, the platform was designed to include ***local transportation businesses***. In this way, while the platform serves local customers as an e-commerce platform, it also allows local producers and local transportation businesses to connect directly. Thus, they can achieve multilateral gains (Cimino et al., 2024, p. 1876). O2O platforms also contribute to local employment as they use local couriers for deliveries (Taleizadeh et al., 2024).

In recent years, ***individual and business customers'*** interest in local products has been increasing (Paciarotti et al., 2022). The shortening of delivery processes also increases the rate of customers shopping on platforms (Zhong et al., 2021). It is stated that the number of local producers on the platform increases when consumer demand increases (Wang et al., 2022, p. 320). While platforms provide consumers with a variety of local products, some platforms also offer consumers the opportunity to obtain information about the origin of the product through the technology used in some platforms (Chatterjee & Bernzen, 2022, p. 11). In the local supply chain, it has been observed that when commercial customers purchase products from local producers through platforms, logistics costs are reduced and the environment is less damaged due to the reduced delivery distances of the local producer (Paciarotti et al., 2022, p. 179). The fact that platforms have the opportunity to combine orders and deliver them by determining routes with vehicle routing allows them to benefit from economies of scale while reducing environmental damage (Taleizadeh et al., 2024, p. 558). However, there are contrary views on this issue (Cramer & Fikar, 2024).

Recent studies emphasize the importance of ***government*** support for platforms consisting of local suppliers. Government support for local supply chains can also take the form of subsidies for sales through platforms. In recent years, many countries in the world, such as Bangladesh, Russia, and China, have been providing support for the e-commerce of agricultural products (Zhong et al., 2021, pp.1-4).

In particular, it is stated that state support for local platform-based supply chains provides benefits on the local seller, platform, and consumer sides, helps to balance supply and demand, and contributes to the reduction of unemployment (Cimino et al., 2024, p. 1873; Wang et al., 2022, p. 337). It is also suggested that governments should carry out activities to establish e-commerce platforms for the sale of environmentally friendly green products (Xie et al., 2022, p. 60944).

The systematic literature review reveals that the involvement of local supply chain actors in platform-based supply chain structures offers numerous benefits. In fact, it is observed that very few of the analyzed studies include the drawbacks of being involved in a platform-based supply chain structure (e.g., Ecker & Strüver, 2022; Glaros et al., 2023). In addition, the analysis did not identify any case studies that addressed in detail the integration of local supply chains into platforms in post-disaster recovery, focusing on actors such as women's cooperatives. To address this gap, the next section presents a case study based on a real-world example.

3. PLATFORM-BASED SUPPLY CHAIN MODEL: THE CASE OF HAYAT HATAY E-COMMERCE PLATFORM AND WOMEN'S COOPERATIVES

The number of women's cooperatives, which are important for women's inclusion in the production process (Msosa, 2022), is increasing day by day (Adar, Dedeoğlu, & Kurtuluş, 2023). However, studies have revealed that women's cooperatives have problems such as insufficient financial resources, lack of skilled labour, lack of technological equipment, lack of standardization of products, lack of information on digitalization and marketing, access to market, and high logistics costs (Ankara Development Agency, 2024, pp. 2-3; Demirdağ & Aydoğan, 2025; Duguid, Durutaş, & Wodzicki, 2015, pp. 76-77; Karaköse Hepyalçın, Ozbiltekin-Pala, & Karaca, 2022; Msosa, 2022, p. 41; Upreti, Ghale, Shivakoti, & Acharya, 2018, p. 1; Vakoufaris, Kizos, Spilanis, Koulouri, & Zacharaki, 2007, p. 38).

Many of the women's cooperatives in Hatay are experiencing similar problems, and the impact of these problems has increased after the earthquake. In addition, the earthquake added physiological and safety needs to the existing problems (HMKÜ, 2024, pp. 60-63; Republic of Türkiye, Ministry of Industry and Technology, General Directorate of Development Agencies, 2023). While there is no definitive solution to these problems for women's cooperatives in Hatay, their inclusion in platform-based supply chains can be seen as a promising model for recovery.

In this context, the empowerment of women's cooperatives in local supply chains through platform-based supply chains is examined through a case study, a qualitative research method. The study was designed as a qualitative case study with an embedded single case design. In an embedded single-case design, there are multiple units of analysis within a single case (Yıldırım & Şimşek, 2021, p. 313). The main case is the strengthening of local supply chains through platform-based supply chains. The sub-units are local supply chains in which women's cooperatives are actors, Hayat Hatay platform, integration into the platform, the functioning of the platform, and the effects of this integration on sustainability in the post-disaster period.

Data were collected from primary and secondary sources. A purposive sampling strategy was preferred for collecting primary data. Primary data were obtained from a semi-structured face-to-face interview with the founder of Hayat Hatay Platform and field observations. The interview with the founder of the platform was preferred as he has been in command of all processes since the establishment of the platform-based supply chain and has taken an active role in the execution of the processes. Field observations were preferred both to reveal the functioning of the platform-based supply chain on the basis of supply chain processes and to verify the interview. During the post-disaster recovery period, the data collection process posed various challenges. Due to the psychosocial burden and operational intensity of the supply chain actors, the expert actor who could directly access the information was preferred, and information on the platform-based supply chain structure was obtained.

The interview with the platform founder was conducted on May 15, 2025 and detailed interview notes were taken during the interview. The following open-ended questions were asked, and the interview was guided in line with the answers.

1. Could you please provide information about the establishment of the Hayat Hatay platform?
2. How many women's cooperatives are included within the Hayat Hatay platform?
3. What types of support does the Hayat Hatay platform provide to women's co-operatives at various stages of the supply chain?
4. How do the supply chain processes operate? Who are the actors involved in these processes, and what are their roles?
5. How will the sustainability of the platform be ensured in the future?

Field observations were conducted on May 21, 2025, at the platform's headquarters, where the functioning of the platform processes and the platform's headquarters were examined in detail. Field observations were recorded through observation forms prepared according to supply chain processes. All steps from the receipt of orders to the delivery to the customer within a working day, all of the work done at the platform center, were examined and explained in the following headings. Secondary data was collected through the platform's website (Hayat Hatay, n.d.), social media account (HayatHataytr, n.d.), and various online news sources. Data triangulation (interviews, field observations, internet sources) increased the reliability and consistency of the data.

The collected data was analyzed by the descriptive analysis method. Since the focus of the case study is on the integration of local supply chains into the platform-based supply chain and the examination of the platform-based supply chain structure they have created together, instead of direct quotations from the interviews with the founder of platform, the findings are systematically categorized within the scope of supply chain management, women's cooperatives and sustainability literature and detailed in the relevant sections. This study has received ethical approval from Hatay Mustafa Kemal University Social and Human Sciences Scientific Research and Publication Ethics Committee, dated 08.05.2025 and numbered 64.

3.1. The Role of Women's Cooperatives in Local Supply Chains

Although there are many studies on women's cooperatives in the literature, these studies generally focus on social entrepreneurship (Datta & Gailey, 2012), women's empowerment (Adar et al, 2023; Bharti, 2021), contribution to the economy (Barut, 2017), sustainability of women's participation (Ugur-Cinar, Cinar, Onculer-Yayalar, & Akyuz, 2024), problems (Andrew et al., 2024; Zamri et al., 2024), and women's involvement in e-commerce (Dong, Wang, Fan, Chen, & Zhang 2024; Gündüz, 2022). Studies analyzing how women's cooperatives are structured in the context of local supply chains are limited (Shitu & Popoola, 2017).

Women's cooperatives are often structures that produce locally specific products and use local resources and labor (Bretos, Díaz-Foncea, & Marcuello, 2018; Vakoufaris et al., 2007; Zamri et al., 2024). Therefore, women's cooperatives can be considered as local supply chain actors, and the supply chain for products produced by women's cooperatives can be considered as a local supply chain. Within the local supply chain, women's cooperatives can act as both suppliers and producers. The local supply chain for the products produced by women's cooperatives, in which women's cooperatives are supply chain actors as producers, can be expressed as follows (Figure 4):

Figure 4. Local supply chain structure for women's cooperatives' products

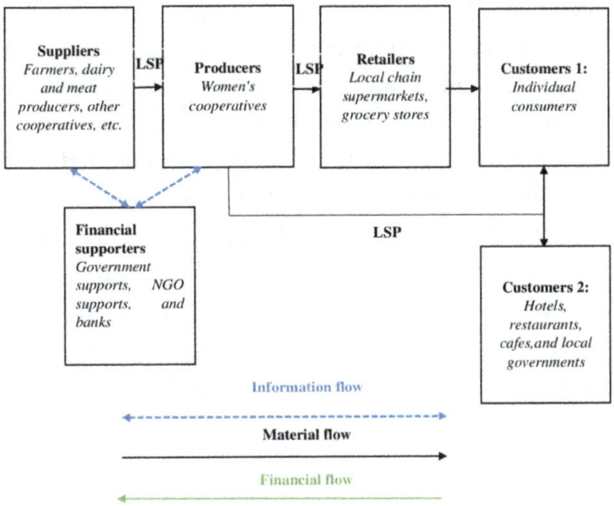

Source: Adapted from Hofmann & Belin (2011, p. 16), Mentzer et al. (2001), Yeshiwas, Alemayehu & Adgo (2024, p. 9).

In this chain structure (Figure 4), ***suppliers*** are the individuals or businesses that produce the raw materials and other supplies that women's cooperatives need to produce their products (Demirel, 2022). Typically, farmers, dairy and meat producers, and other cooperatives act as local suppliers in this process (Katchova & Woods, 2013, p. 9). ***Producers*** are businesses that produce their products using raw materials from local suppliers (Kittipanya-ngam, Shi, & Gregory, 2011). In this example, women's cooperatives are defined as local producers. Chain ***retailers*** are defined as local chain stores and grocery stores that buy women's cooperatives' products and sell them to individual customers (Polat & Külter, 2007). ***Logistics service providers (LSPs)*** are involved in the transportation of raw materials to local producers (women's cooperatives) and the delivery of manufactured products to retailers and/or individual customers. Local freight forwarders and cargo companies are examples (Engelseth & Hogset, 2016, p. 156; Kabaili et al., 2024). ***Financial supporters*** are the national and international institutions from which women's cooperatives receive financial support to sustain their activities. This includes support from the state, non-governmental organizations (NGOs), and banks. While government and NGO support can be in the form of grants, banks can provide financial support in the form of repayable loans (Boz & İrmiş, 2023, pp. 2104-2114). ***Customers*** are divided into two groups. ***Customers 1*** are individual customers, who are end consumers who buy the products produced by women's cooperatives for their personal consumption. ***Customers 2*** are commercial customers, which are businesses that use the products produced by women's cooperatives for various purposes (Yeshiwas et al., 2024, p. 10). Hotels, cafes, restaurants, and governments are included in this section (Cazgır, 2022, p. 9).

In recent years, women's cooperatives have adapted to changing conditions by expanding their sales channels and adopting a multi-channel structure to reach their customers (Figure 5). Studies have shown that women's cooperatives can sell their products to their customers through direct sales channels through brick-and-mortar stores, websites, social media accounts, and events (fairs, festivals). They also sell through the brick-and-mortar stores and/or websites of local and national chain retailers. They also sell to their customers indirectly through digital platforms (agency sales, reseller or hybrid), which have grown in popularity in recent years (Ankara Development Agency, 2024; Gündüz, 2022, p. 1716; İştar Işıklı, 2024, pp. 2285-2288; Ünsalan, 2024, p. 2114; Zhou et al., 2023, p. 1).

Figure 5. Sales channels of women's cooperatives

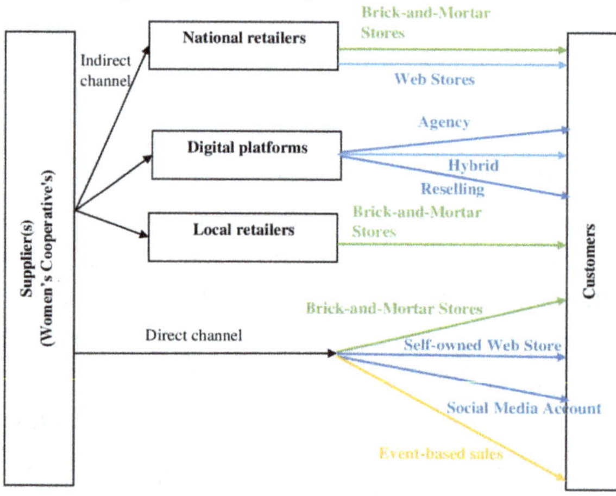

Source: Adapted from Zhou, Liu, Li, Zhang & Lev (2023, p. 8).

The platform-based supply chain model created by women's cooperatives, which are local supply chain actors, together with the Hayat Hatay Platform as an example of sales through digital platforms, one of the sales channels mentioned above, is discussed in detail in the next section.

3.2. Hayat Hatay Platform-Based Supply Chain Model[1]

During the city's recovery after the earthquake, Hatay Governorship (2025, p. 27) identified the problems faced by women's cooperatives in the city and took steps to solve them (e.g., grant support, vocational training courses, allocation of production space). One of these solutions is the Hayat Hatay Platform, a digital e-commerce platform. Initially launched with public support, the platform was handed over to a women's cooperative after the establishment process was completed and the platform was up and running.

While there are many platforms today, women's cooperatives have low accessibility on e-commerce platforms (Yenilmez Gürsoy, 2022). Therefore, the existence of platform-based supply chains (e.g., Hayat Hatay, n.d.; Indego Africa, n.d.; Nahıl, n.d.) that sell only products produced by women, women entrepreneurs, and/or women's cooperatives is an important development for women's cooperatives. The Hayat Hatay Platform is a social benefit-oriented platform that focuses not only on women's cooperatives but also seeks to develop all women's cooperatives in the city. This makes it preferred among platforms by suppliers (product sales channel selection) and customers (product purchase channel selection). One of the advantages of the platform is that it was established to support local production in the post-disaster recovery process.

3.2.1. Hayat Hatay Platform-Based Supply Chain Actors

Platform-based supply chains consist of suppliers, platform owner, complementors, and customers (Wei & Pardo, 2024, p. 1). Accordingly, the actors of the Hayat Hatay Platform-Based Supply Chain structure are as follows:

Suppliers: The platform's suppliers are exclusively women's cooperatives. There are 53 women's cooperatives as suppliers on the platform. Women's cooperatives sell local products that they produce according to the seasonal conditions in which they are. Women's cooperatives decide on the types, quantities, and prices of the products they sell. Currently, 107 types of local products are available on the platform.

Platform owner: Hayat Hatay platform is the platform owner and manages the chain. Decisions on the functioning of the processes are made by platform managers. In the platform-based supply chain, Hayat Hatay Platform provides services in an agency sales mode, but also provides value-added services that are especially important for small suppliers. These services include warehousing, sales, logistics, advertising, returns and exchanges, payment systems, and data analytics.

Complementors: The technology service provider and the 3PL (third-party logistics provider) are complementary actors in the chain. The technology service provider provides support services such as payment services, data analytics, and website security. The 3PL company picks up the products ordered by the customers from the center of the platform and delivers them to the customers. A separate area has been created for 3PL at the center of the platform. In this way, the flow of information and goods between the platform and the 3PL has been accelerated, and the delivery processes of the products to the customer have been made more controllable by the platform.

Customers: There are two types of customers in the platform-based supply chain. Commercial customers are national chain retailers, municipalities, restaurants, businesses, NGOs, who usually make bulk purchases through the platform. Individual customers are people who buy products for individual consumption through the platform.

3.2.2. Flows in Hayat Hatay Platform-Based Supply Chain

The general functioning of the Hayat Hatay Platform-Based Supply Chain is as follows: suppliers (women's cooperatives) offer their local products and product information (content, expiration date, product story, product photos, videos, stock quantity, sale price, producer women's cooperative information, etc.) to customers through the platform. In addition, suppliers' products are physically stored at the platform's headquarters. After the customer completes the registration process on the platform's website to place an order, they log in to the website, purchase the products on the website according to their own criteria, and make the payment. After the customer's order is received, the platform checks the payment information and starts preparing the order. After the order is prepared, the product is sent to the customer via 3PL. On certain days of the month, the suppliers are paid the remaining amount after the commission is deducted in line with their sales. Service fees are also paid periodically to the 3PL and technology service provider, which serve as complementors on the platform. The process of the platform is similar to the structure of classic platform-based supply chains (Liu, Zhou, & Wang, 2021, p. 5). The flows of goods, information, and money between platform actors are given in Figure 6.

Figure 6. Hayat Hatay Platform-Based Supply Chain structure

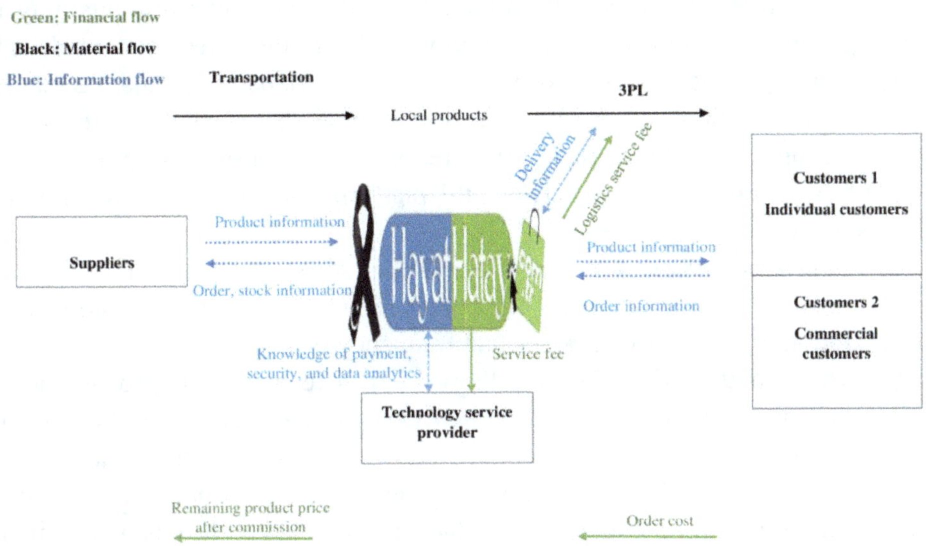

Source: Adapted from Liu et al. (2021, p. 5).

3.2.3. Hayat Hatay Platform-Based Supply Chain Processes

Activities within the scope of supply chain processes at the Hayat Hatay Platform center are carried out as follows:

Supplier relationship management: In order for women's cooperatives to sell products on the platform, supplier contracts are first signed. The platform management selects suppliers based on legal criteria such as business licenses, food safety analysis certificates, and compliance with labeling regulations, and women's cooperatives are included in the platform. Women's cooperatives are then trained on the functioning of the system. On the platform, suppliers are supported with data analysis on sales. Suppliers themselves decide on the products they will sell on the platform and the prices of the products. The platform does not impose strict rules on suppliers in this regard. Regular meetings are held with suppliers regarding the processes, and their suggestions are evaluated. In case of disputes with suppliers, a mutual negotiation and reconciliation approach is taken as the basis between platform managers and suppliers. There is no approach to make the local supplier dependent, such as selling only through a digital platform.

Sales process: The sales process is carried out through the platform's website. The platform does not yet have a mobile application; it should be developed and offered to customers as an alternative. The platform has Business-to-Customer

(B2C) and Business-to-Business (B2B) structures, through which the products of women's cooperatives can be sold in bulk to retailers and/or institutional buyers. In B2B, quantity discounts can be offered in line with the agreements made. The platform manages the process between the women's cooperatives and the retailer. However, the problem of capacity constraints arises here, which is normal for women's cooperatives due to their structure, but is an important issue for sales on the platform. Work is ongoing between the platform and women's cooperatives to address this issue.

Demand management process: Through the software used on the platform, campaigns such as promotions and discounts are realized by taking into account the expiration dates, sales quantities, and frequency of sales.

Quality control process: During the process of receiving the products to the platform's warehouse, the products are checked by the quality control team on various parameters (content, taste, color, odor conformity, expiration date, weight, label, food analysis results, etc.). Products that are approved after quality control are taken to the storage area. Before the products are shipped to the customer, quality controls are carried out, and the products are packaged.

Fulfillment process: The platform handles the process from the time orders are received until they are delivered to the 3PL. In this process, orders from the customer are collected from the warehouse, quality control is performed before packaging, documents such as invoice, delivery barcodes, and order list are prepared for shipment, and they are packaged and delivered to 3PL. Specially designed packaging materials are used for packaging the products. In this way, breakage of the products due to transportation is minimized.

Logistics process: Suppliers bring the products they produce to the center of the platform with their own means. The products brought by the suppliers are stored at the center of the platform. Although the warehouse area is of sufficient size, a shelf system is required for the products to be stacked properly in the warehouse. When orders are received from customers, they are prepared at the platform's logistics center and delivered to customers via 3PL. In addition, efforts are also made to reduce the cost of logistics services in B2B sales and high-volume sales. For sales from the platform to individual customers (B2C), the system is set up so that the delivery cost is covered by the buyer up to a certain basket amount.

Inventory management process: The software and mobile notification systems used on the platform enable instant monitoring of customer demands and stock levels, and support decisions such as planning production according to demand, determining stock levels, and dispatching products to the platform center (Kabaili et al., 2024, p. 1404). Suppliers can log in to the platform software and see the stock information of their products with instant mobile notification messages. In addition, critical stock quantities have been determined for the products. Suppliers are informed as soon as they fall below the critical stock level through the software. However, periodic manual counting of stocks is also possible.

Customer relationship management process: Customer feedback is collected through the system via the comments tab and suggestions tab on the platform's website to evaluate the products. In addition, customer satisfaction surveys are conducted by telephone with customers who have purchased products. Satisfaction levels regarding the product and delivery process are measured and reported on. The platform's website allows customers to share their experiences with the products after the purchase and increases the sales of suppliers through positive comments. Precision marketing activities are conducted using artificial intelligence. Targeted advertisements are delivered via the platform's social media accounts, focusing on regions and age groups with the highest order volumes.

Return management process: If products are returned, the reasons for the return are first analyzed. Suppliers are then informed, and improvement efforts are carried out together with the platform. This process is not carried out by recording on forms such as customer complaint forms and using corrective-preventive action forms for solutions. In the future, the platform may establish an ISO 9001 Quality Management System to ensure that this process is carried out more professionally.

Finance and payment services process: Payments for customer orders are realized through the digital payment infrastructure provided by the technology service provider. Processes such as delivery note entries, invoice issuance, VAT payments, and the issuance of contracts for the sale of products to customers electronically (for example, the platform's website includes a customer contract, Hayat Hatay, n.d.) are carried out by the platform in accordance with Turkish legislation and payments are tracked. Platform suppliers are paid by deducting a certain percentage of commission from their sales on a certain day of the month. Since the platform was established for social purposes, commissions are set at a minimum level to cover only the platform's expenses. The commission fee is set equally for all suppliers and is specified in the contracts made with the supplier during the process of including the supplier in the platform, taking into account the principle of transparency.

Data analytics and decision support processes: Data analytics processes are carried out on the platform, and data-based decisions are made. The data generated as a result of the activities carried out within the scope of the platform-based supply chain are collected in the platform's software. While data ownership belongs to the platform administrators, access rights and control of data are limited to the platform administrators and the technology service provider. Suppliers can access only their own product-related inventory, sales, and payment data using a unique username and password. This restricted access ensures the protection of other suppliers' and customers' information and aligns with principles of fairness and transparency in data governance. Additionally, considering the limited digital literacy among some suppliers and the post-earthquake disruptions to internet infrastructure in the city (HMKÜ, 2024, p. 49), mobile notifications are also used to provide necessary updates.

The data collected through the platform is analyzed and then shared with suppliers to support their production planning, inventory management, and customer relationship management processes. However, a high level of dependence on the technology service provider has been observed in the data analysis process. The analyses are carried out by the technology provider and then communicated to the platform managers, who subsequently disseminate the findings to suppliers. One contributing factor to this dependency is that platform managers are encountering such systems for the first time and lack sufficient technical knowledge about data-related processes. To address this issue, the technology provider offers training to platform managers, who in turn make efforts to understand and manage the processes themselves. As the platform grows, recruiting staff with experience in data analytics may help reduce this reliance. There are policies and protocols in place to ensure data security on the client side of the platform, and clients can access this information through the platform's website. However, the lack of data exchange between different platforms in the country and the lack of common use of information on products cause local suppliers selling on different digital platforms to have limited control over data. To address this issue, the use of open standards across platforms and integration with decentralized data storage and sharing tools are recommended (Glaros et al., 2023, pp. 2-11).

Performance evaluation process: At present, only a limited set of performance indicators is being monitored. For instance, the delivery time of a product to the customer is used as a performance metric; the delivery process is tracked, and products are delivered within a maximum of three days. Additionally, the breakage rate of cargo packages during delivery is monitored, with the reported rate being below 0.004%. Another performance indicator used on the platform is the growth rate in the number of suppliers. Supplier performance is also tracked on a product-specific basis, such as identifying which products from which suppliers are sold the most. However, it is essential to expand these performance metrics to cover the entire range of operational processes more comprehensively.

3.2.4. Assessment of Hayat Hatay Platform-Based Supply Chain and Strategic Directions

Under this section, the current status of the Hayat Hatay Platform-Based Supply Chain is evaluated in light of the relevant literature, along with its strategic directions.

The Hayat Hatay Platform-Based Supply Chain has projects for growth. Shared supply chain, one of the popular topics in supply chain in recent years (Sun, Xu, & Liu, 2021, p. 1), is one of these projects. Although small in scale, the platform, which plays the role of a shared supply chain, already provides packaging for pulses produced in the city by women's cooperatives. At the center of the platform, a production line is being established to enable local producers to process olives and produce olive oil. With the inclusion of a shared supply chain structure, the income generated from local producers for the use of the production facility will contribute to the financial sustainability of the platform.

In ensuring financial sustainability, large-scale platform-based supply chains operating in the country should be taken into account. Characteristics such as the number and loyalty of customers, number of suppliers, diversity of products, mobile accessibility, services provided to suppliers, and payment security increase the attractiveness of such large-scale platforms, while their bargaining power can reduce transaction costs (O'Brien, 2014; Tang & Lei, 2016, pp. 70–72; Yu-Wei, Gui-Hua, & Chen, 2024, p. 11). In such cases, it is essential to emphasize social responsibility and to enhance collaborations with local governments, ministries, NGOs, other cooperatives acting as suppliers, B2B retailers, and other digital platforms.

In order to ensure geographical scalability of the platform, the model is being promoted at the national level, and its applicability is being discussed. It is envisaged that the platform can be replicated if the public provides support and incentives during the initial phase of the platform, such as conducting the integration process and allowing the use of idle facilities in cities as storage centers. The main reasons for the need for such support and incentives are the lack of knowledge of local suppliers about digital processes and the lack of financial means. As a matter of fact, Paciarotti et al. (2022, p. 178) also mentioned similar inadequacies and stated that the support of policymakers is needed to establish local food hubs and ensure their sustainability. For this reason, the platform should be supported by public resources until it becomes operational, and these resources should not be cut. It is possible to establish such platform-based supply chains through public and private sector cooperation. For example, during the pandemic, an e-commerce platform was established in China for agricultural and food products, where the products of local producers were sold through both private and public initiatives, and local administrators supported local producers by participating in product promotion activities (Guo et al., 2021, p. 446). In addition, the study found that the utilization of state funds is among the most effective solutions to increase the number of local suppliers on the platform and to ensure the growth of the supply chain, thereby benefiting consumers shopping on the platform, and ensuring the platform's sustainability even if state support is withdrawn (Wang et al., 2022, p. 337).

In the future, new technologies can be incorporated into the processes to develop and incentivize the platform and local suppliers, and improve the performance of the platform-based supply chain (Cimino et al., 2024). For example, when blockchain technology is included in the platform, transparency and traceability of the supply chain can be ensured. The customer can track the product produced, production conditions, the labor of women in the chain, the working conditions of women, and the fairness of trade (Xu, He, Zhou, & Cheng, 2023; Zhao, 2025, pp. 613-615). These advantages of blockchain technology are also an indicator of an ethical supply chain (Ko, Lee, Park, & Ryu, 2023). This indicator can become a decisive factor for customers to choose the platform. Today, large-scale platforms such as Amazon, JD. Com and local platforms such as Taobao villages have applications such as uploading product videos to the platform and live streaming services (Tian, Wang, & Xu, 2024, p. 351). In Hayat Hatay Platform-Based Supply Chain, videos of local suppliers' production stages are available on the platform's social media account (HayatHataytr, n.d.). Similar technologies can be incorporated into the platform. By monitoring the production sites of local suppliers and the center of the platform with virtual reality technology, both sides will be able to make gains. The sales of the local supplier through different sales channels may also increase due to the interesting technology. The virtual marketplace of the platform can be created, allowing the customer to visit the area where local products are located. Augmented reality technology can provide information about the products (Ma, Li, Hu, & Wang, 2024, p. 2; Zhang, Li, & Tayi, 2023, p. 1). However, in order to incorporate these technologies into the platform-based supply chain, training and infrastructure support are required to improve the technical capacity of both platform managers and local suppliers.

One of the issues to be addressed in order to ensure the sustainability of the Hayat Hatay Platform-Based Supply Chain structure is to determine key performance indicators for monitoring the effectiveness of the processes of the platform and to make measurements accordingly. Otherwise, it will not be possible to measure the real success of the platform and express it with indicators. Using the Balanced Scorecard approach (Kaplan & Norton, 1996), performance measurements can be made monthly, semi-annually, and annually. The indicators to be determined in this way will enable the measurement of the efficiency of the platform and will be a preliminary preparation for the transition to the ISO 9001 Quality Management System. Similarly, a supplier performance evaluation process should be defined, and performance criteria should be established for local suppliers.

3.3. Integration Process of Women's Cooperatives into Hayat Hatay Platform-Based Supply Chain

Supply chain integration is defined as the cooperative and simultaneous execution of chain activities by actors in a supply chain (Qi, Wang, Zhang, & Wang, 2023, p. 477). The integration of women's cooperatives into the Hayat Hatay platform was coordinated and supported by the Hatay Governorship. The integration was carried out by taking into account the existing problems of women's cooperatives in Hatay. Since the integration process would also affect the sustainability of the platform, the process was carried out meticulously.

The first step in the integration process was comprehensive ***training*** for women's cooperatives. In this process, which can be defined as supplier development, women's cooperatives were trained on the functioning of the platform, legal and commercial requirements for selling on the platform (e.g. preparing delivery notes, invoicing, etc.), sales and marketing, product development, standardization in production, food analysis, quality control, and the preparation of product labels and barcodes. The platform also continues to provide periodic training to support the development of women's cooperatives.

After the first phase, women's cooperatives were asked to prepare ***their products*** for sale on the platform. They were asked to determine the products they would offer for sale on the platform and the prices of these products. One of the factors affecting the sale of local products is the lack of standards for the products produced. In order to sell products through the platform, the packaging, volume, quality, and content of the products must be standardized, and food products must comply with food safety and regulations (Abatekassa & Peterson, 2011, p. 74). In this process, many women's cooperatives were found to be lacking. For this reason, women's cooperatives and the platform have taken steps to standardize their products. However, a review of the platform's website (Hayat Hatay, n.d.) shows that standardization in packaging has not been fully achieved. An agreement was made with the university in the city for food analysis, and food analysis was carried out on behalf of the women's cooperatives. Women's cooperatives were asked to prepare the labels of their products in accordance with the Turkish Food Codex Regulation on Food Labeling and Consumer Information (Republic of Türkiye, Presidency of the Legislation Information System, 2017), and after the legal procedures were completed, the products were transferred to the platform's digital environment. In this step, it was observed that many of the women's cooperatives did not have sufficient information, so a studio was set up to take photos and promotional videos of the products and transfer them to the digital platform. The same visual materials are used for social media accounts and advertising. In addition, women's cooperatives were taught how to add new products to the platform in the future.

The third phase focused on the integration of logistics processes. In the inbound logistics phase, women's cooperatives were initially supported to transport the products to be sold to the platform's headquarters, and the transportation processes were carried out by the platform. Later on, women's cooperatives started to bring the products to the center of the platform with their own resources. While the location of the platform's logistics center is advantageous for the platform management and women's cooperatives, as it is an exit point from the city and is within the borders of the same city compared to other platforms, the logistics center is far away for women's cooperatives located in different districts. Although the highway infrastructure in the city was damaged after the disaster, the fact that the main connection roads are mostly used in the city causes traffic congestion (HMKÜ, 2024, pp. 49-50), which results in a longer transportation time for products to the logistics center. This can lead to higher transportation costs and lower income for women's cooperatives as suppliers. As a solution, it is recommended that women's cooperatives base their stock quantities on sales figures, or that the platform periodically transports products to the platform's headquarters.

In the final stage of the process, women's cooperatives were integrated into the platform's software infrastructure, and the integration process was completed. At this stage, there were cases where there were problems with digital literacy and the use of technology, and women's cooperatives were provided with training. Although there are some cases where integrating into platforms poses challenges for suppliers, from the perspective of women's cooperatives, integration into digital platforms allows them to benefit from the platform's competencies by outsourcing rather than transferring technology or knowledge to specialize in processes (Culotta et al., 2024, p. 464). In addition, the power of the platform (advertising, fairs, festivals, support for social responsibility projects by local governments and corporate firms, implementation and support by the public sector, bargaining power, and operation by women's cooperatives) is thought to have motivated suppliers to join the platform.

However, in the long run, the risk of dependency associated with this utilization should not be ignored (Sindakis, Showkat, & Su, 2023). At this stage, an analysis of the women's cooperatives on the platform (Hayat Hatay, n.d.) shows that they use different sales channels other than the platform (e.g., Hazek Women's Cooperative, n.d.; Samandağ Women's Cooperative, n.d.) and carry out activities to develop their internal capacities. The presence of faith and gastronomy tourism in the city (Çetinsöz, Polat, & Yakıcı, 2019) provides the opportunity to sell through physical sales channels. They should continue to diversify their sales channels and carry out promotional activities outside the platform against problems that may occur on the platform (for example, increasing commission rates, forgetting the effects of the earthquake, ending public support, etc.). Utilizing the knowledge and skills acquired during the platform integration process to continue selling their products through various sales channels—such as other digital platforms and their own websites—is an important strategy for ensuring their sustainability and reducing dependency on the platform.

3.4. Contributions of Integration to Sustainability

Women are seen as "active actors" in ensuring sustainability (Women Watch, 2012). The integration of women's cooperatives as local supply chain actors into platform-based supply chains in the post-disaster period has many contributions in terms of sustainability dimensions. The following paragraphs outline these contributions in terms of sustainability dimensions.

Economic losses caused by disasters include different components (Hallegatte & Przyluski, 2010), and the economic well-being of the affected group affects the recovery process (Markhvida, Walsh, Hallegatte, & Baker, 2020, p. 538). Women in vulnerable and disadvantaged groups with low economic welfare levels may suffer large-scale losses after disasters and may take longer to recover. Therefore, it is recommended to support women's cooperative activities in post-disaster recovery processes (Enarson, 2000, pp. 9-30; International Labour Organization, 2021). In Hatay, women's cooperatives contributed to the revitalization of the city's economy during post-earthquake recovery (Yıldızcan, 2023, p. 26). Integration into the platform supported families to generate income after the earthquake. Some cooperatives may cease their activities due to financial problems, and participation in the platform supply chain will contribute to the economic sustainability of their activities as it will expand their market share (İştar Işıklı, 2024, p. 2281; Ünsalan, 2024, p. 2107). In Hatay, there are women's cooperatives that were newly established after the earthquake (Hatay Governorship, 2025). Promotional mechanisms are needed to ensure the visibility of these cooperatives. As the platform is also a social responsibility project, there are public relations efforts to publicize the platform (e.g., Munyar, 2024). If the platform's website includes products and comments on the products, this would reduce the cost of publicity for women's cooperatives.

The platform's warehousing, cheaper last-mile delivery, and the obligation to standardize products benefit the economic sustainability of cooperatives. Integration into such platforms allows not only local producers but also local suppliers and transporters in the city to continue their activities, especially in the post-disaster recovery process. At the same time, integration also serves sustainable development goals such as ending poverty, ending hunger, decent work and economic growth, responsible production and consumption (Suseno & Nataliningsih, 2023, p. 51; Tümer, 2024, p. 1030; Yang, Jiang, Chen, & Jia, 2024, p. 114).

The integration into the platform supports women in acquiring vocational skills by utilizing their handicrafts and local knowledge (UNDP Türkiye, 2022, p. 16.), contributes to increasing female employment, reducing inequality, and advancing sustainable development goals such as gender equality (Ozdemir, 2013, p. 300; Tümer, 2024, p. 1030). It also contributes to the individual development of women with the training provided during the realization of the platform. Since the establishment of the platform, the number of women included in production has increased, and 600 women's cooperative partners have been brought into production. At the outset, 40 women's cooperatives were involved in the platform (Munyar, 2024), and as of the date of the interview, 53 women's cooperatives were involved as sellers. Furthermore, in B2B bulk purchases, women's cooperatives employ temporary female workers to fulfill orders. Integration enables women to participate in various stages of the supply chain (e.g., order preparation within the platform). Thereby contributing to individual and community recovery, both psychologically and economically.

During post-disaster recovery, many individuals and businesses in the country purchase the products of women's cooperatives through the platform, contributing to social solidarity and the economic sustainability of the cooperatives (Suhara & Muttaqien, 2025). Collaborating with the platform within the scope of B2B supports the social sustainability of retailers and corporate customers. At the same time, it contributes to the sustainability of the city itself and the continuity of its traditional products and culture (Işıklı İştar, 2024, p. 2285).

Through the platform, products purchased by the same customer from different women's cooperatives are sent to the customer in a single transportation activity. Compared to the dropshipping model, this helps to reduce energy consumption and CO_2 emissions, while reducing the amount of packaging materials to be used for packaging (Edwards, McKinnon, & Cullinane, 2010, p.115; Pilati & Tronconi, 2024). However, there are environmental problems arising from the individual delivery of orders one by one (Sindakis et al., 2023, p. 6). In order to reduce the number and costs of transportation, the 3PL's orders are delivered in bulk from the platform center at certain periods. In addition, the packaging used by the platform is selected from recyclable cardboard and plastic packaging materials suitable for food transportation, which ensures that the orders reach the customer without damage. In the future, the use of biodegradable packaging instead of plastic packaging materials will contribute to environmental sustainability (Ecker & Strüver, 2022).

At the center of the platform, in line with the shared supply chain structure, some products are directed directly from production to sales, thereby reducing the environmental damage caused by transportation. With the increasing demand for local and organic products in recent years, products can be delivered to a wide variety of locations through platforms (Wang et al., 2022). This contributes to environmentally friendly production activities (Egiad Yarın, 2024). The platform supply chain also serves sustainable development goals such as sustainable cities and communities (Tümer, 2024, p. 1030). Although there is no measurement mechanism on how much the platform reduces energy consumption or CO_2 emissions, opinions have been expressed within the scope of the observations made and the functioning of the process. In order to institutionalize the platform and measure its environmental performance, it is recommended that the platform align its processes with the ISO

14001 Environmental Management System and obtain the management system certificate. At this stage, it will not be sufficient for the platform to act in accordance with the system. Local suppliers should also be made aware of environmental issues, and training on circular economy, sustainability, waste management and recycling should be provided to suppliers.

4. CONCLUSION AND RECOMMENDATIONS

In this study, the strengthening of local supply chains through platform-based supply chains is discussed through a systematic literature review and case study. The study reveals that a platform-based supply chain, initially established with government support in the post-disaster recovery process, can strengthen local supply chains in a disaster-affected city. This strengthening begins from the integration process into the platform and contributes to all three dimensions of sustainability. Based on the findings, the following recommendations are made for policymakers, platform-based supply chains, local supply chains, women's cooperatives, private sector managers, and academics:

In the post-disaster period, sustaining and strengthening the activities of local supply chains plays a crucial role in the development of the disaster-affected area (Özenir, 2024, p. 142). In response to disasters and/or to support local production, policymakers are advised to develop and implement platform-based supply chain models such as Hayat Hatay. This approach helps local suppliers and producers adapt to the digital age and establish alternative sales channels. Incentives should be provided not only to women's cooperatives but also to local producers in other sectors to facilitate their integration into the platforms. Alternatively, as in the case of Hayat Hatay, the integration process can be managed by the state. It is also recommended that local producers be informed about national and international support available for purchasing technology and equipment to enhance their capabilities, that the procedures to access such support be simplified, and that assistance be provided during the preparatory phase.

One of the important issues here is to ensure the sustainability of platforms such as the Hayat Hatay platform. Instead of being a temporary solution in the post-disaster recovery process, activities should be carried out to ensure the permanence of the platform. Platform managers are recommended to carry out activities to encourage participation in the platform. It is also recommended to support local development by minimizing the commission rates received from local producers on the platforms. This will also contribute to the social sustainability of the platform. It is also recommended that such platforms, where sellers are composed of local suppliers, should be supported by the private sector, and these platforms should be preferred as suppliers (Paciarotti et al., 2022, p. 176). It is stated that platforms have the ability to make the chain resilient against interruptions that may occur in the supply chain and to direct suppliers to ensure quality and safety in the supply chain (Suali et al., 2024, p. 586).

When the literature is examined, it is revealed that despite increasing efforts toward women's empowerment across different regions (Andrew et al., 2024), there have been some developments in women's inclusion in the supply chain from past years to the present; however, sufficient progress has not been achieved. It has been observed that the challenges faced by women living in underdeveloped and developing countries have remained largely unchanged (Ankara Development Agency, 2024). Women's cooperatives, as one of the key actors in local supply chains, should continue to receive support from the state, citizens, the private sector, the public sector, and other cooperatives to ensure social sustainability in society. Moreover, state support aimed at including women in the production process should be increased.

In terms of local supply chains where women's cooperatives are producers, it is clear that more professional steps need to be taken throughout the supply chain, from production to distribution. While this may disrupt the traditional nature of the production of local products, it is necessary to mitigate the effects of this and ensure that local supply chains develop towards digitalization and increased production capacity. At this stage, based on the quantity of orders and the quality of production of women's cooperatives, batch production can be introduced for certain products. The managers of women's cooperatives should provide their members with continuous training on digital literacy, cost calculation, and ensuring the quality and standardization of production, and women and women's cooperatives should be encouraged to improve themselves in these areas. Women and women's cooperatives also need to improve themselves in areas such as determining stock quantities and demand forecasting, taking into account the ability to sell on different platforms. If digital literacy and technical infrastructure support are provided, a decentralized e-commerce platform can be established with the participation of women's cooperatives.

Academics should also be involved in the integration process and provide training to local producers on digitalization, food safety, and selling on e-commerce platforms to ensure local development. It is recommended that they take part in the process of adapting emerging technologies such as blockchain to platform-based supply chains consisting of local suppliers and observe in the field what can be done to solve the problems that local suppliers or small-scale platforms may face in adapting to these technologies. In addition, in future studies, it is recommended to determine the factors that determine the decision of local supply chain actors whether or not to participate in platform-based supply chains, to determine the selection criteria of platform-based supply chains, to conduct performance evaluations of platforms, to conduct literature reviews including other databases, to make detailed comparisons of similar platforms in different countries, to conduct risk analysis of platform-based supply chains, to analyze the balance of power between platforms and within the structure of the platform itself in platform-based supply chains.

The study has several limitations due to its subject and methodology. The literature review process was conducted by one researcher. The literature review is limited to the WoS and Scopus databases. In addition, the study has geographical limitations. The study is limited to the Hayat Hatay e-commerce platform, the women's cooperatives in Hatay province, and the geographical region they serve. Another limitation is the relatively small number of key performance indicators available on the platform included in the case study, and the lack of analysis on their measurement. Due to the platform's novelty and the platform managers' focus on its development and promotion, many performance indicators have yet to be defined, which has limited access to complete numerical data. For this reason, not all of the numerical data could be accessed. Due to these limitations, the generalizability of the results to other countries, regions, different cooperatives, and/or women's cooperatives is limited. Nevertheless, despite these limitations, the case study is expected to be beneficial for future research.

ACKNOWLEDGMENTS

Following the February 6 Kahramanmaraş Earthquake, I would like to thank the Governor of Hatay, Mr. Mustafa Masatlı, who introduced the "Hayat Hatay" project to the city and granted me the necessary permission to conduct this study. I also extend my gratitude to Mr. Mustafa Örgüt, Provincial Planning and Coordination Manager of Hatay Governorship, who generously shared his knowledge and experience about the platform's operations and supports the women who continue to live in the city.

REFERENCES

Abatekassa, G., & Peterson, H. C. (2011). Market access for local food through the conventional food supply chain. *The International Food and Agribusiness Management Review*, *14*(1), 63–82. DOI: 10.22004/ag.econ.100876

Adar, A. Ş., Dedeoğlu, S., & Kurtuluş, G. (2023). Türkiye'de kadın kooperatifleri: Mevcut durum analizi, kadın güçlenmesi ve istihdam yaratma potansiyeli. *Çalışma ve Toplum, 2*(77), 1174-1208. DOI: 10.54752/ct.1280833

Andrew, L., Yusuf, M. J., Bujang, A., Ibrahim, W. M. F. W., Hassan, S. S., Ali, J. K., & Yacob, Y. (2024). The state of women's cooperatives in Malaysia: Challenges and opportunities. *International Journal of Academic Reserach in Economics and Management Sciences*, *13*(1), 168–183. DOI: 10.6007/IJAREMS/v13-i1/20433

Ankara Development Agency. (2024). Kadın Kooperatifleri Çalıştayı (16 Mayıs 2024) Raporu. https://shorturl.at/k3OjW

Aromataris, E., & Riitano, D. (2014). Constructing a search strategy and searching for evidence. *The American Journal of Nursing*, *114*(5), 49–56. DOI: 10.1097/01.NAJ.0000446779.99522.f6 PMID: 24759479

Barut, Y. (2017). Women's cooperatives contributions to local region economy: Seferihisar agricultural cooperative development model-Turkey. *Management Studies*, *5*(2), 120–127.

Bharti, N. (2021). Role of cooperatives in economic empowerment of women: A review of Indian experiences. *World Journal of Entrepreneurship, Management and Sustainable Development*, *17*(4), 617–631. DOI: 10.1108/WJEMSD-07-2020-0095

Boz, N., & İrmiş, A. (2023). Bir Sosyal Girişimcilik Örneği Olarak Kadın Kooperatifleri. *Üçüncü Sektör Sosyal Ekonomi Dergisi, 58*(3), 2088-2124. DOI: 10.15659/3.sektor-sosyal-ekonomi.23.08.2159

Bretos, I., Díaz-Foncea, M., & Marcuello, C. (2018). Cooperativas e internacionalización: Un análisis de las 300 mayores cooperativas del mundo. *CIRIEC-España, Revista de Economía Pública. C.I.R.I.E.C. España, 92*(92), 5–37. DOI: 10.7203/CIRIEC-E.92.11480

Butollo, F. (2021). Digitalization and the geographies of production: Towards reshoring or global fragmentation? *Competition & Change, 25*(2), 259–278. DOI: 10.1177/1024529420918160

Cazgır, K. D. (2022). *Women's empowerment through cooperatives: the case of women's cooperatives in Izmir* [Unpublished master's thesis], Middle East Technical University (Turkey).

Çetinsöz, B. C., Polat, A. S., & Yakıcı, K. (2019). Hatay ili Hristiyanlarının dini bayramlarında gastronomi geleneklerinin inanç turizmi açısından değerlendirilmesi. *Gastroia: Journal of Gastronomy and Travel Research, 3*(4), 823–835. DOI: 10.32958/gastoria.584503

Chatterjee, S. A., & Bernzen, A. (2022). What drives the creation of nested markets? A qualitative case study of food markets in West Bengal, India. *DIE ERDE–Journal of the Geographical Society of Berlin, 153*(1), 1–14. DOI: 10.12854/erde-2022-571

Cimino, A., Longo, F., Solina, V., & Verteramo, S. (2024). A multi-actor ICT platform for increasing sustainability and resilience of small-scale farmers after pandemic crisis. *British Food Journal, 126*(5), 1870–1886. DOI: 10.1108/BFJ-01-2023-0049

Cramer, F., & Fikar, C. (2024). Investigating crowd logistics platform operations for local food distribution. *International Journal of Retail & Distribution Management, 52*(9), 836–855. DOI: 10.1108/IJRDM-10-2022-0400

Culotta, C., Blome, C., & Henke, M. (2024). Theories of digital platforms for supply chain management: A systematic literature review. *International Journal of Physical Distribution & Logistics Management, 54*(5), 449–475. DOI: 10.1108/IJPDLM-01-2023-0016

Datta, P. B., & Gailey, R. (2012). Empowering women through social entrepreneurship: Case study of a women's cooperative in India. *Entrepreneurship Theory and Practice, 36*(3), 569–587. DOI: 10.1111/j.1540-6520.2012.00505.x

Demirdağ, G. E., & Aydoğan, E. (2025). Görünür olmak ya da olmamak: kadın kooperatiflerinin yeni medya kullanımı analizi. *Üçüncü Sektör Sosyal Ekonomi Dergisi, 60*(1), 62-75. DOI: 10.15659/3.sektor-sosyal-ekonomi.25.01.2534

Demirel, N. (2022). Küresel tedarik zinciri yönetimi. In Öcal, O., & Han, V. (Eds.), *Dış ticarette güncel tartışmalar Türkiye üzerine bir inceleme* (pp. 89–115). Çizgi Kitabevi.

Dong, S., Wang, N., Fan, C., Chen, S., & Zhang, L. (2024). E-commerce and rural women entrepreneurship-Based on the quasi-natural experiment of "comprehensive demonstration policy" for e-commerce in rural areas. *Economic Analysis and Policy, 83*, 749–765. DOI: 10.1016/j.eap.2024.07.018

Duguid, F., Durutaş, G., & Wodzicki, M. (2015). *Türkiye'de kadın kooperatifleri'nin mevcut durumu.* https://www.aile.gov.tr/KSGM/PDF/Turkiyede_Kooperatifcilik_Kadin.pdf

Ecker, Y., & Strüver, A. (2022). Towards alternative platform futures in post-pandemic cities? A case study on platformization and changing socio-spatial relations in on-demand food delivery. *Digital Geography and Society, 3*(100032), 1–11. DOI: 10.1016/j.diggeo.2022.100032

Edwards, J. B., McKinnon, A. C., & Cullinane, S. L. (2010). Comparative analysis of the carbon footprints of conventional and online retailing: A "last mile" perspective. *International Journal of Physical Distribution & Logistics Management, 40*(1/2), 103–123. DOI: 10.1108/09600031011018055

Yarın, E. (2024, September 11). Bölgesel kalkınmanın güç kaynağı, kadın kooperatifleri. https://egiadyarin.com/bolgesel-kalkinmanin-guc-kaynagi-kadin-kooperatifleri/

Enarson, E. (2000). *Gender and Natural Disasters.* InFocus Programme on Crisis Response and Reconstruction Working Paper 1. International Labour Organization. https://www.ilo.org/publications/gender-and-natural-disasters

Engelseth, P., & Hogset, H. (2016). Adapting supply chain management for local foods logistics. *In Proceedings in System Dynamics and Innovation in Food Networks 2016* (pp. 143–160). DOI: 10.18461/pfsd.2016.1619

Gawer, A. (2021). Digital platforms' boundaries: The interplay of firm scope, platform sides, and digital interfaces. *Long Range Planning, 54*(5), 102045. DOI: 10.1016/j.lrp.2020.102045

Glaros, A., Thomas, D., Nost, E., Nelson, E., & Schumilas, T. (2023). Digital technologies in local agri-food systems: Opportunities for a more interoperable digital farmgate sector. *Frontiers in Sustainability*, *4*(1073873), 1–14. DOI: 10.3389/frsus.2023.1073873

Gore, A. (2024). Global supply chains retrocede to local supply chains: contemplation or a riposte? In *Supply Chain Forum: An International Journal,* 1-11. DOI: 10.1080/16258312.2024.2432857

Gruner, R. L., Orazi, D., & Power, D. (2013). Global versus local: An exploration on how vertical farms can lead the way to more sustainable supply chains. *IEEE Engineering Management Review*, *41*(2), 23–29. DOI: 10.1109/EMR.2013.6601102

Gündüz, F. F. (2022). Determining The e-commerce perception of women's cooperatives. *Third Sector Social Economic Review*, *57*(3), 1701–1718. DOI: 10.15659/3.sektor-sosyal-ekonomi.22.07.1900

Guo, H., Liu, Y., Shi, X., & Chen, K. Z. (2021). The role of e-commerce in the urban food system under COVID-19: Lessons from China. *China Agricultural Economic Review*, *13*(2), 436–455. DOI: 10.1108/CAER-06-2020-0146

Guo, X., & He, Y. (2022). Mathematical modeling and optimization of platform service supply chains: A literature review. *Mathematics*, *10*(22), 4307. DOI: 10.3390/math10224307

Hallegatte, S., & Przyluski, V. (2010). The economics of natural disasters: concepts and methods. *World Bank policy research working paper No. 5507.* Available at *SSRN*: https://ssrn.com/abstract=1732386

Hatay Governorship. (2025). *Deprem Sonrası Hatay Faaliyet Raporu.* http://hatay.gov.tr/kurumlar/hatay.gov.tr/pop-up/faaliyet_raporu_17032025.pdf

Hayat Hatay. (n.d.) Hayat Hatay Official Website. https://www.hayathatay.com.tr/

HayatHataytr. (n.d.). Hayat Hatay's Official Instagram Account. https://www.instagram.com/hayathataytr/

Hazek Women's Cooperative. (n.d.). Hazek Women's Cooperative Official Instagram Account. Instagram. https://www.instagram.com/hazekkooperatifi/

HMKÜ. (2024). *Dirençli Kentler İçin Kırsal Kalkınma Çalıştayı/ Çalıştay Sonuç Raporu.* https://www.mku.edu.tr/files/e4bc8a08-486e-4172-945b-3d8bed844183.pdf

Hofmann, E., & Belin, O. (2011). *Supply chain finance solutions.* Springer Briefs in Business. DOI: 10.1007/978-3-642-17566-4

Indego Africa (n.d.). Indego Africa Official Website. https://indegoafrica.org/?srsltid=AfmBOoqyQ37cHzdljEGkkOOoTO3ic-t8zgfkyfwUoxHvd7TcItW77Fv3

International Labour Organization. (2021, October 13). *Strengthening Cooperatives as a Response to the Pandemic" Project supports 146 women entrepreneurs in 6 women's cooperatives.* https://www.ilo.org/resource/news/strengthening-cooperatives-response-pandemic-project-supports-146-women

İştar Işıklı, E. (2024). Dijital çağda geleneksel zanaatların yaşatılmasında kadın kooperatifleri. *Motif Academy Journal of Folklore, 17*(48), 2281–2299. DOI: 10.12981/mahder.1533928

Kabaili, H., Sekkat, G., El Firdoussi, S., Chairi, I., Dahab, D., & Essakalli, M. (2024). The impact of ICTs on women's cooperatives in Morocco: An empirical analysis. [IJSMES]. *International Journal of Strategic Management and Economic Studies, 3*(4), 1402–1426. DOI: 10.5281/zenodo.13765137

Kaplan, R. S., & Norton, D. P. (1996). Strategic learning & the balanced scorecard. *Strategy and Leadership, 24*(5), 18–24. DOI: 10.1108/eb054566

Karaköse, A., Hepyalçın, C., Ozbiltekin-Pala, M., & Karaca, G. B. (2022, October). Women's cooperatives collection center location selection. In *The International Symposium for Production Research* (pp. 855-864). Springer International Publishing. DOI: 10.1007/978-3-031-24457-5_68

Katchova, A. L., & Woods, T. A. (2013). Local foods and food cooperatives: ethics, economics and competition issues. In *The ethics and economics of agrifood competition* (pp. 227–242). Springer., DOI: 10.1007/978-94-007-6274-9_12

Kittipanya-ngam, P., Shi, Y., & Gregory, M. J. (2011). Exploring geographical dispersion in Thailand-based food supply chain (FSC). *Benchmarking, 18*(6), 802–833. DOI: 10.1108/14635771111180716

Ko, T., Lee, J., Park, D., & Ryu, D. (2023). Supply chain transparency as a signal of ethical production. *MDE. Managerial and Decision Economics, 44*(3), 1565–1573. DOI: 10.1002/mde.3765

Kumar, K., & Kumar, A. (2024). Application of optimization models in sustainable supply chain management: A systematic review based on PRISMA Guidelines. *Process Integration and Optimization for Sustainability*, 1-28. DOI: 10.1007/s41660-024-00455-x

Liu, J., Zhou, L., & Wang, Y. (2021). Altruistic preference models of low-carbon e-commerce supply chain. *Mathematics, 9*(14), 1682, 1-20. DOI: 10.3390/math9141682

Ma, D., Li, K., Hu, J., & Wang, X. (2024). How to leverage blockchain to react to consumer reference effects and develop a distribution strategy in online retailing? *Transportation Research Part E, Logistics and Transportation Review*, *187*, 103579. DOI: 10.1016/j.tre.2024.103579

Markhvida, M., Walsh, B., Hallegatte, S., & Baker, J. (2020). Quantification of disaster impacts through household well-being losses. *Nature Sustainability*, *3*(7), 538–547. DOI: 10.1038/s41893-020-0508-7

McDougall, N., & Davis, A. (2024). The local supply chain during disruption: Establishing resilient networks for the future. *Journal of Cleaner Production*, *462*, 142743. DOI: 10.1016/j.jclepro.2024.142743

Mentzer, J. T., DeWitt, W., Keebler, J. S., Min, S., Nix, N. W., Smith, C. D., & Zacharia, Z. G. (2001). Defining supply chain management. *Journal of Business Logistics*, *22*(2), 1–25. DOI: 10.1002/j.2158-1592.2001.tb00001.x

Msosa, S. K. (2022). Challenges facing women cooperatives in accessing markets for agricultural products: A systematic literature review. *International Review of Management and Marketing*, *12*(6), 37–43. DOI: 10.32479/irmm.13420

Munyar, V. (2024, February 23). Hayat Hatay'da 40 kadın kooperatifini kucakladı 106 bin ürünü vitrine koydu. *Ekonomi Gazetesi*. https://www.ekonomigazetesi.com/kose-yazisi/hayathatayda-40-kadin-kooperatifini-kucakladi-106-bin-urunu-vitrine-koydu-31304

Nahıl. (n.d.). Nahıl Official Website. https://www.nahil.com.tr/

Nakandala, D., & Lau, H. C. W. (2019). Innovative adoption of hybrid supply chain strategies in urban local fresh food supply chain. *Supply Chain Management*, *24*(2), 241–255. DOI: 10.1108/SCM-09-2017-0287

O'Brien, M. (2014, March, 17). Competition, growth and customer experience from the Amazon perspective. Multi Channel Merchant. https://multichannelmerchant.com/must-reads/competition-growth-customer-experience-amazon-perspective/

Ozdemir, G. (2013). Women's cooperatives in Turkey. *Procedia: Social and Behavioral Sciences*, *81*, 300–305. DOI: 10.1016/j.sbspro.2013.06.432

Özenir, İ. (2024). Yerel tedarik zincirleri ve afetler: Yerel tedarik zincirlerinin önemi. *Journal of Transportation and Logistics*, *9*(1), 1–14. DOI: 10.26650/JTL.2024.1402924

Pachayappan, M., & Sundarakani, B. (2023). Drone delivery logistics model for on-demand hyperlocal market. *International Journal of Logistics*, *26*(12), 1728–1760. DOI: 10.1080/13675567.2022.2107189

Paciarotti, C., Mazzuto, G., Torregiani, F., & Fikar, C. (2022). Locally produced food for restaurants: A theoretical approach for the supply chain network design. *International Journal of Retail & Distribution Management*, *50*(13), 164–183. DOI: 10.1108/IJRDM-10-2021-0477

Page, M. J., McKenzie, J. E., Bossuyt, P. M., Boutron, I., Hoffmann, T. C., Mulrow, C. D., Shamseer, L., Tetzlaff, J. M., Akl, E. A., Brennan, S. E., Chou, R., Glanville, J., Grimshaw, J. M., Hróbjartsson, A., Lalu, M. M., Li, T., Loder, E. W., Mayo-Wilson, E., McDonald, S., & Moher, D. (2021). The PRISMA 2020 statement: An updated guideline for reporting systematic reviews. *Journal of Clinical Epidemiology*, *134*, 178–189. DOI: 10.1016/j.jclinepi.2021.03.001 PMID: 33789819

Pilati, F., & Tronconi, R. (2024). Tri-objective vehicle routing problem to optimize the distribution process of sustainable local e-commerce platforms. *Sustainability (Basel)*, *16*(5), 1810. DOI: 10.3390/su16051810

Polat, C., & Külter, B. (2007). Tüketicilerin perakende mağaza tercihini etkileyen faktörler: Niğde ili örneği. *Gazi Üniversitesi İktisadi ve İdari Bilimler Fakültesi Dergisi*, *9*(3), 109–126.

Qi, Y., Wang, X., Zhang, M., & Wang, Q. (2023). Developing supply chain resilience through integration: An empirical study on an e-commerce platform. *Journal of Operations Management*, *69*(3), 477–496. DOI: 10.1002/joom.1226

Republic of Türkiye, Ministry of Industry and Technology, General Directorate of Development Agencies. (2023). *Hatay Saha Raporu* https://seeco.org.tr/upload_cms/files/rapor/Hatay_Saha_Durum_Raporu-Nisan-2023_R1.pdf

Republic of Türkiye, Presidency of the Legislation Information System. (2017). *Turkish Food Codex Regulation on Food Labelling and Consumer Information*. https://www.mevzuat.gov.tr/mevzuat?MevzuatNo=23282&MevzuatTur=7&MevzuatTertip=5

Samandağ Women's Cooperative. (n.d.) Samandağ Women's Cooperative Official Instagram Account. Instagram. https://www.instagram.com/samandagkadinkooperatifi/?hl=tr

Shitu, S., & Popoola, O. M. J. (2017). Investigating socially sustainable behaviour of local actors in the supply chain of shea butter in Rural Borgu, Nigeria: From a role theory perspective. *Indian-Pacific Journal of Accounting and Finance, 1*(1), 3–16. DOI: 10.52962/ipjaf.2017.1.1.2

Sindakis, S., Showkat, S., & Su, J. (2023). Unveiling the influence: Exploring the impact of interrelationships among e-commerce supply chain members on supply chain sustainability. *Sustainability, 15*(24), 16642, 1-27. DOI: 10.3390/su152416642

Suali, A. S., Srai, J. S., & Tsolakis, N. (2024). The role of digital platforms in e-commerce food supply chain resilience under exogenous disruptions. *Supply Chain Management, 29*(3), 573–601. DOI: 10.1108/SCM-02-2023-0064

Suhara, A., & Muttaqien, Z. (2025). Adaptation strategies in supply chains based on local wisdom in the manufacturing industry. *International Journal for Science Review, 2*(4), 1–13. DOI: 10.71364/ijfsr.v2i4.29

Sun, Z., Xu, Q., & Liu, J. (2021). Pricing and replenishment decisions for seasonal and nonseasonal products in a shared supply chain. *International Journal of Production Economics, 233*(108011), 1–17. DOI: 10.1016/j.ijpe.2020.108011

Suseno, G. P., & Natalinigsih, N. (2023). The role of women's cooperatives as an institution to empower women in supporting the achievement of the SDGs. *International Journal of Ethno-Sciences and Education Research, 3*(2), 45–53. DOI: 10.46336/ijeer.v3i2.418

Taleizadeh, A. A., Darabi, E. S., & Thaichon, P. (2024). Optimizing online selling through an online-to-offline platform: Strategic ramifications for local n stores. *Journal of Revenue and Pricing Management, 23*(6), 554–574. DOI: 10.1057/s41272-023-00448-4

Tang, D., & Lei, J. (2016). B2C e-commerce platform competition strategy research based on the network externality. *Management Science and Engineering, 10*(2), 70–76. DOI: 10.3968/8420 PMID: 35886272

Thilmany, D., Canales, E., Low, S. A., & Boys, K. (2021). Local food supply chain dynamics and resilience during COVID-19. *Applied Economic Perspectives and Policy, 43*(1), 86–104. DOI: 10.1002/aepp.13121

Tian, X., Wang, M., & Xu, Y. (2024). Investment strategies of information-provision technology in the platform-based supply chain. [NRL]. *Naval Research Logistics, 71*(3), 351–364. DOI: 10.1002/nav.22155

Trabucchi, D., & Buganza, T. (2020). Fostering digital platform innovation: From two to multi-sided platforms. *Creativity and Innovation Management, 29*(2), 345–358. DOI: 10.1111/caim.12320

Tümer, Ö. (2024). Niğde kadın emek pazarı projesi ve Niğde'de yerel yönetimlerin kadın girişimcilere verdiği desteklerin sürdürülebilir kalkınma amaçları kapsamında değerlendirilmesi. *Ömer Halisdemir Üniversitesi İktisadi Ve İdari Bilimler Fakültesi Dergisi, 17*(4), 1020-1035. DOI: 10.25287/ohuiibf.1539913

Türkiye, U. N. D. P. (2022). *Kadın Dostu Dijital Yol: Türkiye'de Turizm ve İmalat Sektörlerinde Dijitalleşme ve Kadın İş Gücü Araştırması Raporu.* https://www.undp.org/tr/turkiye/publications/kadin-dostu-dijital-yol

Ünsalan, E. (2024). Kadın girişimci kooperatiflerinin pazarlama stratejilerinin belirlenmesi: Amasya ili örneği. *Third Sector Social Economic Review, 59*(4), 2104–2129. DOI: 10.15659/3.sektor-sosyal-ekonomi.24.10.2476

Ugur-Cinar, M., Cinar, K., Onculer-Yayalar, E., & Akyuz, S. (2024). The political economy of women's cooperatives in Turkey: A social reproduction perspective. *Gender, Work and Organization, 31*(4), 1268–1289. DOI: 10.1111/gwao.12925

Upreti, B. R., Ghale, Y., Shivakoti, S., & Acharya, S. (2018). Feminization of agriculture in the Eastern Hills of Nepal: A study of women in cardamom and ginger farming. *SAGE Open, 8*(4), 2158244018817124. DOI: 10.1177/2158244018817124

Vakoufaris, H., Kizos, T., Spilanis, I., Koulouri, M., & Zacharaki, A. (2007). Women's cooperatives and their contribution to the local development of the North Aegean Region, Greece. *Journal of Rural Cooperation, 35*(1), 19–41. DOI: 10.22004/ag.econ.58681

Wang, L., Rabinovich, E., & Richards, T. J. (2022). Scalability in platforms for local groceries: An examination of indirect network economies. *Production and Operations Management, 31*(1), 318–340. DOI: 10.1111/poms.13536

Wang, P., Du, S., Hu, L., & Tang, W. (2023). Logistics choices in a platform supply chain: A co-opetitive perspective. *The Journal of the Operational Research Society, 74*(1), 93–111. DOI: 10.1080/01605682.2021.2023675

Wei, R., & Pardo, C. (2024). Looking for resource density in the platform supply network. *Journal of Purchasing and Supply Management, 30*(3), 100938. DOI: 10.1016/j.pursup.2024.100938

Wei, Y. D., Lin, J., & Zhang, L. (2020). E-commerce, Taobao villages and regional development in China. *Geographical Review, 110*(3), 380–405. DOI: 10.1111/gere.12367

Women Watch. (2012). *Empower Rural Women-End Hunger and Poverty*. https://www.un.org/womenwatch/feature/ruralwomen/index.html

Xie, N., He, H., & Tong, Y. (2022). Game model of green financial supply chain based on government subsidy analysis. *IEEE Access : Practical Innovations, Open Solutions, 10*, 60929–60945. DOI: 10.1109/ACCESS.2022.3181044

Xu, H., Zhang, Y., & Liu, Y. (2024). Digital platform-based domestic supply chain: International comptitiveness and government regulation. *Economic Theory and Business Management, 44*(2), 133.

Xu, S., Govindan, K., Wang, W., & Yang, W. (2024). Supply chain management under cap-and-trade regulation: A literature review and research opportunities. *International Journal of Production Economics, 109199*, 109199. Advance online publication. DOI: 10.1016/j.ijpe.2024.109199

Xu, X., He, P., Zhou, L., & Cheng, T. C. E. (2023). Coordination of a platform-based supply chain in the marketplace or reselling mode considering cross-channel effect and blockchain technology. *European Journal of Operational Research, 309*(1), 170–187. DOI: 10.1016/j.ejor.2023.01.057

Yang, X., Jiang, Y., Chen, X., & Jia, F. (2024). ICT-empowered rural e-commerce development in China: An adaptive structuration perspective. *International Journal of Technology Management, 95*(1-2), 90–119. DOI: 10.1504/IJTM.2024.137016

Yenilmez Gürsoy, Ö. B. (2022). Türkiye'de e-ticaret platformlarında satış yapan kırsal kadın kooperatiflerinin incelemesi. *Beykoz Akademi Dergisi, 10*(1), 265–279. DOI: 10.14514/BYK.m.26515393.2022.10/1.265-279

Yeshiwas, Y., Alemayehu, M., & Adgo, E. (2024). Strategic mapping of onion supply chains: A comprehensive analysis of production and post-harvest processes in Northwest Ethiopia. *Frontiers in Sustainability, 5*(1387907), 1–13. DOI: 10.3389/frsus.2024.1387907

Yıldırım, A., & Şimşek, H. (2021). *Sosyal bilimlerde nitel araştırma yöntemleri*. Seçkin Yayıncılık.

Yıldızcan, C. (2023). Deprem sonrasında afetin ve tarımın politik ekolojisi: Hatay örneği. *PRAKSIS, 63*, 11–28.

Yu-Wei, L., Gui-Hua, L., & Chen, P. (2024). Settlement selection strategic analysis for self-operated e-commerce platforms under market competition. *Systems, 12*(8), 293, 1-22. DOI: 10.3390/systems12080293

Zamri, N. M., Andrew, L., Borhani, N., Latiff, S. F. A., & Mathew, D. H. (2024). Women's cooperatives in Turkey: Challenges, success stories, and policy recommendations. *International Journal of Academic Reserach in Economics and Management Sciences, 13*(3). Advance online publication. DOI: 10.6007/IJAREMS/v13-i3/22853

Zhang, T., Li, G., & Tayi, G. K. (2023). A strategic analysis of virtual showrooms deployment in online retail platforms. *Omega, 117*(102824), 1–12. DOI: 10.1016/j.omega.2022.102824

Zhang, Z., Garimella, A., & Fan, M. (2024). Leveraging the social fabric to ımprove rural e-commerce access. *Production and Operations Management*, 1–16. DOI: 10.1177/10591478231224974

Zhao, Y. (2025). Platform economy and global supply chain integration: A technology-driven approach to business model innovation. H. Luo, T. Yao, W. Cui & H. Li (Eds.), *Advances in Economics, Business and Management Research* (ss. 612–620). Atlantis Press. DOI: 10.2991/978-94-6463-642-0_65

Zhong, Y., Lai, I. K. W., Guo, F., & Tang, H. (2021). Research on government subsidy strategies for the development of agricultural products E-commerce. *Agriculture, 11*(11), 1152. DOI: 10.3390/agriculture11111152

Zhou, X., Liu, H., Li, J., Zhang, K., & Lev, B. (2023). Channel strategies when digital platforms emerge: A systematic literature review. *Omega, 120*(102919), 1–13. DOI: 10.1016/j.omega.2023.102919

ADDITIONAL READING

Abatekassa, G., & Peterson, H. C. (2011). Market access for local food through the conventional food supply chain. *The International Food and Agribusiness Management Review, 14*(1), 63–82. DOI: 10.22004/ag.econ.100876

Kabaili, H., Sekkat, G., El Firdoussi, S., Chairi, I., Dahab, D., & Essakalli, M. (2024). The impact of ICTs on women's cooperatives in Morocco: An empirical analysis. [IJSMES]. *International Journal of Strategic Management and Economic Studies, 3*(4), 1402–1426. DOI: 10.5281/zenodo.13765137

McDougall, N., & Davis, A. (2024). The local supply chain during disruption: Establishing resilient networks for the future. *Journal of Cleaner Production, 462*, 142743. DOI: 10.1016/j.jclepro.2024.142743

Mentzer, J. T., DeWitt, W., Keebler, J. S., Min, S., Nix, N. W., Smith, C. D., & Zacharia, Z. G. (2001). Defining supply chain management. *Journal of Business Logistics*, *22*(2), 1–25. DOI: 10.1002/j.2158-1592.2001.tb00001.x

Nakandala, D., & Lau, H. C. W. (2019). Innovative adoption of hybrid supply chain strategies in urban local fresh food supply chain. *Supply Chain Management*, *24*(2), 241–255. DOI: 10.1108/SCM-09-2017-0287

Paciarotti, C., Mazzuto, G., Torregiani, F., & Fikar, C. (2022). Locally produced food for restaurants: A theoretical approach for the supply chain network design. *International Journal of Retail & Distribution Management*, *50*(13), 164–183. DOI: 10.1108/IJRDM-10-2021-0477

KEY TERMS AND DEFINITIONS

Agency platform: A type of platform-based supply chain where the ownership of the supplier's goods is with the supplier and the platform serves as a marketplace.

Business-to-Business (B2B): The sale of goods or services from one business to another.

Business-to-Customer (B2C): The sale of goods or services from the business to the customer.

Local supply chain: Small-scale supply chains that are established, operate and serve within one or a few cities, villages or rural areas.

Platform-based supply chain: A supply chain model in which supply chain processes are carried out through a digital platform and the supplier delivers its products to its customers through a digital platform.

Reseller platform: A type of platform-based supply chain where the platform buys the supplier's products in bulk and sells them on its behalf.

Women's cooperatives: Structures whose founders and members are exclusively women, where women come together and organize for social and economic purposes.

ENDNOTE

1 The information presented from this section through to the conclusion and recommendations is based on an interview conducted with Mr. Mustafa Örgüt, the Provincial Planning and Coordination Manager of the Hatay Governorship, on May 15 2025 at the Hatay Governorship; field observations carried out at the platform headquarters on May 21 2025; as well as secondary data obtained

from the platform's website, its social media account, and various online news sources. All secondary data sources are listed in the references.

Chapter 12
Enhancing Environmental Performance Through Green Supply Chain Management Practices: An Empirical Study From Morocco

Imad Ait Lhassan
https://orcid.org/0000-0002-5429-8557
Laboratory of Organizational and Territorial Governance (LARGOT), National School of Applied Sciences of Tetouan, Abdelmalek Essaadi University, Morocco

Rihab Ezzaher
Higher School of Technology of Fez, Sidi Mohamed Ben Abdellah University, Morocco

ABSTRACT

The ecological impact of supply chains, especially greenhouse gas emissions and resource waste, is a growing concern for companies (Sanders, 2011). Global warming and pollution from supply chain activities pose risks to human health and the environment. As a result, companies are encouraged to adopt green supply chain management (GSCM) practices to meet regulations and improve performance. This study evaluates the role of GSCM in enhancing the environmental performance of industrial firms in Morocco. Based on data from 95 companies in the northern region and using structural equation modeling, the findings show that GSCM practices positively influence environmental performance. These practices go beyond compliance, acting as strategic tools to reduce ecological impact and increase competitiveness in environmentally aware markets.

DOI: 10.4018/979-8-3373-3012-9.ch012

1. INTRODUCTION

In today's world, marked by urgent environmental challenges such as climate change, biodiversity loss, and the depletion of natural resources, companies face increasing pressure to adopt sustainable and environmentally responsible strategies. The supply chain plays a pivotal role in this transition toward sustainability, as organizations strive to implement "green" management practices that reduce their environmental impact while sustaining competitiveness.

This study aims to investigate the contribution of green supply chain management (GSCM) practices to enhancing the environmental performance of industrial companies in Morocco. Specifically, it seeks to identify the key drivers and barriers influencing the adoption of these practices. While the benefits of green supply chains have been widely acknowledged, their practical application remains limited, especially in emerging economies. This leads to the central research question: To what extent do green supply chain practices improve the environmental performance of Moroccan industrial firms, and what challenges hinder their effective implementation?

Focusing on six fundamental GSCM practices: eco-design, green purchasing, green manufacturing, waste management, reverse logistics, and green distribution. This research draws on survey data collected from industrial companies in northern Morocco to explore the relationships between these practices and various indicators of environmental performance.

The chapter is structured as follows: The first section presents the theoretical framework, synthesizing key literature on green supply chains and environmental performance. The second section details the research methodology, including questionnaire design and data collection procedures. The third section reports and analyzes the empirical findings. Finally, the chapter concludes with a discussion of the results, study limitations, and suggestions for future research.

2. CONCEPTUAL FRAMEWORK

2.1 Supply Chain

The supply chain is a complex and ever-evolving system that covers every stage in transforming raw materials into finished products for end customers. It begins with the extraction or sourcing of raw materials and moves through various stages, including production, assembly, storage, transportation, and ultimately, distribution. Thus, it involves a wide range of interconnected processes and actors whose coordination is essential to ensure efficiency and customer satisfaction. According to Teng and Jaramillo (2006), the supply chain can be seen as a set of upstream and downstream links between firms particularly suppliers and customers. In other words, these links are not limited to basic transactions but rather represent strategic relationships that influence numerous operational activities, such as sourcing strategies, inventory management, and delivery systems. Consequently, the overall performance of a supply chain largely depends on how well these relationships are managed and synchronized. Furthermore, Mentzer et al. (2001) emphasize that supply chains involve the movement of various resources: products, services, finances, and information. The smooth flow of these elements among suppliers, manufacturers, distributors, and customers is essential for the supply chain to operate effectively. In particular, the flow of information is crucial for supporting real-time decisions, enhancing coordination, and boosting responsiveness to changing market needs.

As a result, in today's rapidly changing business environment, the evaluation of supply chains extends beyond traditional performance indicators such as cost and speed. Instead, organizations are increasingly concerned with agility, resilience, and sustainability (Alla et al., 2022). Therefore, supply chains are now viewed as strategic assets that can drive innovation, enhance competitiveness, and contribute to long-term value creation. Ultimately, the ability to manage the supply chain effectively and in an environmentally responsible way has become a crucial factor for success in modern industries.

2.2 Supply Chain Management (SCM)

Monczka et al. (2015) explain that Supply Chain Management (SCM) seeks to coordinate and manage the flow of materials across the entire chain taking into account different levels of suppliers and customers. It is crucial to manage this in order to ensure smooth operations, reduce expenses and optimize customer satisfaction. According to Carbone et al. (2012), SCM plays an increasingly crucial strategic role, enabling companies to adjust quickly to market changes and consumer expectations, while maximizing the efficiency of available equipment. In addition, effective supply chain management contributes to minimizing lead times and improving service levels, which are essential in highly competitive markets. Moreover, the growing complexity of global supply networks requires enhanced coordination and visibility across all supply chain partners. Consequently, organizations are increasingly adopting advanced technologies like real-time tracking, predictive analytics, and integrated platforms to enhance decision-making and boost their responsiveness (Ezekari et al., 2024).

Furthermore, in light of the current environmental and regulatory pressures, supply chain management is evolving beyond a purely operational function to embrace sustainability objectives. Hence, companies are increasingly incorporating green practices into their supply chain strategies, not only to comply with environmental standards but also to strengthen their brand image and respond to the expectations of socially conscious consumers.

2.3 Green Supply Chain Management (GSCM)

GSCM incorporates environmental responsibility into every stage of the supply chain. Its goal is to reduce the environmental footprint of products from design through to end-of-life. In this regard, according to Wang and Gupta (2011), it is possible to make supply chains truly green if they are integrated into all elements of the value chain. Green Supply Chain Management (GSCM) represents an evolution of traditional Supply Chain Management (SCM), incorporating environmental sensitivity, a focus on green productivity and minimizing environmental impact at every stage of the value chain. Thus, a green and sustainable Supply Chain integrates green initiatives into its various stages, from raw material sourcing, product manufacturing, storage, transportation and distribution. Moreover, as Dawei et al. (2015) point out, Green Supply Chain Management (GSCM) is an innovative approach that aligns with modern social development trends. It weaves together economic performance, environmental responsibility, and resource efficiency across all its activities from sourcing raw materials and components to manufacturing, packaging, distribution, retail, and recycling. Additionally, according to Sharma et al. (2017), GSCM encompasses innovative SCM techniques to reduce environmental impact and maximize economic benefits. Consequently, GSCM not only promotes sustainability but also creates competitive advantages for companies by aligning ecological goals with operational efficiency and long-term profitability.

2.4 Environmental Performance

To understand the term environmental performance, it is essential first to define the broader concept of performance itself. Indeed, this term has been interpreted in various ways since its initial use by managers. Initially, during the 1980s, corporate performance was mainly focused on three key aspects: objectives, resources, and performance (Gibert, 1980). These three aspects namely relevance, effectiveness, and efficiency are explicitly distinguished: effectiveness refers to whether the company is able to achieve its set objectives; relevance assesses whether the company possesses adequate resources to reach these objectives; while efficiency relates to the ability to accomplish these objectives at a reduced cost (Oubaouzine, 2019).

Building on this foundation, Desmazes and Lafontaine (2007) characterize environmental performance as "the quantifiable results of the environmental management system (EMS), in relation to the entity's control of its environmental elements, based on its environmental policy, goals and objectives, ecological balance sheets, green indicators, and environmental financing."

Furthermore, companies have a key role to play in enhancing their environmental performance by embracing green practices at every stage of their supply chains. Zhu et al. (2008) emphasize that implementing Green Supply Chain Management practices not only benefits the environment but also boosts overall company performance, including operational efficiency and financial results, by minimizing waste, increasing productivity, and strengthening the company's reputation.

In this context, Salburg (2000) defines environmental performance indicators (EPIs) as "qualitative and quantitative data that enable the effectiveness of the operational and management system to be evaluated." Similarly, according to ISO (1996), management EPIs measure the efforts of management to improve environmental performance by organizing and planning corrective actions, while operational EPIs evaluate environmental performance in terms of the consumption of materials, energy, and services (company inputs), as well as products, services, waste, emissions (company outputs), physical facilities, and logistics (company operations).

2.5 GSCM Practices

In order to implement GSCM, it is necessary for organizations to adopt a number of GSCM practices, which are guidelines for the environmental management of the Supply Chain. A number of studies have sought to identify in-house Green Supply Chain Management practices include internal environmental and quality management systems. Internal environmental management plays a vital role in enhancing a company's environmental performance (Zhu et al., 2008). Additionally, Zhu and Sarkis (2008) emphasize that quality management helps to streamline the implementation of GSCM.

Organizations can implement structured mechanisms to continuously enhance their environmental performance, notably through the ISO 14001 Environmental Management System (EMS) certification. According to Beamon (1999), Green Supply Chain Management (GSCM) practices support companies in optimizing closed-loop supply chains, which encompass the design, control, and operation of systems that maximize value creation across a product's life cycle.

Research has also addressed external factors, such as customers, suppliers, and even competitors, emphasizing the role of environmental considerations in supply chain practices. These include establishing environmental requirements for suppliers by integrating them into procurement specifications, encouraging the use of eco-friendly packaging (reusable, biodegradable, and safe), and selecting suppliers based on environmental criteria. Audits can be conducted to assess suppliers' internal environmental management systems. Internally, companies can implement green production planning, including the scheduling of production modules and resource allocation aligned with environmental strategies.

Infrastructure also plays a role, with the development of environmentally friendly warehouses and buildings to support sustainable activities. Packaging and related documentation can be optimized to simplify the use and return of reusable packaging. Distribution strategies can be refined by better planning delivery routes for materials and components, while efforts in product resale and market creation for reused items further enhance sustainability. Additionally, promoting the return, reuse, and recycling of packaging, along with the inspection and classification of production and consumption waste, simplifies subsequent operations. Finally, mapping hazardous, toxic, and harmful substances through detailed product listings contributes to safer and more responsible supply chain management.

2.6 Environmental Performance Dashboards and Indicators

The methods for implementing and evaluating environmental performance indicators are developed based on established practices and guidelines recommended by international standards, particularly the ISO 14000 series.

Companies implement Environmental Management Systems (EMS) to structure their environmental management strategies. These systems often rely on the ISO 14001 standard, which guides organizations in formulating an environmental policy, setting objectives, and establishing procedures to achieve those objectives while continuously improving their environmental performance.

To assess environmental performance (EP) within an organization, the voluntary standard NF EN ISO 14031 (AFNOR Group) recommends the use of three types of indicators. Environmental Performance Indicators (EPIs) are divided into two subcategories: Environmental Management Performance Indicators (EMPIs), which measure the initiatives taken by management to improve environmental performance (e.g., the number of staff training hours on environmental topics or the achievement rate of environmental objectives), and Operational Performance Indicators (OPIs), which reflect the direct environmental impact of the organization's activities.

Environmental Condition Indicators (ECIs) provide information on the state of the environment at various levels local, regional, national, or international. These indicators are typically developed by external entities such as governmental agencies or research institutions, offering a broader contextual understanding of environmental issues. Altogether, these indicators serve as a decision-making tool to support the continuous improvement of environmental management within the organization.

These indicators offer insights into the condition of the environment across different scales: local, regional, national, and international, and are generally developed by external organizations such as governmental agencies or research centers. Environmental dashboards consolidate environmental indicators in a concise format to enable effective internal monitoring. These tools allow managers to track performance on a daily basis and to respond promptly in the event of deviations from established objectives.

The environmental performance indicator dashboard presents concrete examples of how these indicators are applied in different companies. It incorporates a variety of daily environmental indicators, covering aspects such as wastewater discharge, water usage, energy consumption, waste, and volatile organic compounds (VOCs).

Below is a dashboard summarizing selected indicators for evaluating environmental performance **(Table 1)**:

Table 1. Environmental Performance Indicators

Category	Indicator	Description	Unit of Measurement
Energy	Total Energy Consumption	Total amount of energy used for manufacturing processes	kWh
	Renewable Energy	Percentage of energy from renewable sources used in production	%
	Energy Efficiency of Equipment	Energy performance of production machines (ratio of energy consumed to quantity produced)	kWh/unit produced
Waste	Total Waste Production	Total amount of waste generated during component manufacturing	Tonnes
	Material Recycling Rate	Percentage of materials recycled or reused in the production process	%
	Hazardous Waste	Amount of hazardous waste produced, including chemicals used in manufacturing	Tonnes
Emissions	CO2 Emissions	Tonnes of CO2 emitted by manufacturing processes	Tonnes of CO2
	Carbon Footprint of Products	Calculation of the carbon footprint over the entire life cycle of the manufactured products	

continued on following page

Table 1. Continued

Category	Indicator	Description	Unit of Measurement
Social Indicators	Employee Awareness of Chemical Risks	Percentage of employees trained in chemical management and environmental risks	%
	Environmental CSR Projects	Number of corporate social responsibility (CSR) projects specific to the microelectronics sector	Number
Regulatory Compliance	Environmental Non-Compliance	Number of environmental incidents or non-compliance cases recorded at production sites	Number
	Environmental Fines	Amount of fines received for environmental non-compliance	MAD

Source: El Maqaddem (2024)

3. THE MOROCCAN CONTEXT

3.1 Sustainable Performance in Morocco

Situated at a strategic crossroads between Europe and Africa, Morocco benefits from a privileged geographical location that facilitates trade and logistical flows. However, this position also presents challenges, notably in managing increasing freight volumes while minimizing environmental impacts (Ezzaher et al., 2024). The dynamic growth observed in key sectors such as automotive, aerospace, and renewable energies has led to heightened demand for environmentally friendly logistics solutions.

Environmental issues, now a national priority, necessitate a transition towards sustainable logistics practices, given the sector's significant share in greenhouse gas emissions. Logistics sustainability thus represents not only a response to global ecological imperatives but also a strategic lever to enhance Morocco's international competitiveness. By adopting eco-responsible logistics approaches, the Kingdom aims to reduce operating costs, optimize logistics performance, and reconcile economic development with environmental protection.

Aware of the urgency to integrate environmental and social dimensions into its development model, Morocco has actively engaged with the international community in combating climate change. During COP22 held in Marrakech, the sustainability of the logistics sector was highlighted, leading to the adoption of the Moroccan Green Logistics Charter in 2016. This public-private initiative aims to limit greenhouse gas emissions associated with logistics activities by enforcing strict environmental standards, including solar panel installations, rainwater harvesting, and low-consumption lighting.

Since the early 2000s, Morocco has pursued a proactive policy of green and inclusive growth, embodied by the National Sustainable Development Strategy (NSDS) launched in 2017, which serves as a framework to embed sustainability principles across all sectoral public policies. This environmental commitment is internationally recognized: in 2021, Morocco ranked 26th on the Green Future Index, first in the MENA region and second in Africa, and stands among global leaders in the Climate Performance Index.

Regarding sustainable governance, 53% of governance-related targets have been achieved, approximately 72% of actions for the green transition are underway, and natural resource preservation efforts have progressed by 76%. Building on these encouraging results, Morocco has raised its climate ambitions, targeting a 45.5% reduction in greenhouse gas emissions by 2030, coupled with investments to improve infrastructure quality and service access. Consequently, Morocco ranks as a leading African country in road and overall infrastructure quality (OECD, 2020; Global Competitiveness Index 2016-2017).

Morocco's regulatory framework, notably the National Charter for Environment and Sustainable Development, provides a favorable foundation for the ecological transition. Nonetheless, effective implementation in the logistics sector faces various challenges, and the systematic integration of sustainability principles into logistics infrastructure ports, transport networks, distribution platforms remains a strategic priority (Bensfia & Aidi, 2025).

The rise of emerging technologies such as the Internet of Things (IoT), artificial intelligence (AI), and blockchain offers new opportunities to optimize the sector's environmental performance. This also necessitates training a skilled workforce capable of adopting and embedding these sustainable practices in logistics processes.

Achieving sustainable logistics in Morocco requires close collaboration among public authorities, businesses, academia, and international partners, alongside a profound cultural shift that places sustainability at the core of all strategic decisions. As Moroccan companies integrate into global value chains, alignment with international sustainability standards becomes imperative.

Despite these advances, Morocco's transport sector remains heavily dependent on fossil fuels gasoline, natural gas, and diesel with emissions calculated via the LEAP model. Globally, biofuels and natural gas still account for a marginal share in road transport fuel consumption (Safaai et al., 2010). Although green innovations have demonstrated potential to reduce CO2 emissions and improve energy efficiency, few

Moroccan logistics service providers have fully adopted these practices (Aronsson & Brodin, 2006).

3.2 Temporal Evolution of GSCM Adoption in Morocco

- Initial Period – Identification of Barriers (2017)

Cherrafi and Elfezazi (2017) conducted a pioneering study identifying 11 barriers to the implementation of green supply chains in Morocco. Using ISM and MICMAC methodologies, they highlighted structural, organizational, and institutional challenges, and proposed a model to help companies prioritize actions.

- Early Empirical Studies – Culture and Awareness (2023)

Souhli and Ennadi (2023) surveyed Moroccan companies and demonstrated the importance of two factors in adopting Green Supply Chain Management (GSCM): leadership in ecological practices and top management commitment. Moreover, another study conducted in 2023 confirmed organizational culture as a critical catalyst for adoption, while national culture was found to be non-significant. Additionally, large firms and foreign subsidiaries exhibited higher adoption rates.

- Transition Toward Sustainable Performance (2024 – Early 2025)

Hmioui et al. (2024), in a literature review, emphasized that green logistics contributes not only to local emission reductions but also to territorial development encompassing economic, social, and environmental dimensions. Similarly, Gourire et al. (2025) analyzed sustainable practices among Casablanca Stock Exchange-listed companies, noting increased attention to sustainable supply chain practices (SSCP), which is driven by regulatory pressure and the desire to improve logistics performance.

- Emphasizing Internal Performance (2025)

Oufkiri et al. (2025) explored the impact of GSCM on the performance of Moroccan industrial firms. They concluded that internal practices such as waste reduction, eco-design, and energy efficiency had a statistically significant positive effect on overall performance, whereas external practices like green purchasing were not yet widely integrated.

3.3 Barriers to the Adoption of GSCM Practices

- Dependence on Fossil Fuels

The Moroccan logistics sector heavily relies on diesel and gasoline, which significantly hinders the transition to low-emission transport modes. This dependence limits the sector's ability to reduce its carbon footprint and adopt greener alternatives (Safaai et al., 2010; Haut-Commissariat au Plan, 2020).

- Lack of Training and Awareness

Knowledge of sustainable practices, including ISO 14001 standards and green logistics, remains limited, particularly among SMEs. This lack of awareness impedes the widespread adoption of environmentally friendly logistics approaches (Aronsson & Brodin, 2006; International Trade Centre, 2017).

- High Cost of Green Technologies

The initial investment required for electric vehicles, solar energy systems, and logistics optimization tools represents a major barrier, especially in the absence of targeted governmental subsidies or financial incentives (Kusrini et al., 2020; Cour des comptes du Royaume du Maroc, 2020).

- Weak Regulatory Framework Enforcement

Although Morocco has established strategic frameworks such as the National Sustainable Development Strategy (SNDD) and the Green Logistics Charter, their implementation suffers from insufficient monitoring, lack of fiscal incentives, and absence of binding operational standards (Ministry of Environment, 2017; CGEM & AMDL, 2016).

- Lack of Multi-Stakeholder Coordination

The absence of integrated governance involving government bodies, local authorities, the private sector, and academia slows down the large-scale deployment of sustainable logistics solutions. Effective collaboration is essential to overcome structural and operational challenges (World Bank, 2021; OECD, 2020).

3.4 Facilitators of GSCM Practices Adoption

- Strategic Geographical Advantage

Morocco's location at the crossroads between Europe and Africa positions it as a key regional logistics platform. This strategic role encourages the adoption of international sustainability standards to meet global trade requirements and enhance competitiveness (AMDL, 2018; MIT Technology Review Insights, 2021).

- Strong Institutional Commitment

Following COP22 in Marrakech and the establishment of the National Sustainable Development Strategy (SNDD), Morocco has firmly committed to integrating sustainability into sectoral policies, including logistics. This political will fosters regulatory frameworks and incentives that support green supply chain initiatives (Ministry of Energy Transition, 2021; Cour des comptes, 2020).

- Modern Logistics Infrastructure

Advanced logistics platforms such as Tanger Med port and integrated logistics zones facilitate the gradual incorporation of green technologies into supply chains, improving efficiency and reducing environmental impact (Global Competitiveness Index, 2016–2017; AMDL, 2019; Bensfia & Aidi, 2025; Bensfia, 2025).

- Massive Deployment of Renewable Energy

With flagship projects like Noor Ouarzazate, Morocco has become a leader in renewable energy in Africa, providing clean power sources that support greener logistics operations (IRENA, 2020; ONEE, 2021).

- High Technological Potential

Emerging technologies such as the Internet of Things (IoT), blockchain, and artificial intelligence (AI) are increasingly integrated in pilot projects, paving the way for smart, sustainable logistics systems (UNCTAD, 2021; McKinnon, 2018; Badr, 2022).

- International Support and Cooperation

Morocco benefits from technical and financial assistance from the European Union, the World Bank, and the African Development Bank to develop and structure green logistics projects, enhancing capacity building and innovation (African Development Bank, 2022; GIZ Morocco, 2021).

4. LITERATURE REVIEW

Green Supply Chain Management (GSCM) encompasses a range of practices, including eco-design, green purchasing, green manufacturing, green distribution, and reverse logistics (Srivastava, 2007). Strong internal environmental management plays a vital role in enhancing a company's environmental performance (Zhu et al., 2008). Zhu and Sarkis (2008) further note that quality management helps streamline the adoption of GSCM practices.

4.1 Eco-Design: A Key Lever for Environmental Performance

Eco-design, defined as the integration of environmental criteria from the product design phase, plays a crucial role in developing more sustainable and energy-efficient products (Gottberg et al., 2006). It promotes the use of non-hazardous materials, facilitates disassembly for recycling purposes, and enables significant modifications regarding product size, lifespan, and recyclability. These practices contribute to waste reduction while enhancing resource efficiency.

More specifically, eco-design impacts several dimensions of the product life cycle, yielding significant improvements in environmental performance. First, it encourages the use of recycled, renewable, and less polluting materials. For example, incorporating bioplastics and recycled polymers in packaging helps reduce the extraction of non-renewable resources and plastic waste volumes (Zhu et al., 2021). Additionally, modular design facilitates repair and disassembly, thereby extending product lifespan and reducing waste generation (González-García et al., 2020).

Second, eco-design contributes to reducing energy consumption. Designing lighter and more energy-efficient products lowers energy use during manufacturing, transportation, and operation. For instance, in the automotive and electronics sectors, eco-design helps curb electricity consumption and greenhouse gas emissions (Luo et al., 2022).

Third, the integration of life cycle assessment (LCA) within the eco-design process is a fundamental tool. It enables the identification of major sources of CO_2 and other pollutant emissions, optimizing technical and material choices. This approach is essential to meet climate targets and reduce the carbon footprint of products (Koffler et al., 2023).

Finally, eco-design supports the development of the circular economy by facilitating product reuse, refurbishment, and recycling at end-of-life stages. By doing so, it contributes to more sustainable resource management and waste reduction (Geissdoerfer et al., 2020).

Therefore, eco-design stands as a strategic lever to enhance product environmental performance while addressing sustainability challenges and resource efficiency.

Based on this, the following hypothesis is proposed:

> H1: Eco-design positively influences environmental performance.

4.2 Green Purchasing: A Strategic Driver of Environmental Performance

Green purchasing refers to procurement practices that prioritize environmental sustainability by acquiring recyclable, reusable, or recycled materials (Zhu et al., 2008). This concept encompasses several essential actions, including the selection of suppliers committed to sustainable and eco-friendly production methods, the preference for eco-friendly materials that are durable, recycled, or less harmful to the environment, and the assurance of environmental compliance through products and suppliers certified by recognized ecological labels.

These procurement practices are interconnected and mutually reinforcing, collectively contributing to the reduction of the environmental footprint across the supply chain. By prioritizing suppliers who integrate sustainability into their production processes, green purchasing supports the transition to cleaner energy sources and more environmentally responsible manufacturing. This synergy amplifies the effectiveness of environmental management practices, fostering significant improvements in overall environmental performance.

Moreover, green purchasing not only reduces waste and emissions directly linked to procurement but also drives suppliers to innovate and improve their environmental practices. The ripple effect extends beyond the immediate supply chain, promoting a culture of sustainability throughout the production network.

Thus, green purchasing stands as a pivotal component in advancing sustainable supply chain management and enhancing environmental outcomes in organizations.

Based on this rationale, the following hypothesis is proposed:

> H2: Green purchasing positively influences environmental performance.

4.3 Green Manufacturing: Enhancing Environmental Performance through Sustainable Production

Green manufacturing is a contemporary production strategy focused on minimizing environmental impact while optimizing resource utilization (Li et al., 2016). As Melnyk and Smith (1996) highlight, it integrates product and process design with production planning and control to achieve sustainable manufacturing outcomes. The core objective of green manufacturing is to identify, measure, evaluate, and manage environmental waste flows in order to reduce or eliminate adverse environmental effects while maximizing resource efficiency.

This approach involves several key practices. Firstly, the adoption of environmentally friendly technologies reduces pollution and improves energy efficiency. Examples include cleaner production techniques, energy-saving machinery, and renewable energy integration. Secondly, waste minimization is pursued through internal recycling processes and more efficient use of raw materials, which both lower waste generation and decrease the demand for virgin resources. Thirdly, the installation of emission control systems such as particulate filters and gas treatment units serves to limit the release of harmful substances into the environment. Lastly, sustainable management of vital resources, especially water and raw materials, prevents their depletion and supports long-term environmental sustainability.

Collectively, these practices position green manufacturing as a vital lever for organizations striving to enhance their environmental performance. By embedding sustainability into production operations, companies can simultaneously reduce environmental burdens and improve operational efficiency, thus fostering competitive advantage in an increasingly eco-conscious market.

Based on these considerations, the following hypothesis is formulated:

H3: Green manufacturing positively influences environmental performance.

4.4 Waste Management: A Crucial Factor for Enhancing Environmental Performance

Inadequate waste management is a major contributor to environmental pollution, affecting water, soil, and air quality, and posing significant risks to public health (Giusti, 2009). Effective waste management encompasses the design and implementation of processes aimed at preventing waste generation and mitigating its harmful impacts throughout the product life cycle. This includes strategies for waste reduction, pollution prevention, and proper disposal (Srivastava, 2007).

Understanding that all products undergo a life cycle from raw material procurement to end-of-life treatment waste management plays a vital role in controlling environmental externalities associated with each stage (Jofre & Morioka, 2005). By integrating waste management principles into production and operational processes, organizations can minimize pollutant emissions, reduce landfill dependency, and promote recycling and recovery activities.

Moreover, sound waste management contributes to resource conservation and fosters a circular economy by encouraging reuse and valorization of materials. This holistic approach not only decreases the environmental footprint but also aligns with regulatory requirements and societal expectations for sustainability.

Consequently, waste management emerges as a fundamental driver for improving environmental performance, enabling firms to balance ecological responsibility with economic efficiency.

Based on this, the following hypothesis is proposed:

H4: Waste management positively influences environmental performance.

4.5 Reverse Logistics: Enhancing Environmental Performance through Product Recovery and Reuse

Reverse logistics primarily involves managing the return flow of products and materials that are suitable for recycling or reuse. This process encompasses several critical steps, including collection, sorting, compacting, temporary processing, transportation, and reintegration into the supply chain (Sarkis, 2020). By effectively coordinating these activities, organizations can minimize waste, reduce resource consumption, and lower the overall environmental footprint of their operations.

Moreover, integrating reverse logistics into supply chain information systems enables better tracking, management, and optimization of returned goods. Chouinard et al. (2005) emphasize that reverse logistics not only facilitates the recovery and management of unused or end-of-life products but also supports the redistribution and reuse of valuable materials within the supply chain. This integration enhances operational efficiency while promoting sustainability.

By closing the loop on product life cycles, reverse logistics contributes significantly to the circular economy, conserving natural resources and reducing pollution associated with raw material extraction and manufacturing. As such, it represents a strategic lever for companies aiming to improve their environmental performance while maintaining competitive advantage.

Based on these insights, the following hypothesis is proposed:

H5: Reverse logistics (recycling or reuse) positively influences environmental performance.

4.6 Green Distribution: Reducing Environmental Impact through Sustainable Logistics Practices

The rapid expansion of international trade has led to logistics chains that span vast geographic areas, resulting in increased consumption of fossil fuels for transportation and a significant rise in carbon dioxide emissions compared to previous decades (Venkat & Wakeland, 2006). This growth has heightened concerns about the environmental consequences of logistics activities, particularly regarding climate change, air pollution, and energy use.

In response, the concept of green distribution has emerged as a vital approach to mitigate these impacts. Green distribution encompasses various practices aimed at optimizing transportation and distribution processes to reduce environmental footprints. Key actions include redesigning distribution networks to improve efficiency, optimizing delivery routes to minimize fuel consumption, assessing and monitoring pollution levels, adopting alternative fuels such as biofuels or electricity, and enhancing packaging design to reduce waste and improve load efficiency (Sharma, 2000).

By integrating these measures, green distribution not only lowers greenhouse gas emissions but also improves overall resource efficiency in the supply chain. This approach supports organizational sustainability goals and contributes to broader efforts in combating climate change.

Based on these considerations, the following hypothesis is proposed:

H6: Green distribution positively influences environmental performance.

5. RESEARCH MODEL

After presenting the various hypotheses of our research, we proceed to the design of the research model (**Figure 1**), which is structured around key elements: Eco-design, Green purchasing, Green manufacturing, Waste management, Reverse logistics, as well as Green distribution as explanatory variables, and environmental performance as the dependent variable.

Figure 1. Conceptual Model

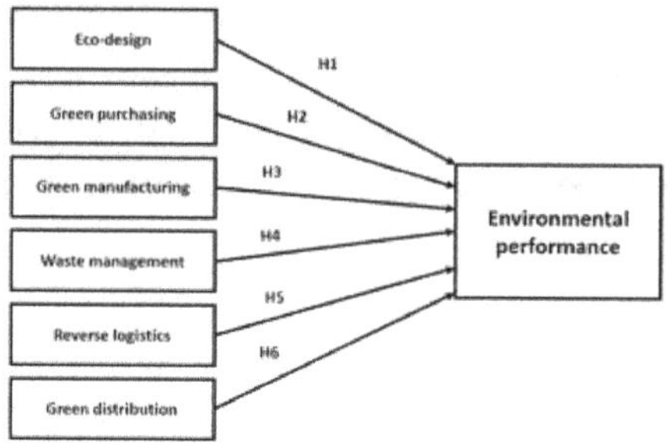

6. METHODOLOGY

6.1 Data Collection

In this study, data were gathered using a structured questionnaire distributed through two main channels: an online survey via Google Forms and direct, face-to-face administration. A total of 250 questionnaires were sent to industrial companies located in northern Morocco. From these, 98 completed responses were returned, representing an initial response rate of 39.2%. After excluding 3 questionnaires due to substantial missing information, the final valid sample consisted of 95 respondents, resulting in an effective response rate of 38.0%. These 95 valid responses formed the foundation for analyzing green supply chain management practices and their effects on environmental performance within the industrial sector of northern Morocco

6.2 Instrument Development

To assess the study variables, a structured questionnaire was designed using a five-point Likert scale, with options ranging from 'strongly disagree' to 'strongly agree.' This format enables respondents to express the intensity of their perceptions and attitudes regarding various green practices and their impact on environmental performance. In order to ensure the reliability and validity of the instrument, the measurement items were adapted from widely recognized and validated scales in the existing literature (**Table 2**). Specifically, Eco-design was measured using 4 items inspired by the works of Afum et al. (2020), Zhu et al. (2008), and Green et al. (2012), capturing practices aimed at reducing environmental impacts through improved product design. Green purchasing was assessed using 6 items, focusing on the organization's efforts to collaborate with suppliers that demonstrate environmental commitment, such as seeking low-energy-consumption suppliers or encouraging the delivery of eco-friendly materials.

Similarly, Green manufacturing was measured with 4 items targeting the integration of environmentally friendly production techniques. Sample items include the use of eco-technological equipment and processes, the reduction of energy consumption during manufacturing and transportation, and the production of goods that minimize material usage during consumption. Waste management was captured through 5 items addressing waste reduction, pollution prevention, and environmentally sound disposal practices. Reverse logistics was measured using 5 items reflecting the company's efforts to manage product returns, recycling, and reuse. Green distribution was evaluated through 5 items focused on optimizing transport systems to minimize carbon emissions and improving packaging design to reduce environmental impact.

Finally, environmental performance was measured with seven items capturing different aspects of ecological outcomes, including reduced air emissions, decreased use of hazardous or toxic materials, and fewer environmental incidents. Together, these measures provide a well-rounded understanding of green supply chain management practices and their impact on environmental sustainability.

Table 2. Measurement instruments

	Number of items	Authors
Eco-design	4	• Afum et al. (2020) • Zhu, Sarkis, et Lai (2008) • Green et al. (2012)
Green purchasing	6	
Green manufacturing	4	
Waste management	5	
Reverse logistics	5	
Green distribution	5	
Environmental performance	7	

6.3 Data Analysis Method

To assess the proposed research model and test the hypotheses, the study employed Partial Least Squares Structural Equation Modeling (PLS-SEM) using SmartPLS 4 software. This method is well-suited for predictive and exploratory research, especially in social sciences where models can be complex and sample sizes relatively small (Hair et al., 2017; Ringle et al., 2015). The analysis was conducted in two main stages. First, the measurement model was evaluated to confirm its reliability and validity. Convergent validity was examined by checking that factor loadings exceeded 0.7, composite reliability (CR) was above 0.7, and average variance extracted (AVE) was greater than 0.5, following established guidelines (Hair et al., 2017). Discriminant validity was verified using the Fornell-Larcker criterion alongside cross-loading analysis. In the second stage, the structural model was tested by analyzing the coefficient of determination (R^2), predictive relevance (Q^2), and the significance of path coefficients, which were assessed through bootstrapping techniques. This comprehensive approach provided a rigorous validation of both the measurement tools and the proposed relationships among the constructs.

6.4 Sample

According to the data presented in **Table 3**, 85% of the surveyed companies are limited liability companies (LLCs), while 15% are public limited companies (PLCs). In terms of workforce size, 44% of the companies employ between 100 and 250 people, 41% have more than 250 employees, and 15% have between 10 and 100 employees. This distribution highlights the predominance of medium to large-sized enterprises within the sample.

With regard to the industrial sector, the automotive industry is the most represented, comprising 74% of the companies surveyed. It is followed by the textile and leather sector (17%) and the aeronautics sector (9%). These figures underscore the strong presence of the automotive sector in northern Morocco's industrial ecosystem.

The sample consists of multinational industrial firms operating in northern Morocco. While the specific nationalities of these companies are not detailed in this table, the focus on strategic, high-value sectors suggests a significant international dimension.

As for the respondents' profiles presented in **Table 4**, 64% are Logistics Managers, while 36% are Logistics Directors. In terms of professional experience, 35% of respondents have between 6 and 10 years of experience, 24% have 3 to 6 years, 19% have more than 10 years, 18% have 1 to 3 years, and 4% have less than 1 year. This indicates that the majority of participants are mid- to senior-level professionals with substantial experience in the logistics domain.

Table 3. Study participants' characteristics

Company characteristics	Frequency	%
Legal form Public limited company (PLC) Limited liability company (LLC)	14 81	15% 85%
Workforce Between 10 and 100 Between 100 and 250 More than 250	14 42 39	15% 44% 41%
Industry sector Automotive Aeronautics Textile and Leather	70 9 16	74% 9% 17%

Table 4. Study participants' characteristics

Respondent characteristics	Frequency	%
Position Logistics Director Logistics Manager	34 61	36% 64%
Number of years of experience 0 – 1 year 1 – 3 years 3 – 6 years 6 – 10 years 10 years and more	4 17 23 33 18	4% 18% 24% 35% 19%

7. RESULTS

7.1 Validity and Reliability Analysis

Table 5 presents the findings from the measurement model assessment. All indicator outer loadings surpass the recommended cutoff of 0.70, confirming a strong link between the observed variables and their respective constructs. Furthermore, the Average Variance Extracted (AVE) values for all constructs are above 0.50, indicating that each construct explains more than half of the variance in its indicators and thus meets the standard for convergent validity (Hair et al., 2017). In addition, both Cronbach's alpha and Composite Reliability values surpass the commonly accepted threshold of 0.70, further supporting the internal consistency and reliability of the measurement model.

Table 5. Results of reliability and convergent validity

	Cronbach's Alpha	rho A	Composite Reliability	Average Variance Extracted (AVE)
Eco-design	0.752	0.804	0.857	0.668
Environmental Performance	0.905	0.908	0.925	0.638
Green Distribution	0.824	0.868	0.896	0.745
Green Manufacturing	0.906	0.931	0.934	0.781
Green Purchasing	0.920	0.924	0.940	0.759
Reverse Logistics	0.888	0.903	0.917	0.690
Waste Management	0.828	0.860	0.896	0.743

Discriminant validity is also firmly established. According to the Fornell-Larcker criterion, each construct shares more variance with its own indicators than with those of other constructs, consistent with the recommendations of Henseler et al. (2016). This is further supported by the cross-loading analysis shown in Table 6, which reveals that each indicator has a higher loading on its own construct than on any other, confirming the model's discriminant validity.

Table 6. Results of discriminant validity

	Eco-design	Environmental Performance	Green Distribution	Green	Green Purchasing	Reverse Logistics	Waste Management
Eco-design	**0.817**						
Environmental Performance	0.559	**0.899**					
Green Distribution	0.481	0.804	**0.863**				
Green Manufacturing	0.379	0.711	0.578	**0.884**			
Green Purchasing	0.537	0.643	0.515	0.386	**0.871**		
Reverse Logistics	0.495	0.794	0.784	0.668	0.469	**0.830**	
Waste Management	0.396	0.704	0.556	0.769	0.335	0.690	**0.862**

7.2 Results of Assessing Structural Model

The evaluation of the inner model focuses on several important criteria, such as the coefficient of determination (R^2), predictive relevance (Q^2), and the goodness-of-fit (GoF) index. As shown in Table 7, the R^2 value for environmental performance is 0.823, reflecting strong explanatory power. This means that the model's independent variables explain about 82.3% of the variation in environmental performance, which reflects a substantial level of determination for this dependent variable.

Table 7. Results of R2

	R Square	R Square Adjusted
Environmental Performance	0.823	0.814

Furthermore, the model's predictive relevance is assessed using the Q^2 value. As illustrated in **Table 8**, the Q^2 value for environmental performance is 0.512, which exceeds the commonly accepted threshold of 0.39. According to Hair et al. (2020), this confirms the model's strong predictive capability and its relevance in forecasting the endogenous construct.

Finally, the GoF index calculated at 0.769 demonstrates an overall strong model fit. This value, based on the approach proposed by Henseler et al. (2009), suggests that the model achieves a robust balance between the measurement and structural components, thereby reinforcing the validity and adequacy of the overall model.

Table 8. Results of Q2

	SSO	SSE	Q^2 (=1-SSE/SSO)
Eco-design	360.000	360.000	
Environmental Performance	840.000	409.709	**0.512**
Green Distribution	360.000	360.000	
Green Manufacturing	480.000	480.000	
Green Purchasing	600.000	600.000	
Reverse Logistics	600.000	600.000	
Waste Management	360.000	360.000	

7.3 Hypotheses Testing

The empirical results presented in **Table 9** and **Figure 2** provide partial support for the research hypotheses. Specifically, the findings reveal that green purchasing has a statistically significant and strong positive effect on environmental performance ($T = 5.600$; $p < 0.001$), confirming Hypothesis H2. Similarly, green distribution also exerts a significant and positive influence on environmental performance (T = 3.529; $p < 0.001$), validating Hypothesis H6.

In addition, waste management and reverse logistics are both found to have significant effects on environmental performance, with T-values of 2.330 ($p = 0.020$) and 2.001 ($p = 0.046$), respectively, thereby supporting Hypotheses H4 and H5. Green manufacturing also shows a significant impact (T = 2.070; $p = 0.039$), supporting Hypothesis H3.

However, the analysis does not provide sufficient evidence to support Hypothesis H1, as the relationship between eco-design and environmental performance is not statistically significant (T = 1.148; $p = 0.252$). This suggests that, within the context of this study, eco-design practices may not directly contribute to improved environmental performance.

Overall, the model confirms the critical role of several green supply chain practices particularly green purchasing, green distribution, and waste-related strategies in enhancing environmental outcomes.

Figure 2. Results of model testing

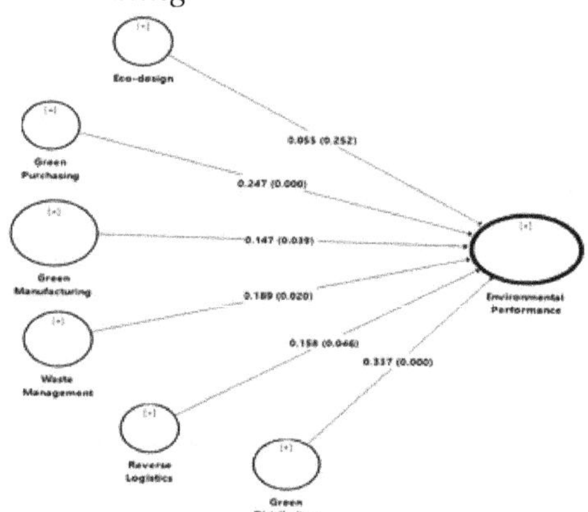

Table 9. Results of hypothesis tests.

Hypothesis	Original Sample	Sample Mean	Standard Deviation	T Statistics	P Values
H1: Eco-design -> Environmental Performance	0.055	0.055	0.048	1.148	0.252
H2: Green Purchasing -> Environmental Performance	0.247	0.253	0.044	5.600	0.000
H3: Green Manufacturing -> Environmental Performance	0.147	0.135	0.071	2.070	0.039

continued on following page

Table 9. Continued

Hypothesis	Original Sample	Sample Mean	Standard Deviation	T Statistics	P Values
H4: Waste Management -> Environmental Performance	0.189	0.195	0.081	2.330	0.020
H5: Reverse Logistics -> Environmental Performance	0.158	0.161	0.079	2.001	0.046
H6: Green Distribution -> Environmental Performance	0.337	0.334	0.096	3.529	0.000

8. DISCUSSION

The findings of this study provide important insights into the environmental performance of industrial firms in northern Morocco and the role that various green supply chain management (GSCM) practices play within this context. Understanding the specific environmental challenges faced by Moroccan companies helps contextualize the urgency and relevance of adopting sustainable practices. Moroccan industries frequently confront sector-specific environmental issues such as excessive waste generation, high emissions, and critical water usage concerns. For example, the textile and manufacturing sectors often struggle with wastewater treatment and emission controls, while food processing firms face significant challenges managing organic waste. These contextual realities reinforce the necessity for effective GSCM adoption tailored to local constraints and priorities.

Our results reveal a complex picture regarding the influence of different green practices on environmental performance. Contrary to expectations and previous international research (e.g., Gottberg et al., 2006), eco-design did not show a statistically significant positive effect in our sample. This finding can be explained by several factors. Firstly, eco-design remains a relatively immature concept in the Moroccan industrial landscape, where limited technical expertise and scarce financial resources constrain its development and implementation. Additionally, regulatory incentives or market pressures encouraging eco-design adoption are still weak compared to more developed economies. From a managerial perspective, this suggests a need for capacity building and technical support programs to enhance firms' eco-design capabilities. Theoretically, it highlights the importance of considering contextual maturity when studying the impact of green innovations in developing country settings.

In contrast, green purchasing, green manufacturing, and waste management practices demonstrated significant positive effects on environmental performance. These findings align well with prior studies emphasizing their foundational role in advancing sustainability within supply chains (Zhu et al., 2008; Li et al., 2016; Srivastava, 2007). Moroccan firms appear to be increasingly recognizing the strategic value of selecting environmentally responsible suppliers, adopting energy-efficient production technologies, and proactively managing waste streams to comply with evolving regulations and reduce costs. These efforts not only mitigate environmental risks but also enhance operational efficiency, creating a virtuous cycle of sustainability and performance.

Beyond these core practices, the discussion is further enriched by recent global trends reshaping supply chain sustainability strategies. Notably, the circular economy model, which emphasizes resource recovery, reuse, and closed-loop systems, is gaining momentum among Moroccan companies. Our findings on reverse logistics support this shift, demonstrating how reclaiming and reintegrating materials contribute to tangible environmental improvements. Furthermore, decarbonization efforts are becoming increasingly relevant as firms respond to mounting pressures to reduce greenhouse gas emissions. International regulations such as the Corporate Sustainability Reporting Directive (CSRD) and the European Green Deal, though primarily targeting developed markets, are influencing Moroccan firms through supply chain linkages and export requirements, pushing for greater transparency and environmental accountability. These developments underscore the dynamic and evolving nature of sustainability strategies in emerging economies.

Finally, green distribution practices, including optimizing transportation routes, switching to cleaner fuels, and enhancing packaging efficiency, also showed positive impacts on environmental outcomes. Given that transportation is a significant contributor to carbon emissions globally, the proactive adoption of these measures by Moroccan firms signals an encouraging responsiveness to environmental challenges and reflects broader global sustainability trends.

Taken together, these results offer a nuanced understanding of the state of environmental management in Moroccan industries. While certain innovative practices like eco-design require targeted support to realize their potential, more established green supply chain practices are being effectively integrated and driving improvements in environmental performance. The incorporation of emerging sustainability trends and international regulatory influences highlights the need for continuous adaptation and capacity building to ensure that Moroccan firms can meet both local environmental challenges and global sustainability expectations.

9. CONCLUSION

The empirical findings of this study underscore the increasing significance of environmental management practices in fostering sustainable performance among industrial firms. While the maturity and effectiveness of various green supply chain initiatives vary, the strong positive effects observed for green purchasing, manufacturing, waste management, reverse logistics, and green distribution indicate that Moroccan companies are progressively integrating environmental goals into their operations.

These results align with existing research but also reveal gaps, particularly in the adoption of eco-design practices, which appear less developed. Addressing this gap presents an opportunity to further enhance the environmental and economic benefits of green supply chains in Morocco.

However, this study faces certain limitations, including its cross-sectional design and the geographic focus on the northern region of Morocco, which may affect the generalizability of the findings. Additionally, the reliance on self-reported data may introduce bias.

To strengthen the practical relevance of future work, it will be important to link statistical findings more explicitly to real-world applications and provide additional sector-specific examples from the Moroccan context. Moreover, a comparative analysis considering company size and sectoral differences could offer deeper insights into the adoption and impact of GSCM practices.

Based on these insights, it is recommended that companies and policymakers focus on building technical capacity and providing financial incentives to support comprehensive GSCM implementation. Encouraging innovation in eco-design and fostering collaboration across the supply chain can also drive further progress.

Further, expanding the discussion on the unexpectedly low impact of eco-design will help clarify barriers and inform targeted managerial and policy interventions.

Finally, future research and practice should integrate emerging global trends such as the circular economy, decarbonization, and international sustainability regulations, which are increasingly influencing Moroccan firms and their environmental strategies.

Future research could expand the scope to other regions and sectors, employ longitudinal approaches to assess changes over time, and explore the role of emerging technologies in strengthening green supply chain practices.

REFERENCES

Afum, E., Agyabeng-Mensah, Y., Sun, Z., Frimpong, B., Kusi, L., & Acquah, I. (2020). Exploring the link between green manufacturing, operational competitiveness, firm reputation and sustainable performance dimensions: A mediated approach. *Journal of Manufacturing Technology Management, 31*(7), 1417–1438. Advance online publication. DOI: 10.1108/JMTM-02-2020-0036

Alla, L., Bentalha, B., & Bouhtati, N. (2022). Assessing supply chain performance in the COVID-19 context: A prospective model. In *2022 14th International Colloquium of Logistics and Supply Chain Management (LOGISTIQUA)* (pp. 1–6). IEEE. DOI: 10.1109/LOGISTIQUA55056.2022.9938083

Aronsson, H., & Brodin, M. (2006). The environmental impact of changing logistics structures. *International Journal of Logistics Management, 9*(4), 307–317. DOI: 10.1108/09574090610717545

Badr, B. (2022). Smart technologies for a sustainable service supply chain: A prospective perspective. *Journal of Environmental Issues and Climate Change, 1*(1), 77–89. DOI: 10.59110/jeicc.v1i1.81

Beamon, B. M. (1999). Designing the green supply chain. *Logistics Information Management, 12*(4), 332–342. DOI: 10.1108/09576059910284159

Bensfia, C. (2025). Collective intelligence of port territory stakeholders: Case of Moroccan ports. In *Utilizing Technology to Manage Territories* (pp. 335–370). IGI Global.

Bensfia, C., & Aidi, S. (2025). Port community systems and organizational performance: A socio-technical approach to evaluation. In *Industrial and Logistics Systems Design and Efficient Operation* (pp. 42–52). https://doi.org/DOI: 10.1007/978-3-031-87309-6_4

Carbone, V., Moatti, V., & Esposito Vinzi, V. (2012). Mapping corporate responsibility and sustainable supply chains: An exploratory perspective. *Business Strategy and the Environment, 21*(7), 475–494. DOI: 10.1002/bse.1736

CGEM & AMDL. (2016). *Charte marocaine de la logistique verte.*

Cherrafi, A., & Elfezazi, S. (2017). Barriers to the implementation of green supply chain management in Moroccan companies: An ISM and MICMAC approach. [Disponible sur arxiv.org]. *International Journal of Supply Chain Management, 6*(4), 260–268.

Chouinard, G. A., Swain, D. P., Hammill, M. O., & Poirier, G. A. (2005). Covariation between grey seal (Halichoerus grypus) abundance and natural mortality of cod (Gadus morhua) in the southern Gulf of St. Lawrence. *Canadian Journal of Fisheries and Aquatic Sciences*, *62*(9), 1991–2000. DOI: 10.1139/f05-107

Cour des comptes. (2020). *Évaluation de la Stratégie Nationale de Développement Durable au Maroc*. Rabat, Maroc.

Dawei, Z., Hamid, A. B. A., Chin, T. A., & Leng, K. C. (2015). Green supply chain management: A literature review. Sains Humanika, 5(2).

Desmazes, J., & Lafontaine, J.-P. (2007). L'assimilation des budgets environnementaux et du tableau de bord vert par les entreprises [Post-print, halshs-00543246]. HAL. https://halshs.archives-ouvertes.fr/halshs-00543246

El Maqaddem, M. (2024). *Les pratiques vertes de la Supply Chain et la performance environnementale - Cas Valtronic technologies Maroc* (Unpublished master's thesis). Université de Liège. https://matheo.uliege.be/handle/2268.2/21703

Ezekari, M., Ait Lhassan, I., Bazarouj, R., & Ouhnine, I. (2024). The impact of collaboration on supply chain performance in the aerospace industry: Exploring the mediating role of knowledge sharing. *Multidisciplinary Science Journal*, *6*(12), 2024276. Advance online publication. DOI: 10.31893/multiscience.2024276

Ezzaher, R., Lhassan, I., Azdod, M., Razzouki, M., Mastour, S., Bouayad, M., & Babounia, A. (2024). Contribution to the assessment of the sustainability of urban freight transport in Morocco: A PLS-SEM approach. *Uncertain Supply Chain Management*, *12*(4), 2651–2660. DOI: 10.5267/j.uscm.2024.5.006

Geissdoerfer, M., Savaget, P., Bocken, N. M. P., & Hultink, E. J. (2020). The circular economy – A new sustainability paradigm? *Journal of Cleaner Production*, *143*, 757–768. DOI: 10.1016/j.jclepro.2016.12.048

Gibert, P. (1980). *Le contrôle de gestion dans les organisations publiques*. Les Éditions d'Organisation.

Giusti, A. (2009). Review of waste management practices and their impact on human health. *Waste Management (New York, N.Y.)*, *29*(8), 2227–2239. DOI: 10.1016/j.wasman.2009.03.028 PMID: 19401266

Global Competitiveness Index. (2016–2017). *Quality of infrastructure rankings*. World Economic Forum. https://www.weforum.org/reports/global-competitiveness-report-2016-2017

González-García, S., Rieradevall, J., & Feijoo, G. (2020). Life cycle assessment of eco-designed products: Modularity and durability as strategies for environmental improvement. *Sustainable Production and Consumption*, *22*, 144–156. DOI: 10.1016/j.spc.2020.03.002

Gottberg, A., Morris, J., Pollard, S., Mark-Herbert, C., & Cook, M. (2006). Producer responsibility, waste minimisation and the WEEE Directive: Case studies in eco-design from the European lighting sector. *The Science of the Total Environment*, *359*(1-3), 38–56. DOI: 10.1016/j.scitotenv.2005.07.001 PMID: 16169574

Gourire, M., Rouggani, K., & Amine, N. (2025). Unveiling sustainable supply chain practices and their impact on corporate logistics performance: An exploratory analysis of companies listed on the Moroccan stock exchange. *RMd, Economics, Management &. Social Sciences*, *2*(1), e202510. DOI: 10.23882/emss25125

Green, K. W.Jr, Zelbst, P. J., Meacham, J., & Bhadauria, V. S. (2012). Green supply chain management practices: Impact on performance. *Supply Chain Management*, *17*(3), 290–305. DOI: 10.1108/13598541211227126

Haut-Commissariat au Plan (HCP). (2020). *Statistiques environnementales et énergie*.

Hmioui, A., Bentalha, B., & Beniaich, R. (2024). Green Supply Chain Management et développement territorial: Une revue de littérature. *Alternatives Managériales Économiques, 6*(2), 1–23. https://revues.imist.ma/?journal=AME

International Trade Centre (ITC). (2017). *Rapport sur la logistique durable pour les PME marocaines*.

IRENA. (2020). *Renewable energy statistics 2020*. International Renewable Energy Agency. Abu Dhabi, Émirats arabes unis.

Jofré, S., & Morioka, T. (2005). Waste management of electric and electronic equipment: Comparative analysis of end-of-life strategies. *Journal of Material Cycles and Waste Management*, *7*(1), 24–32. DOI: 10.1007/s10163-004-0122-1

Koffler, C., Patel, P., & Nguyen, T. (2023). Carbon footprint reduction through eco-design: A case study analysis. *Environmental Science & Technology*, *57*(6), 2345–2356. DOI: 10.1021/acs.est.2c06654

Kusrini, E., Widyastuti, D., & Raharjo, K. (2020). Green supply chain management practices and performance: A literature review. *International Journal of Supply Chain Management*, *9*(3), 155–160.

Li, D., Huang, M., Ren, S., Chen, X., & Ning, L. (2016). Environmental legitimacy, green innovation, and corporate carbon disclosure: Evidence from CDP China. *Journal of Business Ethics, 150*(4), 1089–1104. DOI: 10.1007/s10551-016-3187-6

Luo, Z., Chen, H., & Wang, J. (2022). Energy-saving design in electric vehicles: Impact assessment through life cycle analysis. *Energy Reports, 8*, 1465–1475. DOI: 10.1016/j.egyr.2022.01.015

McKinnon, A. (2018). *Decarbonizing logistics: Distributing goods in a low carbon world*. Kogan Page.

Melnyk, S. A., & Smith, R. T. (1996). *Green manufacturing*. Society for Manufacturing Engineering.

Mentzer, J. D., DeWitt, W., Keebler, J. S., Min, S., Nix, N. W., Smith, C. D., & Zacharia, Z. G. (2001). Defining supply chain management. *Journal of Business Logistics, 22*(2), 1–25. DOI: 10.1002/j.2158-1592.2001.tb00001.x

Monczka, R. M., Handfield, R. B., Giunipero, L. C., & Patterson, J. L. (2015). *Purchasing & supply chain management* (6th ed.). Cengage Learning.

ONEE. (2021). *Rapport annuel sur les énergies renouvelables au Maroc.*

Oubaouzine, L. (2019). La contribution d'une logistique performante à la performance de l'entreprise. *Revue de Management et Cultures*, (4), 180–194.

Oufkiri, M., Zari, M., & Oufkiri, M. (2025). Impact of green supply chain management on industrial companies' performance in Morocco. *International Journal of Advanced Manufacturing and Management, 8*(1), 45–59.

Safaai, S., Sopian, K., & Zaidi, S. (2010). Life cycle assessment of biofuels and fossil fuels for road transport. *Renewable & Sustainable Energy Reviews, 14*(6), 1839–1845.

Salburg, H. (2000). *Green logistics: Improving the environmental sustainability of logistics*. Kogan Page.

Sarkis, J., Cohen, M. J., Dewick, P., & Schröder, P. (2020). A brave new world: Lessons from the COVID-19 pandemic for transitioning to sustainable supply and production. *Resources, Conservation and Recycling, 159*, 104894. DOI: 10.1016/j.resconrec.2020.104894 PMID: 32313383

Sharma, S. (2000). Managerial interpretations and organizational context as predictors of corporate choice of environmental strategy. *Academy of Management Journal, 43*(4), 681–697. DOI: 10.2307/1556361

Sharma, V. K., Chandna, P., & Bhardwaj, A. (2017). GSCM related performance indicators in agro industry: A review. *Journal of Cleaner Production, 141*, 1194–1208. DOI: 10.1016/j.jclepro.2016.09.103

Souhli, K., & Ennadi, A. (2023). Adoption of GSCM practices and sensitivity/ influencing factors: An empirical study at the Moroccan firm level. *International Journal of Industrial Engineering and Management, 14*(3), 214–231. DOI: 10.24867/ IJIEM-2023-3-334

Srivastava, S. K. (2007). Green supply-chain management: A state of the art literature review. *International Journal of Management Reviews, 9*(1), 53–80. DOI: 10.1111/j.1468-2370.2007.00202.x

Technology Review Insights, M. I. T. (2021). *Climate performance index 2021.* Massachusetts Institute of Technology.

Teng, J. T. C., & Jaramillo, P. (2006). Analyzing sustainable supply chain management using system dynamics modeling. *Systems Research and Behavioral Science, 23*(5), 647–661. DOI: 10.1002/sres.793

UNCTAD. (2021). *Emerging technologies in sustainable logistics: Opportunities for developing countries. United Nations Conference on Trade and Development.*

Venkat, K., & Wakeland, W. (2006). Is lean necessarily green? In *Proceedings of the 50th Annual Meeting of the International Society for the Systems Sciences (ISSS).* https://journals.isss.org/index.php/proceedings50th/article/view/401

Wang, H. F., & Gupta, S. M. (2011). *Green supply chain management A product life cycle approach.* McGraw-Hill Education.

Zhu, Q., & Geng, Y. (2001). Integrating environmental issues into supplier selection and management. *Greener Management International, 35*(35), 27–40. DOI: 10.9774/ GLEAF.3062.2001.au.00005

Zhu, Q., Sarkis, J., & Lai, K. H. (2008). GSCM implications for closing the loop. *Transportation Research Part E, Logistics and Transportation Review, 44*(1), 1–18. DOI: 10.1016/j.tre.2006.06.003

Zhu, X., Zhao, L., & Liu, J. (2021). Sustainable packaging design: A review on the development of bio-based and biodegradable materials. *Journal of Cleaner Production, 280*, 124420. DOI: 10.1016/j.jclepro.2020.124420

Chapter 13
Blockchain and Supply Chain Transparency:
Bibliometric and Visualization Analysis

Ghizlene Ouhrir
https://orcid.org/0009-0004-7523-5266
Cadi Ayyad University, Morroco

Nawfal Bahha
Cadi Ayyad University, Morroco

ABSTRACT

In the extremely dynamic environment, the concept of digitalization and supply chain transparency have attracted increasing attention from researchers and practitioners, due to increasing the social, economic and environmental challenges in recent years. To address these challenges, supply chain actors have been obliged to improve their supply chain transparency, both upstream and downstream, to strengthen their proactive and reactive capabilities in this competitive business environment. Due to the complexity of supply chain, digitalization appears as inevitable pillar to improve transparency of flows between different actors involved in this network in real time, by implementing industry 4.0 technologies. Blockchain is one such advanced technology has been discussed in literature to enable supply chain transparency. Our chapter aims to map scientific production of the nexus between blockchain technology and supply chain transparency using bibliometric approach. A total of 745 documents published in the Scopus database from 2017 to 2025 have been analyzed via Vosviewer software.

DOI: 10.4018/979-8-3373-3012-9.ch013

INTRODUCTION

In this competitive era, no organization can guarantee its survival, without transparence of its supply chain operations in real time. With complexity and ambiguity of environment, transparent supply chain has become a prerequisite for building a resilient supply chain and achieving sustainability. Supply chain transparency is defined as "an indicator of the quality, availability, accuracy, accessibility and actuality of supply chain data" (Bastian & Zentes, 2013). Transparent supply chain guarantees the smooth flow of information's and clarity, represent two pivotal elements for decision making (Ahmed & Omar, 2019). Data availability in real and opportune time offers numerous advantages for supply chain actors, including managing supply chain risks, improving supply chain efficiency, and strengthening trust between them (Sodhi & Tang, 2019).

In recent years, the world has experienced a multitude of disruptive events that significantly impacted supply chain management and increase its vulnerability when facing these events. This has prompted researchers and practicians to find methods to improve supply chain transparency to manage both expected and unexpected changes. In this context, the researchers considered digitalization as a catalysator allows organizations to increase the transparency of their supply chain operations, by implementing advanced technologies such as blockchain. Blockchain technology (BCT) is a " digital distributed ledger that stores data in the form of blocks that are linked together using a cryptographic function"(Tripathi et al., 2023). Integrating blockchain into supply chain activities leads to decentralizing the management of the supply chain that contributes to an improvement in the efficiency and security of supply chain (Alabaddi et al., 2023; Queiroz et al., 2020). Numerous studies have been published in recent years investigating the role of blockchain technology for supply chain transparency (Gligor et al., 2022; Xu et al., 2021; Zelbst et al., 2020). Due to increased interest to improve supply chain transparency, we decided to proceed a bibliometric analysis at the nexus between blockchain technology and supply chain transparency. The aim of this study is to answer the questions:

- Q1: Which journals, authors and countries are contributing to evolution of research in the field of blockchain technology and supply chain transparency?
- Q2: What are the main major research themes in the literature on the blockchain technology and supply chain transparency nexus?

The paper is organized as follows. Firstly, we present the literature review and secondly, we describe the methodology used for this study and then we present the findings of this bibliometric analysis.

LITERATURE REVIEW

Blockchain Technology

Blockchain is a distributed ledger technology, it is considered as a key technology of industry 4.0 (Wamba & Queiroz, 2022).This technology refers to a "chain of blocks, serving both as a repository of information and as a secured ledger for transactions. Each block keeps a list of recent transactions, and once the block is complete, then it becomes a part of the permanent blockchain. The database updates with each new block that comes along, allows for secure money, property, and contract transfers"(Kumar et al., 2025) . Blockchain is a peer-to-peer technology, characterized by decentralized structure that record and store data chronologically and employs distributed concord algorithms to produce and upgrade databases (M. Li et al., 2020) . It was applied for the first time in 2009 to support cryptocurrency transactions (M. Wang et al., 2020). The primary application of blockchain was Bitcoin (Iranmanesh et al., 2023). It was linked by the financial sector, it allows to ensure with transparency the financial transactions realized. This technology is based on cryptology techniques to protect access and exchange of data (Iranmanesh et al., 2023) . It is characterized by automated transactions and smart contracts to control and execute data (Z. Li et al., 2018; Rejeb & Rejeb, 2020) . Blockchain is a technology that guarantees the integrity, transparency, and traceability of data (Rejeb & Rejeb, 2020; Wong et al., 2020; Yu et al., 2021). Therefore, the records of transactions carried out are made in an immutable, transparent and secure manner (M. Li et al., 2020; Wong et al., 2020). Moreover, we can distinguish two main types of blockchain: public blockchain, and private blockchain(Agrawal et al., 2021). A public blockchain is open, everyone to have access to the data (Agi & Jha, 2022). This type of blockchain constitutes the backbone in cryptocurrency transactions such a Bitcoin and Ethereum(Kumar et al., 2025) . Contrary to public blockchain, private blockchain is restricted to a limited number of people that must adequate permission to record and access to data (Agi & Jha, 2022). This type of blockchain offers more confidentiality of data and quick transaction processing (Kumar et al., 2025). But both types (public and private) are characterized by implementation of consensus mechanisms to verify, validate the data before being stored in the blockchain (Agi & Jha, 2022).

Blockchain technology has generated great interest from various industries in recent years, when different organizations of different sizes are clamouring to adopt it in their activities to obtain a competitive advantage in this competitive market (Ghobakhloo et al., 2021; Maroufkhani et al., 2022; Morkunas et al., 2019). For example, in the healthcare sector, blockchain allows to strengthen the reproducibility and transparency of clinical trials through the recording of trial data and verification of results (McGhin et al., 2019). In the energy and sustainability sector allows to create an auditable system to control carbon emissions (Kumar et al., 2025) .

In parallel with these advantages offered by blockchain, certain barriers for adopting this technology are highlighted in the literature, such a technological, organizational and environmental context (Kumar Singh et al., 2023).

Blockchain And Supply Chain

In a disruptive environment, an effective management of supply chain needs a transparency of flows and process between actors of chain in a real and opportune time, to anticipate the risks and to quickly react to reduce their impact. The lack of this transparency constitutes a barrier to gain competitive advantages. Improving supply chain transparency allows to better control of supply chain operations (Ada et al., 2021), and strengthen the collaboration between partners (Agi & Jha, 2022). This transparency challenge supply chain actors and compel them to adopt the advanced technologies to achieve sustainable advantages (Badr, 2022). These technologies are advancing supply chains significantly (Bensfia, 2025; Bentalha et al., 2019). As a digital technology, blockchain guarantees the full traceability and improve visibility of downstream and upstream of different supply chain activities(S. Khan et al., 2023), that can reduce the complexity and ambiguity of supply chain processes(Brandon-Jones et al., 2014). As a results, supply chain actors can interact between them, create, verify, validate and record the different information linked a product, such as certificates, localization, records and transactions records (Agi & Jha, 2022). This could help to eliminate supply chain intermediaries, and increase trust between actors (S. Khan et al., 2023). This trust is an indispensable element for optimal collaboration between supply chain actors (Tönnissen & Teuteberg, 2020). Also, through the collaboration and effective communication, supply chain partners can adapt and respond quickly to changes and disruptions. Therefore, improving supply chain transparency can contribute to achieve performance objectives(Agi & Jha, 2022).

RESEARCH METHODOLOGY

Bibliometric analysis is a rigorous technique, utilized to analyze a large volume of scientific publications in a research area (Donthu et al., 2021). This method offers a transparent and systematic presentation of data using indicators such as co-authorship analysis, keywords co-occurrence.

For our analysis bibliometric, the data was extracted from "Scopus" database. This database is a multidisciplinary source of peer-reviewed publications, includes scientific articles, books, chapters and conference proceedings. The search string used was as follows: "("Blockchain" OR "Block chain" OR "BCT" OR "Distributed ledger" OR "Digital ledger" OR "Shared ledger") AND (supply chain*) AND (Transparen*)". This research equation was applied in the title, abstract and keywords. The publication period extended from 2017 to 2025. The field of study was limited to business and management disciplines. 745 documents were identified to match the criteria for bibliometric analysis. About 54,8% of the 745 documents are articles (i.e. 408 articles), followed by book chapter (20,8%), conference papers (17,4%), and review represent 5% of the documents in this study, see figure 1.

Figure 1. Document Types

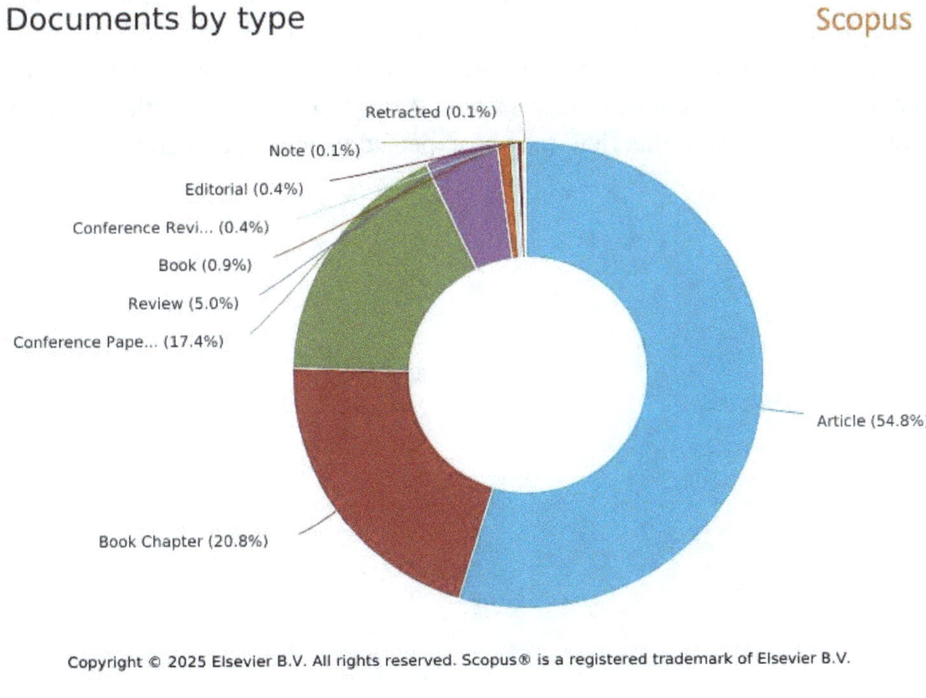

RESULTS

Annual Scientific Production

The annual scientific production is an indicator that reflets the evolution of a scientific community interests regarding a topic. According to the results illustrated in figure 2, we observe an upward trend, especially in the last 3 years. This indicate that the interest in studying the relationship between blockchain technology and supply chain transparency has increased over the years.

The number of publications is growing at an annual rate of 55,77%. In 2017 and 2018, the research characterized by a weak rhythm that does not exceed 10 documents per year. From 2019, we noticed a significant increase to achieve 244 publications in 2024, where the annual rate of this period (2019-May 2025) is 44%. Moreover, it should be noted that 94,76% of the publications were published over the last 5 years, 31.12% were published in the last 3 years.

These results demonstrate that the nexus of blockchain technology and supply chain transparency is an emerging research field due to its increased awareness among researchers and practitioners. In addition to that, the emergence of corona-virus in late 2019 accelerated the rate of adoption of advanced technologies such as blockchain to ensure the availability of data in order to track, to control the supply chain from end to end in real time, to improve the communication and exchange between supply chain actors, which allows to anticipate and to react in an efficient and effective way in the face of disruptions. With the dynamism of the world, we suppose that research in this field may be expected to continue to grow.

Figure 2. Evolution of publications per year

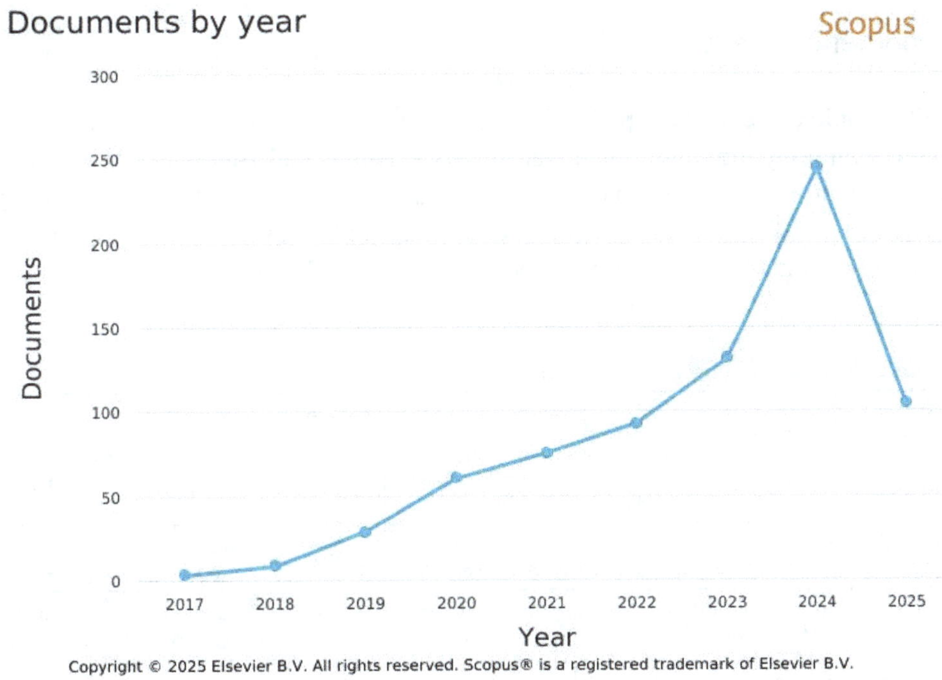

Distribution of Publications By Journal

The 745 documents selected has been extracted from 358 sources. The top 10 sources represent 20% of the total scientific production, with 146 scientific references, see figure 3. The journal "International Journal Of Production Research" is the most productive journal with 30 articles (i.e.4,02% of the references), flowed by "Journal Of Cleaner Production" with 20 articles (i.e.2,68% of the references) and "Logistics" with 15 articles (i.e.2,013% of the references).

Figure 3. Distribution of publications by journal

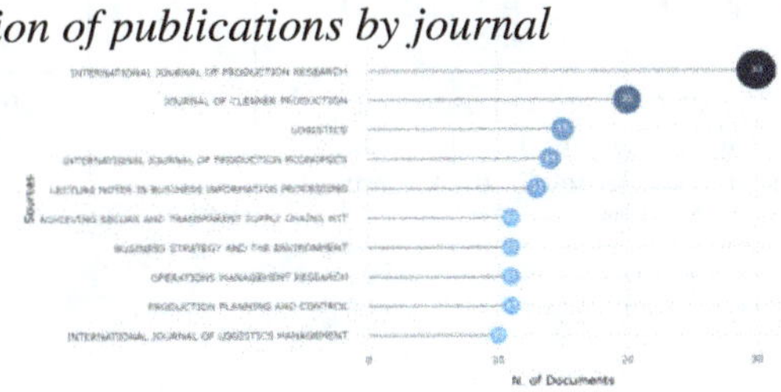

Citation Analysis

Documents Citation

The table presents the top 10 scientific publications cited in the nexus of blockchain technology and supply chain transparency. The article" Blockchain technology and its relationships to sustainable supply chain management", published in 2019 by "Sara Seberi, Mahtab Kouhiyadeh, Joseph Sarkis and Lejia shen" in "International Journal of Production Research" is the most cited article with 2527 citations.

Table 1. Top cited scientific publications

Reference	Title	Citation	Literature Sources
(Saberi et al., 2019)	Blockchain technology and its relashionships to sustainable supply chain management	2527	International Journal of Production Research
(Kshetri, 2018)	1 Blockchain's roles in meeting key supply chain management objectives	1543	International journal of information management
(Dutta et al., 2020)	Blockchain technology in supply chain operations: applications, challenges and research opportunities	1056	Transportation research part e: logistics and transportation review
(Kouhizadeh et al., 2021)	Blockchain technology and the sustainable supply chain: theoretically exploring adoption barriers	878	International Journal of Production economics
(Y. Wang et al., 2018)	Understanding blockchain technology for future supply chains: a systematic literature review and research agenda	867	Supply chain management
(Wüst & Gervais, 2018)	Do you need a blockchain?	751	2018 Crypto Valley Conference on Blockchain Technology (CVCBT)
(S. S. Kamble et al., 2020)	Modeling the blockchain enabled traceability in agriculture supply chain	717	International journal of information management
(Tian,2017)	A supply chain traceability system for food safety based on haccp, blockchain& internet of things	704	14th International Conference on Service Systems and Service Management
(S. Kamble et al., 2019)	Understanding the blockchain technology adoption in supply chains-Indian context	697	International Journal of Production Research
(Pournader et al., 2020)	Blockchain applications in supply chains, transport and logistics: a systematic review of the literature	643	International Journal of Production Research

Source: Authors' own
Sources citation

The network of the most cited sources by the authors is illustrated in Figure 4. After applying the criterion of a minimum of 5 documents per source, only 29 sources were selected from 358, dispersed across 9 clusters. "International Journal Of Production Research" is classified as the most cited source by the authors with 6442 citations and a total link strength of 219, followed by the journal "International Journal of Production Economics" with 2186 citations and a total link strength of 33.

Figure 4. Citation sources

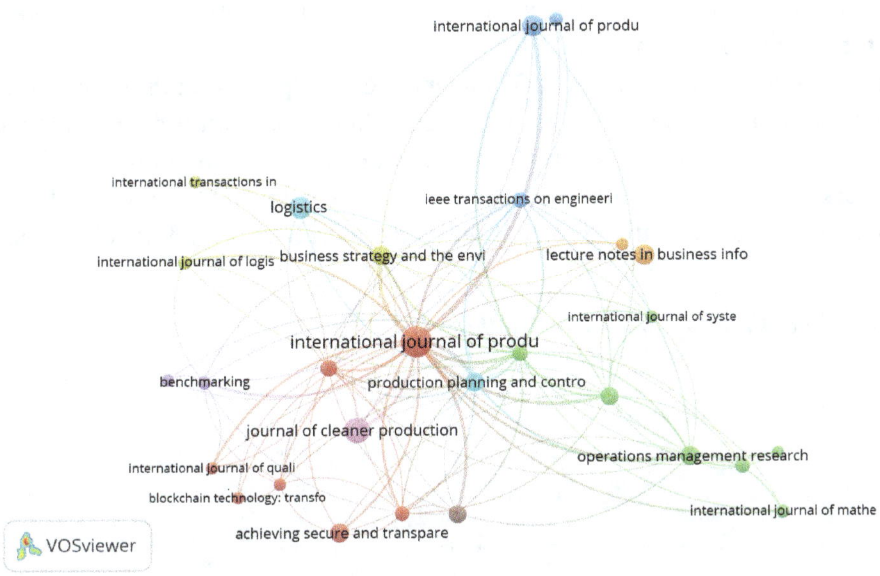

Co-Authorship Analysis

Co-authorship is an indicator that examine the strength of collaboration and co-production in scientific research for authors and countries which are linked to each other according to the number of publications that they jointly wrote.

Authors

Figure 5 illustrates the co-authors network based on the collaboration between the researchers in the nexus of blockchain and supply chain transparency. The big nodes represent the most influential authors in this topic. The links between nodes show the cooperation relationship between the authors.

The co-authors network represents the authors with at least two publications. Out of 2106 authors, 229 authors were selected and dispersed in 85 clusters. The majority of clusters are disconnected. The largest set of connected items consists of 26 items, see figure 5. This suggests that the weak link between the authors is due to the novelty of the subject.

In terms of the number of scientific papers, the most prolific authors are Sarkis Joseph with 10 documents, Flowed by Panghal Anupama with 9 documents and Vern Priyanka with 8 documents. However, authors citations exhibit a different result, Sarkis Joseph has a high number of citations with 4093 citations, flowed by Kouhizadeh Mahtab with 3576 citations and Saberi Sara with 3405 citations.

Figure 5. Co-authorship analysis of authors

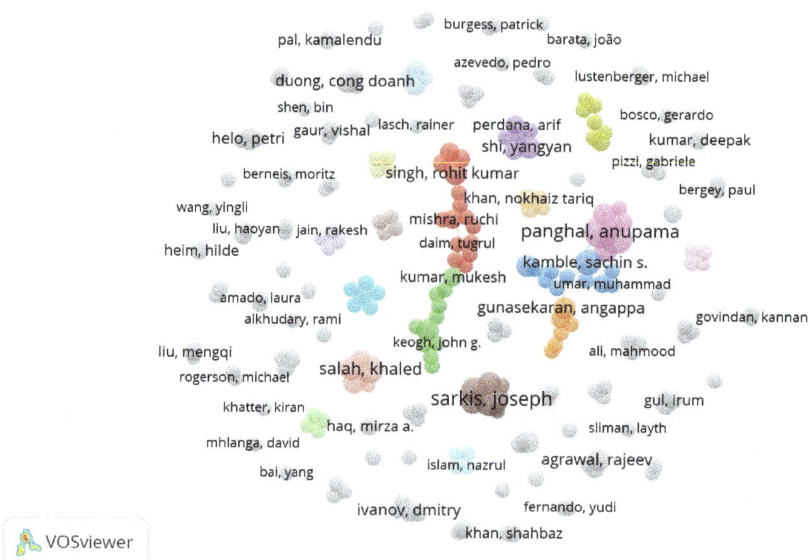

Figure 6. The largest set of connected authors

Co-Authorship Analysis Of Countries

Figure 7 shows collaboration network between countries, with a minimum threshold of 2 per country. 63 countries were selected among 89 countries and grouped in 12 clusters.

According to the number of links, the countries that collaborate most with other countries are: United Kingdom (35 links, total link strength of 137), followed by India (34 links, total link strength of 113), United States (27 links, total link strength of 85), China (25 links, total link strength of 84) and Malaysia (25 links, total link strength of 157).

The major research on blockchain and supply chain transparency is being carried out in the India and United Kingdom. They represent 41,07% of the total scientific production. India contributes with 205 documents (i.e. 27,51%) and United Kingdom with 101 documents (i.e. 13,55%).

Based on this analysis, it can be concluded that research at the nexus of blockchain technology and supply chain transparency is global in scope.

Figure 7. Co-authorship analysis of countries

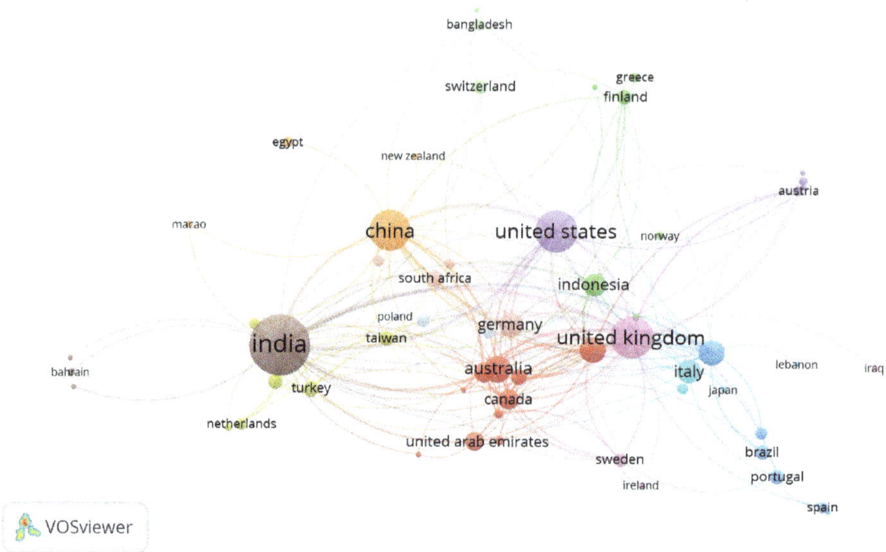

Keywords Analysis

Keywords reflect a precise description of content of publications. The objective of analyzing the co-occurrence of keywords is to identify the research topics trend in blockchain technology and supply chain transparency studies in the current research domain. to map the keywords co-occurrence network, we have fixed the number of occurrences at 5, and we also used thesaurus file to structure and to clean up the data. This analysis allowed us to identify 3 clusters, see figure 8.

Figure 8: Keywords co-occurrence analysis

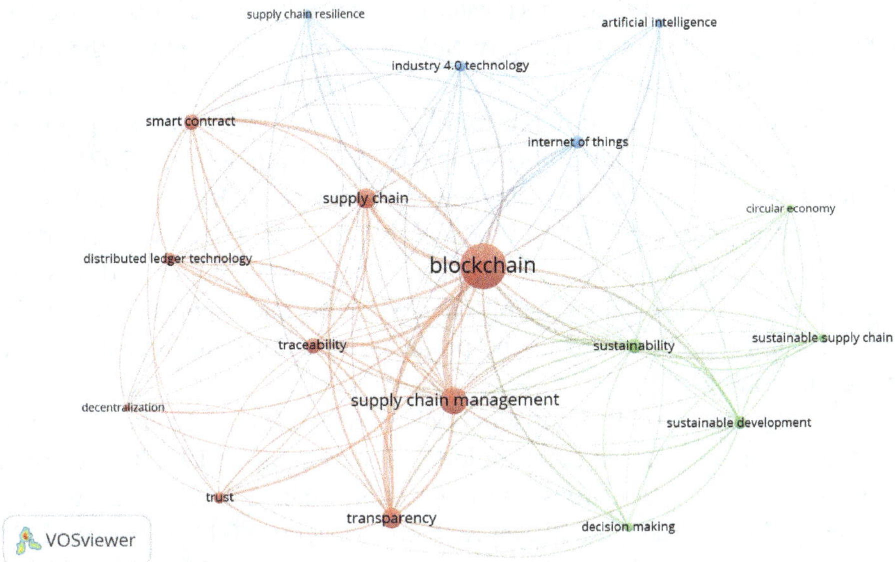

Cluster 1: BCT Improve Transparency And Traceability For Trust In Supply Chain

Cluster 1 is colored red. The first cluster emphasizes the role of transparency and traceability in building trust in supply chain network through blockchain technology.

Blockchain introduces traceability and transparency into supply chain operations through contract smart and its decentralized architecture (Sakib, 2024). Blockchain as a distributed ledger technology allowing transactions to be recorded securely and reliably in blocks (Marpina et al., 2024). Access to these blocks is limited to actors with a cryptography code (Azzi et al., 2019) to prevent modification and suppression of recorded data(Cole et al., 2019) .

Traceability, transparency and security of data favor the collaboration, the exchange of information, resources, and communication between actors which strengthens the trust between them (Yavaprabhas et al., 2024). Dubey et al., (2020) argue that blockchain in the supply chain has a positive impact on supply chain transparency and enhancing swift-trust between supply chain partners.

Cluster 2: Blockchain and Supply Chain Sustainability

Cluster 2 is the green cluster. This cluster is related to the benefits of blockchain technology in supply chain sustainability. Several researchers examined the role of blockchain for sustainability. Khan et al.(2022) indicated that there is a significant indirect effect of blockchain, through supply chain integration and supply chain mapping on supply chain sustainability. Relatedly, S. S. Kamble et al., (2021) examined the intermediate role of supply chain integration between BCT and sustainable supply chain performance. Similarly, Shujaat Mubarik et al., (2023) confirmed that there is an important impact of BCT on supply chain sustainability, and they also showed that the BCT can enhance supply chain visibility and supply chain mapping by offering a transparent record of the products, inventory and transactions for achieving sustainability.

Economically, adopting BCT in supply chain network reduces the number of intermediaries between network stakeholders (Saberi et al., 2019) . This disintermediation leads to cost reductions and transaction times(Saberi et al., 2019), to improve the quality of products and supply chain efficacity (Mukherjee et al., 2022) and higher profit margins(Ghobakhloo et al., 2023).

Blockchain through tractability and transparency of process can promote the integration of circular economy practices in supply chain operations (Ghobakhloo et al., 2023). This improves environmental performance in terms of raw material consumption and energy efficiency (Belhadi et al., 2018).

So, BCT has a significant impact on triple line of supply chain sustainability (economic, social and environmental) as Rejeb & Rejeb, (2020) mentioned in their systematic review.

Cluster 3: BCT, as an Industry 4.0 Technology Enhances Supply Chain Resilient.

Cluster 3 is the blue cluster. It revolves on the contribution of industry 4.0 such as the internet of things, artificial intelligence and blockchain to supply chain resilience.

The contributions of technologies tools to supply chain resilience has been treated in numerous studies. Huang et al., (2023) examined the effect of 16 industry 4,0 technologies on supply chain resilience trough mediating roles of supply chain collaboration and supply chain visibility. Moreover, Balakrishnan & Ramanathan (2021) showed that the use of technologies tools strengthens supply chain practices in case of disruptions, which allows the company to recover quickly and to improve its commercial performance. So, adopting the advanced technologies has a positive potential on supply chain resilience. For example, the study of (Modgil et al., 2022) indicated that artificial intelligence allows to identify the potential risks, to analyze exposure to these risks using hypothetical scenarios and resistance test, to reconfigure the action plans most suited to the perturbations, which guarantees the optimal performance of supply chain in case of disruptions such covid-19. For Blockchain technology, G. Li et al. (2022) demonstrated that novelty and efficiency centered business model design (BMD) based on blockchain technology positively impact agility as a reactive component of supply chain resilient, whereas only blockchain enabled efficiency centered BMD improves supply chain robustness as a proactive supply chain resilient component.

CONCLUSION

The objective of this study is to map the state of research on the relationship between Blockchain technology and supply chain transparency. For this reason, we realized bibliometric analysis to identify scientific production evolution, principal journal, authors, citation analysis, co-authorship network between authors and countries, and principal research clusters. This bibliometric includes 745 documents indexed in the database "Scopus", covering the period from 2017 to 2025.

Chronologic results showed that this research area is emergent, the scientific production has seen an unprecedented increase in the number of papers after covid-19, and the production rate continues to grow to show the advantages of adopting blockchain technology in supply chain in this dynamic world. In terms of productivity, the results showed Sarkis Joseph, Panghal Anupama and Vern Priyanka are the authors with the highest productivity and that the journal "International Journal Of Production Research" is the most productive journal. In terms of collaboration, we concluded that there are weak links between the authors due to the novelty of this topic. This field search attracts worldwide attention, where the productivity is centralized in United Kingdom and India. Finally, keywords occurrence analysis allowed us to identify 3 major research related to this nexus. These clusters constitute a complete overview of current research trends in blockchain technology and supply chain transparency. According to these clusters, we can propose a set of questions for future research:

- How does blockchain contribute to supply chain transparency?
- How does transparency contribute to supply chain resilience and sustainability?
- What are the barriers to adopting blockchain in the Moroccan context?

Nevertheless, several limits are noteworthy in this research. Firstly, an important limitation is related to the exclusive use of "Scopus" as a data source, which could exclude an important amount of literature that could have enriched our analysis.

Secondly, in our bibliometric analysis, we have not explored bibliographic coupling, and co-citation analyses. Therefore, for future developments, it would be primordial to include more database and to exploit the other bibliometric tools such bibliographic coupling, and co-citation analyses.

REFERENCES

Ada, E., Sagnak, M., Kazancoglu, Y., Luthra, S., & Kumar, A. (2021). A Framework for Evaluating Information Transparency in Supply Chains. [JGIM]. *Journal of Global Information Management, 29*(6), 1–22. DOI: 10.4018/JGIM.20211101.oa45

Agi, M. A., & Jha, A. K. (2022). Blockchain technology in the supply chain : An integrated theoretical perspective of organizational adoption. *International Journal of Production Economics, 247*, 108458. DOI: 10.1016/j.ijpe.2022.108458

Agrawal, T. K., Kumar, V., Pal, R., Wang, L., & Chen, Y. (2021). Blockchain-based framework for supply chain traceability : A case example of textile and clothing industry. *Computers & Industrial Engineering, 154*, 107130. DOI: 10.1016/j.cie.2021.107130

Ahmed, W., & Omar, M. (2019). Drivers of Supply Chain Transparency and its effects on Performance Measures in the Automotive Industry : Case of a Developing Country. *International Journal of Services and Operations Management, 33*(2), 159–186. DOI: 10.1504/IJSOM.2019.100291

Alabaddi, Z., Obidat, A., & Alziyadat, Z. (2023). Exploring the effect of blockchain technology on supply chain resilience and transparency : Evidence from the healthcare industry. *Uncertain Supply Chain Management, 11*(2), 787–798. Advance online publication. DOI: 10.5267/j.uscm.2023.1.001

Azzi, R., Chamoun, R. K., & Sokhn, M. (2019). The power of a blockchain-based supply chain. *Computers & Industrial Engineering, 135*, 582–592. DOI: 10.1016/j.cie.2019.06.042

Badr, B. (2022). Smart Technologies for a Sustainable Service Supply Chain : A Prospective Perspective. *Journal of Environmental Issues and Climate Change, 1*(1), 1. Advance online publication. DOI: 10.59110/jeicc.v1i1.81

Balakrishnan, A. S., & Ramanathan, U. (2021). The role of digital technologies in supply chain resilience for emerging markets' automotive sector. *Supply Chain Management, 26*(6), 654–671.

Bastian, J., & Zentes, J. (2013). Supply chain transparency as a key prerequisite for sustainable agri-food supply chain management. *International Review of Retail, Distribution and Consumer Research, 23*(5), 553–570. DOI: 10.1080/09593969.2013.834836

Belhadi, A., Touriki, F. E., & El Fezazi, S. (2018). Benefits of adopting lean production on green performance of SMEs : A case study. *Production Planning and Control, 29*(11), 873–894. DOI: 10.1080/09537287.2018.1490971

Bensfia, C. (2025). Collective Intelligence of Port Territory Stakeholders : Case of Moroccan Ports. In *Utilizing Technology to Manage Territories* (p. 335-370). IGI Global Scientific Publishing. DOI: 10.4018/979-8-3693-6854-1.ch011

Bentalha, B., Hmioui, A., & Alla, L. (2019). The digitalization of the supply chain management of service companies : A prospective approach. *Proceedings of the 4th International Conference on Smart City Applications*, 1-8. DOI: 10.1145/3368756.3369005

Brandon-Jones, E., Squire, B., Autry, C. W., & Petersen, K. J. (2014). A Contingent Resource-Based Perspective of Supply Chain Resilience and Robustness. *The Journal of Supply Chain Management*, *50*(3), 55–73. DOI: 10.1111/jscm.12050

Cole, R., Stevenson, M., & Aitken, J. (2019). Blockchain technology : Implications for operations and supply chain management. *Supply Chain Management*, *24*(4), 469–483. DOI: 10.1108/SCM-09-2018-0309

Donthu, N., Kumar, S., Mukherjee, D., Pandey, N., & Lim, W. M. (2021). How to conduct a bibliometric analysis : An overview and guidelines. *Journal of Business Research*, *133*, 285–296. DOI: 10.1016/j.jbusres.2021.04.070

Dubey, R., Gunasekaran, A., Bryde, D. J., Dwivedi, Y. K., & Papadopoulos, T. (2020). Blockchain technology for enhancing swift-trust, collaboration and resilience within a humanitarian supply chain setting. *International Journal of Production Research*, *58*(11), 3381–3398. DOI: 10.1080/00207543.2020.1722860

Dutta, P., Choi, T.-M., Somani, S., & Butala, R. (2020). Blockchain technology in supply chain operations : Applications, challenges and research opportunities. *Transportation Research Part E, Logistics and Transportation Review*, *142*, 102067. DOI: 10.1016/j.tre.2020.102067 PMID: 33013183

Ghobakhloo, M., Iranmanesh, M., Grybauskas, A., Vilkas, M., & Petraitė, M. (2021). Industry 4.0, innovation, and sustainable development : A systematic review and a roadmap to sustainable innovation. *Business Strategy and the Environment*, *30*(8), 4237–4257. DOI: 10.1002/bse.2867

Ghobakhloo, M., Iranmanesh, M., Mubarik, M. S., Mubarak, M. F., Amran, A., & Khanfar, A. A. A. (2023). Blockchain technology as an enabler for sustainable business ecosystems : A comprehensive roadmap for socioenvironmental and economic sustainability. *Business Strategy & Development*, *7*(1), e319. DOI: 10.1002/bsd2.319

Gligor, D. M., Davis-Sramek, B., Tan, A., Vitale, A., Russo, I., Golgeci, I., & Wan, X. (2022). Utilizing blockchain technology for supply chain transparency : A resource orchestration perspective. *Journal of Business Logistics*, *43*(1), 140–159. DOI: 10.1111/jbl.12287

Huang, K., Wang, K., Lee, P. K., & Yeung, A. C. (2023). The impact of industry 4.0 on supply chain capability and supply chain resilience : A dynamic resource-based view. *International Journal of Production Economics*, *262*, 108913. DOI: 10.1016/j.ijpe.2023.108913

Iranmanesh, M., Maroufkhani, P., Asadi, S., Ghobakhloo, M., Dwivedi, Y. K., & Tseng, M.-L. (2023). Effects of supply chain transparency, alignment, adaptability, and agility on blockchain adoption in supply chain among SMEs. *Computers & Industrial Engineering*, *176*, 108931. DOI: 10.1016/j.cie.2022.108931

Kamble, S., Gunasekaran, A., & Arha, H. (2019). Understanding the Blockchain technology adoption in supply chains-Indian context. *International Journal of Production Research*, *57*(7), 2009–2033. DOI: 10.1080/00207543.2018.1518610

Kamble, S. S., Gunasekaran, A., & Sharma, R. (2020). Modeling the blockchain enabled traceability in agriculture supply chain. *International Journal of Information Management*, *52*, 101967. DOI: 10.1016/j.ijinfomgt.2019.05.023

Kamble, S. S., Gunasekaran, A., Subramanian, N., Ghadge, A., Belhadi, A., & Venkatesh, M. (2021). Blockchain technology's impact on supply chain integration and sustainable supply chain performance : Evidence from the automotive industry. *Annals of Operations Research*, *327*(1), 575–600. DOI: 10.1007/s10479-021-04129-6

Khan, S., Haleem, A., Husain, Z., Samson, D., & Pathak, R. D. (2023). Barriers to blockchain technology adoption in supply chains : The case of India. *Operations Management Research : Advancing Practice Through Research*, *16*(2), 668–683. DOI: 10.1007/s12063-023-00358-z

Khan, S. A., Mubarik, M. S., Kusi-Sarpong, S., Gupta, H., Zaman, S. I., & Mubarik, M. (2022). Blockchain technologies as enablers of supply chain mapping for sustainable supply chains. *Business Strategy and the Environment*, *31*(8), 3742–3756. DOI: 10.1002/bse.3029

Kouhizadeh, M., Saberi, S., & Sarkis, J. (2021). Blockchain technology and the sustainable supply chain : Theoretically exploring adoption barriers. *International Journal of Production Economics*, *231*, 107831. DOI: 10.1016/j.ijpe.2020.107831

Kshetri, N. (2018). 1 Blockchain's roles in meeting key supply chain management objectives. *International Journal of Information Management, 39*, 80–89. DOI: 10.1016/j.ijinfomgt.2017.12.005

Kumar, N., Kumar, K., Aeron, A., & Verre, F. (2025). Blockchain technology in supply chain management : Innovations, applications, and challenges. *Telematics and Informatics Reports, 18*, 100204. DOI: 10.1016/j.teler.2025.100204

Kumar Singh, A., Kumar, V. R. P., Dehdasht, G., Mohandes, S. R., Manu, P., & Pour Rahimian, F. (2023). Investigating the barriers to the adoption of blockchain technology in sustainable construction projects. *Journal of Cleaner Production, 403*, 136840. DOI: 10.1016/j.jclepro.2023.136840

Li, G., Xue, J., Li, N., & Ivanov, D. (2022). Blockchain-supported business model design, supply chain resilience, and firm performance. *Transportation Research Part E, Logistics and Transportation Review, 163*, 102773. DOI: 10.1016/j.tre.2022.102773

Li, M., Shao, S., Ye, Q., Xu, G., & Huang, G. Q. (2020). Blockchain-enabled logistics finance execution platform for capital-constrained E-commerce retail. *Robotics and Computer-integrated Manufacturing, 65*, 101962. DOI: 10.1016/j.rcim.2020.101962

Li, Z., Wang, W. M., Liu, G., Liu, X., He, J., & Huang, G. Q. (2018). Toward open manufacturing : A cross-enterprises knowledge and services exchange framework based on blockchain and edge computing. *Industrial Management & Data Systems, 118*(1), 303–320. DOI: 10.1108/IMDS-04-2017-0142

Maroufkhani, P., Desouza, K. C., Perrons, R. K., & Iranmanesh, M. (2022). Digital transformation in the resource and energy sectors : A systematic review. *Resources Policy, 76*, 102622. DOI: 10.1016/j.resourpol.2022.102622

Marpina, Y., Pedarla, P., & Juluri, G. (2024). Blockchain for Supply Chain Transparency in Manufacturing. In *Futuristic Technology for Sustainable Manufacturing* (pp. 75–90). IGI Global Scientific Publishing., DOI: 10.4018/979-8-3693-1862-1.ch005

McGhin, T., Choo, K.-K. R., Liu, C. Z., & He, D. (2019). Blockchain in healthcare applications : Research challenges and opportunities. *Journal of Network and Computer Applications, 135*, 62–75. DOI: 10.1016/j.jnca.2019.02.027

Modgil, S., Gupta, S., Stekelorum, R., & Laguir, I. (2022). AI technologies and their impact on supply chain resilience during COVID-19. *International Journal of Physical Distribution & Logistics Management, 52*(2), 130–149. DOI: 10.1108/IJPDLM-12-2020-0434

Morkunas, V., Paschen, J., & Boon, E. (2019). How blockchain technologies impact your business model. *Business Horizons, 62*(3), 295–306. Advance online publication. DOI: 10.1016/j.bushor.2019.01.009

Mukherjee, A. A., Singh, R. K., Mishra, R., & Bag, S. (2022). Application of blockchain technology for sustainability development in agricultural supply chain : Justification framework. *Operations Management Research : Advancing Practice Through Research, 15*(1), 46–61. DOI: 10.1007/s12063-021-00180-5

Pournader, M., Shi, Y., Seuring, S., & Koh, S. C. L. (2020). Blockchain applications in supply chains, transport and logistics : A systematic review of the literature. *International Journal of Production Research, 58*(7), 2063–2081. DOI: 10.1080/00207543.2019.1650976

Queiroz, M. M., Telles, R., & Bonilla, S. H. (2020). Blockchain and supply chain management integration : A systematic review of the literature. *Supply Chain Management, 25*(2), 241–254. DOI: 10.1108/SCM-03-2018-0143

Rejeb, A., & Rejeb, K. (2020). Blockchain and supply chain sustainability. *Logforum, 16*(3). https://yadda.icm.edu.pl/baztech/element/bwmeta1.element.baztech-3e949e5f-8a67-4d5a-ad4f-342c28a6cb32

Saberi, S., Kouhizadeh, M., Sarkis, J., & Shen, L. (2019a). Blockchain technology and its relationships to sustainable supply chain management. *International Journal of Production Research, 57*(7), 2117–2135. DOI: 10.1080/00207543.2018.1533261

Saberi, S., Kouhizadeh, M., Sarkis, J., & Shen, L. (2019b). Blockchain technology and its relationships to sustainable supply chain management. *International Journal of Production Research, 57*(7), 2117–2135. DOI: 10.1080/00207543.2018.1533261

Sakib, S. M. N. (2024). Blockchain Technology for Smart Contracts : Enhancing Trust, Transparency, and Efficiency in Supply Chain Management. In *Achieving Secure and Transparent Supply Chains With Blockchain Technology* (p. 246-266). IGI Global Scientific Publishing. DOI: 10.4018/979-8-3693-0482-2.ch014

Shujaat Mubarik, M., Ahmed Khan, S., Kusi-Sarpong, S., & Mubarik, M. (2023). Supply chain sustainability in VUCA : Role of BCT-driven SC mapping and 'Visiceability'. *International Journal of Logistics, 0*(0), 1–19. DOI: 10.1080/13675567.2023.2222660

Sodhi, M. S., & Tang, C. S. (2019). Research Opportunities in Supply Chain Transparency. *Production and Operations Management, 28*(12), 2946–2959. DOI: 10.1111/poms.13115

Tönnissen, S., & Teuteberg, F. (2020). Analysing the impact of blockchain-technology for operations and supply chain management : An explanatory model drawn from multiple case studies. *International Journal of Information Management, 52*, 101953. DOI: 10.1016/j.ijinfomgt.2019.05.009

Tripathi, G., Ahad, M. A., & Casalino, G. (2023). A comprehensive review of blockchain technology : Underlying principles and historical background with future challenges. *Decision Analytics Journal, 9*, 100344. DOI: 10.1016/j.dajour.2023.100344

Wamba, S. F., & Queiroz, M. M. (2022). Industry 4.0 and the supply chain digitalisation : A blockchain diffusion perspective. *Production Planning and Control, 33*(2-3), 193–210. DOI: 10.1080/09537287.2020.1810756

Wang, M., Wu, Y., Chen, B., & Evans, M. (2020). Blockchain and supply chain management : A new paradigm for supply chain integration and collaboration. *Operations and Supply Chain Management: An International Journal, 14*(1), 111–122. DOI: 10.31387/oscm0440290

Wang, Y., Han, J. H., & Beynon-Davies, P. (2018). Understanding blockchain technology for future supply chains : A systematic literature review and research agenda. *Supply Chain Management, 24*(1), 62–84. Advance online publication. DOI: 10.1108/SCM-03-2018-0148

Wong, L.-W., Leong, L.-Y., Hew, J.-J., Tan, G. W.-H., & Ooi, K.-B. (2020). Time to seize the digital evolution : Adoption of blockchain in operations and supply chain management among Malaysian SMEs. *International Journal of Information Management, 52*, 101997. DOI: 10.1016/j.ijinfomgt.2019.08.005

Xu, P., Lee, J., Barth, J. R., & Richey, R. G. (2021). Blockchain as supply chain technology : Considering transparency and security. *International Journal of Physical Distribution & Logistics Management, 51*(3), 305–324. DOI: 10.1108/IJPDLM-08-2019-0234

Yavaprabhas, K., Pournader, Mehrdokht, & and Seuring, S. (2024). Blockchain and trust in supply chains : A bibliometric analysis and trust transfer perspective. *International Journal of Production Research, 0*(0), 1–28. DOI: 10.1080/00207543.2024.2389544

Yu, Y., Huang, G., & Guo, X. (2021). Financing strategy analysis for a multi-sided platform with blockchain technology. *International Journal of Production Research, 59*(15), 4513–4532. DOI: 10.1080/00207543.2020.1766718

Zelbst, P. J., Green, K. W., Sower, V. E., & Bond, P. L. (2020). The impact of RFID, IIoT, and Blockchain technologies on supply chain transparency. In *Journal of Manufacturing Technology Management* (Vol. 31, Numéro 3, p. 441-457). Emerald Group Holdings Ltd. DOI: 10.1108/JMTM-03-2019-0118

Chapter 14
An Exploratory Study on the Impact of Sustainability on Supply Chain Resilience:
The Case of the Pharmaceutical Sector in the Souss Massa Region

Imad Terrah
https://orcid.org/0009-0008-2489-0761
Faculty of Legal, Economic, and Social Sciences, Ibn Zohr University, Agadir, Morocco

Amina Kada
https://orcid.org/0009-0008-1487-5031
Polydisciplinary Faculty of Taroudant, Ibn Zohr University, Morocco

Youssef Elliasmine
https://orcid.org/0009-0005-8698-7292
Faculty of Legal, Economic, and Social Sciences, Ibn Zohr University, Agadir, Morocco

ABSTRACT

This study explores how sustainability practices influence supply chain resilience in the pharmaceutical sector of Morocco's Souss Massa region. Amid increasing global uncertainty and post-COVID disruptions, the research investigates the extent to which environmental, social, and economic sustainability initiatives support the robustness and adaptability of pharmaceutical logistics. Drawing on an interpretivist epistemological approach, the study employs a qualitative design based on seven

DOI: 10.4018/979-8-3373-3012-9.ch014

semi-structured interviews with key supply chain actors including manufacturers, distributors, and pharmacies. The findings reveal a mutual reinforcement between sustainability and resilience, with practices such as local sourcing, digitalization, and waste reduction enhancing operational continuity. However, implementation disparities and contextual constraints persist, particularly among smaller actors. The study contributes to a deeper understanding of sustainability-resilience interfaces by offering empirically grounded insights specific to regional supply chains in developing economies.

INTRODUCTION

In 2023 alone, over 300 supply chain disruptions were recorded globally, ranging from health emergencies to climate-related events and political unrest (Resilinc, 2023). These figures highlight a troubling reality: supply chains, once considered engines of efficiency, are now increasingly viewed through the lens of fragility and risk. Among the sectors most acutely affected stands the pharmaceutical industry, where even short-term interruptions in the availability of medicines and medical equipment can have life-threatening consequences—particularly in low- and middle-income regions where logistical flexibility is limited.

The importance of this research lies in its contribution to advancing understanding in the field of supply chain resilience within the context of sustainable development. By exploring the pharmaceutical sector in the Souss Massa region, this study aims to address the gap in knowledge surrounding how sustainability practices influence a supply chain's capacity to withstand and recover from disruptions. This investigation unfolds against the backdrop of a growing interest in understanding the interplay between environmental responsibility, operational robustness, and healthcare continuity.

Over the past decade, the concept of sustainability has evolved. Initially framed as an ethical or ecological imperative, it has become a central pillar in supply chain strategy. Today, sustainability encompasses a multidimensional framework that includes environmental stewardship, social equity, and economic viability. Increasingly, it is not merely seen as a moral obligation, but as a proactive mechanism for enhancing resilience.

Previous studies explored this relationship in various sectors, yet few have focused on pharmaceutical supply chains in emerging economies. The research questions guiding this inquiry are designed to shed light on how sustainability-oriented practices contribute to the resilience of supply chains operating in a region marked by logistical constraints and structural vulnerabilities.

This study aims to contribute to filling this gap by exploring how sustainability initiatives impact the resilience of pharmaceutical supply chains in Morocco's Souss Massa region. Guided by an interpretivist epistemology, this research adopts a qualitative design centered on semi-structured interviews with stakeholders from various stages of the pharmaceutical supply chain, including manufacturers, wholesalers, and retail pharmacists. This methodological choice enables an in-depth, contextually grounded understanding of how sustainability and resilience are conceptualized and enacted by supply chain actors in real-world operational settings.

The structure of this article unfolds as follows: The first section lays out the theoretical underpinnings by defining and linking the core concepts of sustainability and resilience within supply chains. The second section introduces the research setting and methodological approach. The third section presents the empirical findings and situates them in relation to existing literature. The final section offers concluding reflections, acknowledges the study's limitations, and outlines directions for future inquiry.

1. THEORETICAL BACKGROUND

1.1. Sustainability in Supply Chain Management

Sustainability has emerged as a pivotal, multifaceted concept at the crossroads of academia, industry, and policy-making, driven by the intensification of environmental crises, social disparities, and widespread vulnerabilities in global supply chains (Bentalha, 2022). At its essence, sustainability denotes the capacity to fulfill present needs without impairing the ability of future generations to satisfy theirs, as first articulated in the seminal Brundtland Report (WCED, 1987). Building upon this foundational perspective, Elkington (1998) introduced the influential "triple bottom line" model, which proposes that sustainability must concurrently balance economic viability, environmental responsibility, and social justice. This tripartite framework has since become central to sustainable development paradigms and is increasingly reflected in corporate and supply chain strategies.

Within the field of supply chain management (SCM), sustainability is broadly understood as the strategic incorporation of environmental, social, and economic objectives into the coordination of inter-organizational activities (Carter & Rogers, 2008; Seuring & Müller, 2008).

Recent research highlights the strategic value of embedding sustainability into supply chains, not only for enhancing corporate performance but also for reinforcing the resilience of supply networks as a whole (Touboulic & Walker, 2015; Zhu, 2024). Rather than serving purely ethical or reputational purposes, sustainability initiatives are increasingly viewed as mechanisms to mitigate operational risks, foster innovation, and respond to the growing volatility of global markets (Bouklata & Bensfia, 2020; Bentalha, Hmioui, and Alla, 2019; Bensfia, 2025).

From an environmental perspective, scholars have long emphasized the role of supply chains in either contributing to or mitigating ecological harm. Seuring and Müller (2008) and Sarkis (2012), for example, argue that sustainable SCM practices – such as reducing carbon emissions, optimizing energy consumption, and adopting cleaner production methods-can significantly lower a firm's ecological footprint. More recent studies have added practical examples, including the deployment of renewable energy in transport and storage, the use of biodegradable packaging, and the development of reverse logistics systems. While these measures aim to comply with increasingly strict environmental regulations, they also offer operational benefits in terms of cost efficiency and resource optimization.

The social dimension of sustainability, by contrast, remains less uniformly implemented across supply chains. While Carter and Rogers (2008) laid early groundwork for integrating social criteria—such as fair labor practices, workplace safety, and stakeholder inclusion-subsequent research has questioned the depth of such integration in practice. Morais (2019) and Khokhar (2020), for instance, highlight persistent gaps between corporate sustainability commitments and actual labor conditions within complex, multi-tiered supply networks. Critics argue that existing frameworks often oversimplify social responsibility by focusing on compliance metrics rather than systemic change.

Although historically underemphasized relative to environmental and economic concerns, social sustainability has gained prominence due to rising awareness of systemic injustices and increased stakeholder scrutiny. As highlighted by Khokhar (2020) and Saidi et al. (2024), prioritizing social equity can lead to enhanced transparency, stronger stakeholder trust, and more robust long-term collaborations across the supply chain.

Economically, sustainability within supply chains emphasizes the need to maintain financial health while adhering to responsible and inclusive practices. This dimension includes efforts toward cost efficiency, sustainable innovation, value generation, and the pursuit of profitability without undermining social or environmental responsibilities (Pagell & Wu, 2009; Lazar, 2021; Bensfia & Bouklata, 2022; Asif, 2024). In today's globalized and ethically aware market, focusing solely on short-term profits is no longer tenable. Instead, businesses are expected to integrate considerations such as job creation, inclusive economic growth, and regional development into their operational strategies.

As supply chain management continues to evolve, the concept of sustainable logistics has gained increased attention. It refers to the design and optimization of logistics functions, including transportation, warehousing, and distribution, through a sustainability lens (Sbihi & Eglese, 2007; Saidi et al., 2024). Examples of sustainable logistics practices include low-carbon transportation modes, energy-efficient infrastructure, green warehouses, and the adoption of circular economy principles. These efforts contribute to environmental protection while simultaneously reinforcing supply chain resilience and improving brand reputation, particularly in industries under intense regulatory and public scrutiny such as pharmaceuticals.

A genuinely sustainable supply chain extends beyond reducing negative externalities; it constitutes a dynamic and ethically governed ecosystem capable of adapting to change while delivering shared value across environmental, social, and economic fronts. As defined by Carter and Rogers (2008, p. 368), such a supply chain involves "the strategic, transparent integration and achievement of an organization's social, environmental, and economic goals in the systemic coordination of key inter-organizational business processes for improving the long-term economic performance of the individual company and its supply chains." This definition underscores the critical role of collaboration, accountability, and performance measurement in achieving comprehensive sustainability.

Recent literature further reinforces the need to evaluate sustainability from a life-cycle perspective, spanning from raw material sourcing to end-of-life management (Shashi, 2022; Silva et al., 2023; Zhu, 2024). Integrating environmental and social concerns allows companies to move beyond short-term compliance and toward more resilient, adaptable structures (Hmioui et al., 2020). Yet this transformation is far from uniform. It requires rethinking business models, investing in innovation, and fostering cooperation across stakeholders—challenges that are particularly acute in sectors like pharmaceuticals, where disruptions can carry critical societal consequences.

1.2. Supply Chain Resilience

This literature review critically examines how the concept of resilience has evolved within the domain of supply chain management. Originally viewed through a narrow operational lens, resilience was primarily defined as a system's capacity to absorb shocks and return to a pre-disruption state. Early research emphasized tactical responses such as inventory buffers, supply base diversification, or backup capacity to manage external turbulence (Rice & Caniato, 2003; Silva, Martins, & Pinto, 2023). While these mechanisms offered short-term protection, critics argue that such redundancy-based approaches fail to address the structural adaptability necessary for long-term resilience.

Over time, the literature has shifted toward a more nuanced and systemic understanding. Ivanov (2022) redefines resilience as a dynamic capability embedded throughout the entire supply network, capable of responding to complex, multi-layered disruptions. This reconceptualization reflects the increasing interdependence and fragility of global logistics, especially in critical sectors like pharmaceuticals (El Bakkouri, 2021).

Several scholars have advanced integrated frameworks that expand the temporal and functional dimensions of resilience. For instance, Gružauskas and Burinskienė (2022) propose a five-stage model—prevention, resistance, response, recovery, and learning—that captures both anticipatory and adaptive behaviors. Contrary to earlier reactive paradigms, this model encourages a proactive posture rooted in continuous improvement and systemic learning.

Building on these developments, Pettit, Fiksel, and Croxton (2010) introduced the concept of the "Zone of Resilience," where the interplay between a firm's vulnerabilities and capabilities determines its adaptive strength. This systems-based approach suggests that resilience does not reside in any single function or asset but emerges from the organization's holistic ability to balance internal robustness with external flexibility.

Ivanov (2018) furthers this discourse by highlighting the role of control systems, structural agility, and visibility in shaping resilient behavior. According to his view, resilience is not an intrinsic attribute but a composite of interdependent capabilities that evolve with contextual volatility.

More recently, the concept of supply chain ambidexterity has attracted scholarly attention. Aslam et al. (2020) define it as the dual ability to maintain operational efficiency under stable conditions while adapting swiftly in times of crisis. Their research emphasizes that agility acts as a mediating force that translates structural preparedness into effective real-time responses.

Alternative typologies have also emerged to capture the layered nature of resilience. Adobor and McMullen (2018) distinguish between engineering, ecological, and evolutionary resilience. While engineering resilience focuses on rapid restoration, ecological resilience highlights absorptive and reorganizational capacities, and evolutionary resilience emphasizes learning and transformation. This typology provides a comprehensive lens to understand resilience across different phases of disruption.

Other scholars have drawn attention to the distinction between absorptive and adaptive capacities. While absorptive capacity allows supply chains to buffer shocks, adaptive capacity reflects the ability to redesign processes, reconfigure networks, or innovate in response to evolving challenges (Bloem, 2022; Silva et al., 2023). This distinction underscores that effective resilience requires not only resistance to stress but also strategic reorientation in its aftermath.

Increasingly, researchers argue that resilience should be seen not just as survival but as an engine for transformation and value creation. Mishra (2024) and Odulaja (2023) suggest that resilient firms are those that do not merely endure disruption but use it as a catalyst for innovation and competitive repositioning. Golan, Jernegan, and Linkov (2021) support this view by framing resilience as a strategic function that empowers organizations to sense, anticipate, and capitalize on environmental volatility.

An emerging strand of research aligns resilience with sustainability objectives. Bamia (2025) and other proponents of this view contend that resilience cannot be divorced from environmental and social considerations. Practices such as circular economy integration, emissions reduction, and stakeholder inclusion are now seen as essential components of a sustainable-resilient supply chain. While the existing literature emphasizes operational continuity, this new wave highlights legitimacy, ethical accountability, and long-term viability.

Finally, the shift toward a network-centric perspective marks another turning point in resilience theory. Wieland and Durach (2021) and Liu et al. (2023) argue that resilience is not solely the responsibility of individual firms but a property of interconnected ecosystems. Effective resilience, in this view, requires inter-firm coordination, transparent information flows, and collaborative risk governance (Radebe & Chipangamate, 2024).

Despite these conceptual advancements, several gaps persist. One key issue is the absence of robust, operationalizable metrics for assessing resilience in complex settings. As Zou (2024) and Faggioni et al. (2024) note, without measurable indicators and clear definitions, translating theory into actionable strategy remains elusive.

1.3. The Influence of Sustainability on Supply Chain Resilience

Recent scholarship increasingly acknowledges the interdependence between sustainability and supply chain resilience, especially in the face of mounting environmental, sanitary, and geopolitical disruptions. Rather than viewing these as distinct objectives, many authors now frame them as mutually reinforcing pillars of modern supply chain design (Negri, 2021; Michel-Villarreal, 2022; Alla, Bentalha and Bouhtati 2022). This conceptual shift is particularly relevant to sectors vulnerable to systemic shocks, such as pharmaceuticals and agri-food, where the capacity to withstand crises is closely tied to ethical, environmental, and social practices (Carissimi, 2023).

This relationship finds strong theoretical grounding in the Triple Bottom Line (TBL) framework, which promotes the integration of environmental sustainability, social responsibility, and economic performance (Elkington, 1998). When applied to supply chain management, the TBL suggests that resilience depends not only on operational efficiency, but also on a system's ability to sustain social legitimacy and environmental compliance. For example, a financially robust supply chain that ignores labor conditions or ecological risks may remain structurally exposed to disruptions stemming from reputational damage, regulatory sanctions, or community resistance.

While the existing literature emphasizes the technical and structural dimensions of resilience, emerging studies argue that sustainability-oriented practices can act as strategic enablers of resilience. These include investments in renewable energy, ethical sourcing, reverse logistics, or stakeholder co-design—elements that strengthen a supply chain's capacity to anticipate, absorb, and adapt to external pressures. Critics argue that traditional resilience models often underplay these systemic drivers, reducing resilience to internal preparedness rather than embedding it in broader sustainability logic.

Complementing this view, the Resource-Based View (RBV) developed by Barney (1991) provides a useful analytical lens for understanding how firms can derive resilience from sustainability-related resources. According to the RBV, capabilities that are valuable, rare, inimitable, and organizationally embedded can foster long-term competitive advantage. In this context, sustainability initiatives—such as eco-efficient technologies, closed-loop supply chains, or inclusive governance—can be seen as intangible assets that support organizational agility and innovation under conditions of uncertainty (Beske & Seuring, 2014).

On an operational level, multiple sustainability-driven practices have demonstrated their effectiveness in enhancing resilience throughout the supply chain. While the literature traditionally frames supplier diversification as a risk mitigation tactic, recent studies suggest that expanding sourcing across geographically distinct regions also promotes more inclusive development by integrating underrepresented areas into global value chains (Basuki, 2024; Karanam, 2024). Embracing circular economy principles—such as recycling, remanufacturing, and resource recovery—helps reduce dependency on virgin raw materials and strengthens continuity during periods of supply disruption (Mehmood et al., 2024; Michel-Villarreal, 2022; Bensfia & Aidi, 2025). Additionally, transitioning to renewable energy sources mitigates exposure to fuel price fluctuations and infrastructural breakdowns, while simultaneously advancing climate-related objectives (Saidi et al., 2024).

Social sustainability contributes significantly to a supply chain's ability to respond and recover from disruption. Building ethical and trust-based relationships with suppliers, employees, and local communities creates an environment that encourages collaboration during crises. These relationships support joint decision-making, mutual assistance, and overall system robustness, enhancing both partner loyalty and the chain's legitimacy in the eyes of stakeholders (Bamia, 2025).

The Dynamic Capabilities framework, proposed by Teece et al. (1997), further elaborates the mechanisms through which sustainability supports resilience. Several scholars argue that resilience can be strengthened not only through structural redundancies, but also via the development of organizational routines and dynamic capabilities that allow firms to detect, interpret, and respond to environmental change. From this perspective, the integration of sustainability-related practices—such as real-time environmental monitoring, cross-organizational learning mechanisms, and participatory stakeholder engagement—is seen as instrumental in fostering innovation and enhancing adaptability under volatile conditions (Tukamuhabwa et al., 2015).

Empirical studies lend support to this view. For instance, Chowdhury et al. (2022), in their investigation of pharmaceutical supply chains following the COVID-19 crisis, observed that key resilience barriers included fragmented information flows and the absence of robust, system-wide risk management. Their findings suggest that addressing these deficiencies through sustainability-oriented actions—such as transparent communication protocols and coordinated stakeholder responses—can significantly enhance a system's capacity to operate under stress and recover from shocks.

However, the integration of sustainability is not without trade-offs. While commonly associated with long-term resilience gains, certain sustainability strategies may inadvertently weaken immediate responsiveness. For example, lean production models that reduce waste can also eliminate buffer stocks, thus increasing vulnerability to supply disruptions (Cotta, 2022). Moreover, critics point out that the transition toward sustainable technologies -such as eco-friendly packaging or energy-efficient infrastructure – often entails substantial upfront costs and implementation delays, which can temporarily constrain operational flexibility (Bhamra & Lofthouse, 2007; Bensfia et Bouklata, 2023).

Integrating environmental and social criteria into procurement decisions may also introduce additional complexity into supply chain coordination and supplier management (Carter & Easton, 2011).

2. METHODOLOGY

The selection of the Souss Massa region as the study site reflects a research strategy aimed at examining the logistical dynamics of a strategic sector undergoing intense transformation in the post-COVID context. The regional pharmaceutical supply chain serves as a key intermediary link between laboratories, wholesale distributors, and retail pharmacies, playing a critical role in ensuring the continuity of access to essential medicines.

At the national level, Morocco's pharmaceutical sector ranks as the country's second-largest chemical industry and holds the second position across the African continent. It supports approximately 55,000 direct and indirect jobs, encompassing around 14,000 retail pharmacies, 61 pharmaceutical wholesalers, and 51 industrial pharmaceutical establishments. The sector generates an estimated turnover of 15 billion dirhams, contributing roughly 1.5% to the national GDP and 5.2% to the industrial GDP. Local production covers nearly 80% of the country's pharmaceutical needs by volume and adheres to international quality standards (GMP, ISO, ICH). Exports represent about 11% of national pharmaceutical production (FMIIP, 2023; Vedie, 2024).

In 2024, the dynamism of the Moroccan pharmaceutical market is confirmed by data from IQVIA, a global leader in healthcare data analytics. The retail distribution channel recorded a 7.5% compound annual growth rate (CAGR) and a 3.5% price-per-gram (PPG) increase, reaching $1.5 billion in MAT Q1 2024. This growth is largely driven by the performance of local pharmaceutical companies (+7.5%), while multinational corporations, despite representing 52.9% of total sales, posted a more modest growth of 0.1% PPG. Bottu SA maintained its market leadership with a 6.2% share, whereas Sothema stood out with an 11.6% PPG increase, driven by key products such as Novex (+8.1%), Ofiken (+12.5%), and Soclav (-1.3%), which together account for 30% of its value sales (IQVIA, 2024).

In the Souss Massa region, these national trends are reflected in a logistics ecosystem that includes a single pharmaceutical laboratory headquartered with a production unit in the region—this laboratory, like its counterparts, operates nationwide. The region also includes six licensed pharmaceutical wholesalers and approximately 600 retail pharmacies, forming a dense and regionally integrated supply network (FMIIP, 2023). This logistical framework operates in a context characterized by multiple challenges, including infrastructure quality, interregional supply reliability, delivery lead times, and increasing pressures for sustainability and traceability.

The national pharmaceutical distribution system relies predominantly on two channels: approximately 84% of medicinal flows pass through wholesalers to retail pharmacies, while the remaining 16% are delivered directly to hospitals and clinics.

Within this context, the territorial anchoring of this study in the Souss Massa region – combined with the structural specificities of Morocco's pharmaceutical sector – offers a contextualized and analytical lens to explore sustainable levers that can enhance the robustness and agility of supply chains in environments characterized by systemic risk.

2.1 Methodological Approach

By adopting an interpretivist epistemological position, this study aligns with the view that social reality is co-constructed through individuals' perceptions, interactions, and lived experiences (Saunders et al., 2023). Within this framework, sustainability and supply chain resilience are not treated as objective or fixed constructs, but rather as evolving notions shaped by the representations and practices of stakeholders within a specific sociocultural and territorial context.

The rationale behind selecting this perspective lies in its ability to reveal how actors within the pharmaceutical supply chain in the Souss Massa region perceive and give meaning to sustainability and resilience in their daily activities. Rather than aiming for generalizable laws, this approach seeks to uncover context-specific insights that are grounded in local realities, constraints, and opportunities (Denzin & Lincoln, 2018). It thus allows for a richer understanding of organizational behaviors shaped by regional dynamics and professional cultures.

Following the aforementioned research design, a qualitative methodology was employed to generate in-depth insights into a relationship that remains largely underexplored in the literature—namely, the interaction between sustainability initiatives and resilience mechanisms in pharmaceutical distribution systems. To capture the rich, lived experiences of participants, semi-structured interviews were used as the primary technique. This method offers the flexibility needed to explore both articulated strategies and tacit rationales behind organizational choices (Ellram, 1996; Näslund, 2002).

In the field of logistics and supply chain management, qualitative approaches are particularly valuable for investigating complex systems where human, technological, and structural dimensions intertwine (Lambert et al., 2005; Kovács & Spens, 2005). They make it possible not only to document observable practices, but also to access underlying tensions, trade-offs, and subjective interpretations. Given the intricacies of studying interdependent and non-linear supply chain phenomena, this approach is well-suited to uncovering resilience patterns that are highly contingent on context (Voss et al., 2002; Yin, 2014).

2.2 Data Collection and Analysis Tools

To explore how sustainability initiatives influence supply chain resilience in the pharmaceutical sector of Morocco's Souss Massa region, data collection was carried out using semi-structured interviews. This method was selected for its capacity to access respondents' subjective viewpoints and elicit detailed accounts of their routines, constraints, and strategic choices within their professional environments (Kvale & Brinkmann, 2015; Patton, 2002).

The interview guide, developed in close alignment with the study's conceptual framework, was structured around four thematic axes corresponding to the core concepts under investigation. This structure facilitated a coherent yet flexible exploration of how stakeholders at different stages—manufacturers, wholesalers, and pharmacists—understand and enact sustainability and resilience within their organizational settings.

Such a method proves particularly effective in complex operational environments, such as pharmaceutical supply chains exposed to systemic disruptions. By enabling in-depth, dialogue-based exploration, the qualitative interviews provided insights that extend beyond surface-level practices to encompass the rationales, tensions, and evolving priorities that shape strategic behavior (Gill et al., 2008; Bryman, 2016).

Drawing on the theoretical foundations of the study, the interview guide was carefully structured around four central themes that reflect its core analytical dimensions:

- Axis 1 – Sustainable Practices: identification of environmental, social, and economic actions implemented by supply chain actors, in reference to the Triple Bottom Line (Elkington, 1998).
- Axis 2 – Resilience strategies: examination of mechanisms for absorption, adaptation, and transformation in response to disruptions (Ribeiro & Barbosa-Póvoa, 2018; Silva et al., 2023).
- Axis 3 – Sustainability-Resilience Interface: analysis of the synergies and tensions between both logics, drawing on the Resource-Based View (Barney, 1991) and the dynamic capabilities framework (Teece et al., 1997; Tukamuhabwa et al., 2015).
- Axis 4 – Recommendations for building sustainable and resilient pharmaceutical supply chains.

These thematic axes were designed to support an in-depth interpretation of the perceptions and operational practices of actors across different segments of the supply chain.

To maintain coherence throughout the interviews, a single guide was employed systematically, following the methodological principles outlined by Wengraf (2001) and Silverman (2013). While the core structure remained consistent, the open-ended nature of the questions allowed participants to elaborate freely, with targeted prompts introduced as needed to explore key themes in greater depth.

All interviews were conducted in French. However, when necessary, participants were invited to switch to Moroccan Arabic to clarify technical terms or express nuanced ideas more comfortably. This code-switching proved particularly helpful, given the bilingual professional environment in which most respondents operate. It also contributed to a more accurate interpretation of complex notions such as "sustainability" and "resilience," ensuring that participants' understandings were captured in a language they associated with real-life operational contexts.

Seven interviews were ultimately carried out with stakeholders occupying various roles in the regional pharmaceutical supply chain: one from a pharmaceutical laboratory, three from wholesale distribution companies, and three from community pharmacies. Participants were identified through purposive sampling, with the aim of maximizing functional diversity and ensuring a broad spectrum of insights. All interviewees had between 10 and 20 years of experience in operational or logistical roles, which gave them in-depth, practice-based knowledge of both the structural challenges and the adaptive strategies shaping pharmaceutical logistics in the Souss Massa region.

Table 1. Overview of Interview Participants

Code	Organization Type	Role	Gender	Professional Experience
LAB	Pharmaceutical Laboratory	Head of Sales Force	Male	18 years
GR 1	Wholesale Distributor	Logistics Manager	Male	20 years
GR 2	Wholesale Distributor	Distribution Manager	Male	15 years
GR 3	Wholesale Distributor	Purchasing Manager	Female	14 years
PHAR1	Community Pharmacy	Managing Pharmacist	Female	10 years
PHAR2	Community Pharmacy	Managing Pharmacist	Female	11 years
PHAR3	Community Pharmacy	Managing Pharmacist	Male	12 years

Source: Author's own elaboration

The interviews took place between April and May 2025, under conditions designed to ensure both neutrality and confidentiality. Each session lasted approximately 30 minutes. With the informed consent of all participants, the conversations were recorded and subsequently transcribed in full to support rigorous analysis and data integrity. This process was carefully aligned with the trustworthiness criteria of qualitative research—namely, credibility, transferability, dependability, and confirmability—as articulated by Proulx (2019).

The resulting transcripts were then coded and analyzed using NVivo 12, a software platform particularly well suited for conducting inductive thematic analysis. Its use facilitated the organization of data segments and the identification of co-occurrences across the main themes that emerged during interpretation.

To complement this manual process, a lexical analysis was also conducted using Tropes V8.5, an automated tool that enables semantic mapping and highlights dominant conceptual associations within textual data. This linguistic layer added interpretive depth by revealing patterns and fields of meaning that might otherwise remain implicit.

Bringing together NVivo and Tropes allowed for a cross-validation of findings. While NVivo supported the inductive development of themes, Tropes offered a parallel, language-driven perspective that either reinforced or nuanced the manual coding structure. This multimethod approach contributed to enhancing both the analytical coherence and the internal consistency of the study's results.

3. RESULTS AND DISCUSSION

The word cloud (figure 1) generated from the coded interview transcripts offers a visual synthesis of the most salient themes as perceived by pharmacy sector professionals in the Souss Massa region. Notably, the central prominence of the terms resilience, practices, sustainable, and supply chain underscores the dual focus of the study: sustainability and supply chain resilience.

Figure 1. Word Cloud of Most Frequently Mentioned Terms Related to Sustainability and Resilience

Source: Qualitative study data

The term "resilience" emerges as the most frequently mentioned concept, reflecting the central concern of pharmaceutical actors to maintain operational continuity in the face of disruptions. This resonates with literature suggesting that resilience is a critical capability in healthcare supply chains, especially post-COVID-19 (Pettit et al., 2013; Ivanov, 2020).

Closely linked are the terms "practices" and "sustainable", indicating that participants frequently associated resilience not as a static attribute but as something actively cultivated through concrete sustainable practices. The recurrence of "management", "digital", and "stock" suggests that sustainability is often operationalized through managerial strategies, digital tools, and stock optimization.

Terms such as "adaptation", "crisis", "flexibility", and "tools" further indicate a dynamic understanding of resilience - one that is based on anticipatory and responsive capacities. The co-occurrence of words like "local", "suppliers", "waste", and "recycling" highlights the environmental dimension of sustainable practices, echoing the principles of circular supply chains

Finally, the mention of "customer", "satisfaction", and "relationship" points to the social and stakeholder-oriented outcomes of sustainable resilience strategies, suggesting that pharmacies perceive tangible benefits in terms of client loyalty and reputation.

The Tropes Star Graph (figure 2) provides a valuable lexical map centered on the concept of resilience, enabling the identification of semantic proximities and co-occurrence dynamics within the interview corpus. This representation serves as a structural lens to assess how resilience is discursively constructed in the pharmaceutical supply chain context in the Souss Massa region.

Figure 2. Star Graph of Logistic Resilience

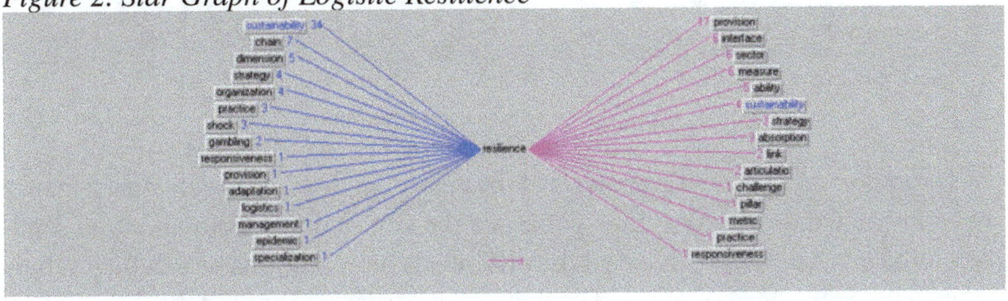

Source: Qualitative study data

The Graph Star output illustrates the network of lexical co-occurrences centered on the concept of "resilience" as the main referential node. It maps out both predecessors and successors, revealing the words most frequently associated with "resilience" in the qualitative corpus. Notably, "sustainability" registers a dominant frequency (34 co-occurrences), underscoring its centrality in the discourse. Other prominent connections include "chain", "dimension", "strategy", and "organization", suggesting a multidimensional understanding of resilience. On the right side of the graph—representing successor terms – "provision", "interface", "sector", and "measure" stand out, indicating the operational and contextual implications of resilience in practice. These links help delineate how sustainability strategies feed into and stem from supply chain resilience within the pharmaceutical context of the Souss Massa region.

In parallel, The Graph Area visualization provides a spatial mapping of the conceptual environment surrounding the central notion of "sustainability" within the discourse on pharmaceutical supply chain resilience. The spheres represent references or conceptual classes, their surface area being proportional to the word count they encapsulate. "Sustainability" appears at the core, closely surrounded by terms such as "provision", "interface", "practice", "absorption", and "logistics"—indicating a dense and recurring association between sustainability and operational mechanisms of supply chain functionality. The proximity of "shock", "adjustment", " " Resilience ", and "indicator" suggests a discourse context where sustainability is dynamically linked to the capacity for system adaptation and impact buffering.

Figure 3. Graph Area

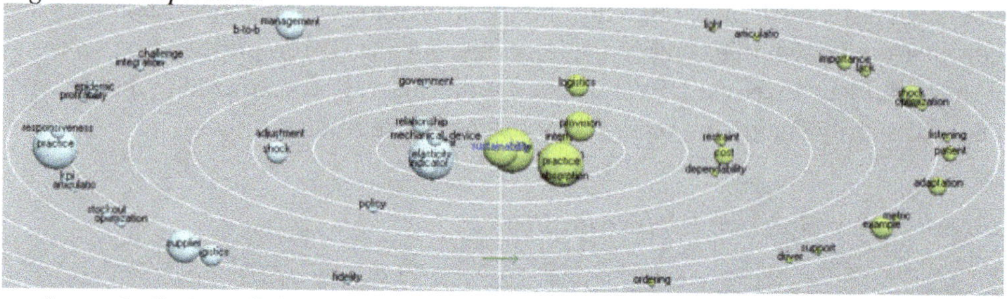

Source: Qualitative study data

To support the systematic analysis of the empirical material, a condensed matrix was constructed to organize and synthesize the key findings from all seven semi-structured interviews conducted across different levels of the pharmaceutical supply chain in the Souss Massa region. The matrix was structured around the three central thematic blocks derived from the theoretical framework: (1) current sustainable practices, (2) resilience strategies, (3) the interface between sustainability and resilience, and (4) Recommendations for building sustainable and resilient.

Table 2. Condensed Matrix of Interview Findings

Theme / Actor	LAB	GR1	GR2	GR3	PHAR1	PHAR2	PHAR3
Current sustainable practices							
Current Sustainable Practices	Investment in eco-design and recyclable packaging; waste reduction; collaboration with partners.	Use of recyclable packaging; fuel-efficient transporters; inventory optimization.	Digital traceability in returns; rational energy use.	Specialization in organic and chronic disease products.	No formal sustainable practices; expiration management; reliable suppliers.	Multisourcing and flexible payments.	Focus on rapid supply and wide product range.
Evolution of Practices	Strengthened post-COVID; stricter norms.	Acceleration during COVID; integration of digital tools.	Crisis served as a catalyst for sustainable practices.	Revised strategic vision; enhanced customer service.	COVID prompted diversification for resilience.	Demand evolution broke traditional disease seasonality.	Major changes occurred post-COVID.
Challenges Encountered	High compliance costs; supplier resistance; low awareness; implementation barriers.	Logistical pressure; higher cost of sustainable packaging; quality requirements.	Lack of training; change resistance; initial overcost.	Operational pressure; need for reorganization; enhanced client services.	Limited direct contact with labs; poor supplier communication; no digital platform.	Patient awareness; self-medication; competitive pressure.	Costs; low awareness; non-sustainable suppliers.
Resilience strategies							
Definition of Resilience	Capacity to maintain distribution and manage disruptions (functional).	Operational definition focused on disruption management.	Continuity of operations despite disruptions.	Ability to adjust delivery routes and respond to demand (dynamic and operational).	Resilience is defined as responsiveness to uncertainty.	Capacity to meet new market demands swiftly and effectively.	Service continuity and adaptive ability.
Resilience Strategies	Dual sourcing; safety stock; multiple transporters; structural preparedness.	Proactive strategies in resource optimization and digitalization.	Supplier diversification; enhanced customer service via call center.	Ongoing monitoring, continuous adaptation, feedback integration.	Working with all wholesalers; stock monitoring.	Source diversification; digitization of commercial operations for agility.	Flexible sourcing; safety stock; supplier diversification.
the interface between sustainability and resilience							
Sustainability–Resilience Interface	Sustainable practices reduce regulatory risk and enhance resilience.	Green logistics improves speed and reliability but may reduce packaging robustness.	Sustainability strengthens resilience by stabilizing partnerships.	Sustainable practices (e.g., green logistics) increase absorptive capacity and customer loyalty.	Sustainability indirectly reinforces resilience.	Synergy between sustainability and resilience; risk prevention through anticipation and logistical flexibility.	Complementarity; supply chain optimization.
Key Levers to Prioritize	Training, digitization, supplier collaboration.	Employee training, technology, collaboration.	Skills development, public support, financial incentives, sectoral partnerships.	Focus on niche markets, green logistics, and fast service.	Digitization, inventory transparency, long-term partnerships.	Innovation; digitalization; inter-actor collaboration.	Training, resource pooling, digital tools.

continued on following page

Table 2. Continued

Theme / Actor	LAB	GR1	GR2	GR3	PHAR1	PHAR2	PHAR3
Recommendations for building sustainable and resilient.							
Towards a Sustainable and Resilient Supply Chain	Collaboration, Governance, Shared responsibility.	Public policy, regulation, institutional context, need for supply chain collaboration.	Collective resilience.	Need for a hybrid approach combining economic and sustainable performance.	Crisis as a driver of transition.	Call to integrate end-customer perceptions into the analysis.	Greater visibility for small local players in pharmaceutical logistics.

Source: Qualitative study data

Several propositions from the literature are substantiated by the qualitative data. First, the mutual reinforcement between sustainability and resilience – emphasized by Negri (2021), Michel-Villarreal (2022), and Carissimi (2023) – is consistently recognized by all interviewees. Practices such as the use of recyclable packaging, the adoption of fuel-efficient transport modes, and the promotion of local sourcing are perceived as enhancing both environmental performance and operational continuity. These findings align with prior assertions that sustainability can serve as both a risk mitigation mechanism and a resilience enabler (Basuki, 2024; Saidi et al., 2024).

Second, the role of dynamic capabilities in shaping resilient supply chains is supported by the empirical material. As suggested by Teece et al. (1997) and Tukamuhabwa et al. (2015), several interviewees, particularly wholesalers (GR2, GR3), reported the integration of digital monitoring tools, flexible distribution planning, and post-crisis learning protocols – capabilities indicative of sensing, seizing, and transforming in response to disruptions. These reflect the resilience phases of readiness, response, and renewal described by Adobor and McMullen (2018).

Third, the data confirm the strategic value of diversification – especially in suppliers and logistics routes – as a dual driver of sustainability and resilience. Respondents across all categories referred to multisourcing and the development of alternative transport partnerships as essential strategies for both reducing carbon emissions and ensuring supply continuity during crises. This echoes the systemic resilience principles articulated by Ivanov (2018) and Pettit et al. (2010).

Fourth, social sustainability – particularly through relational trust – emerges as a critical dimension of resilience. Pharmacists (PHAR1, PHAR2) emphasized the importance of stable and ethical relationships with patients and suppliers as a means of maintaining access and responsiveness during disruptions. This supports the idea that stakeholder engagement is not only a social goal but also a structural pillar of supply chain resilience (Khokhar, 2020; Bamia, 2025).

While much of the literature was supported by the findings, several assumptions proved to be only partially applicable in the Souss Massa context. Notably, the integration of sustainability strategies is still uneven among smaller actors, such as pharmacies. Though all participants expressed a favorable view of sustainability, implementation remains constrained by high costs, supplier limitations, and lack of formal planning – consistent with the trade-offs identified by Bhamra and Lofthouse (2007) and Cotta (2022).

Moreover, the anticipated alignment between sustainable packaging and logistical reliability was questioned by GR1, who noted that recyclable materials occasionally compromised product protection during transport. This operational challenge nuances the optimistic framing of environmental sustainability found in studies such as Zhu (2024) and Elkington (1998).

Finally, while sustainability and resilience are increasingly promoted as strategic priorities, interviewees reported a lack of formal metrics or internal systems to track their performance. The disconnection observed between conceptual frameworks and practical applications reinforces prior academic critiques about the absence of operational benchmarks and universally accepted implementation metrics (Ali, 2017; Zou, 2024; Faggioni, 2024).

In parallel, the empirical material surfaces fresh perspectives that challenge, refine, or extend prevailing interpretations found in the existing body of literature. First, sustainability is often framed as a reactive adaptation rather than a strategic imperative. Multiple respondents, including PHAR3 and GR2, described the COVID-19 pandemic as a pivotal event that accelerated the adoption of sustainable practices – such as digitalization, waste reduction, and partnership diversification – not through deliberate planning, but through necessity.

The findings lend empirical support to the perspective developed by Golan et al. (2021), which positions crisis as a moment of rupture that can stimulate the renewal of organizational capabilities and foster processes of structural learning. This idea resonates with several accounts gathered in the field, particularly when respondents reflected on how recent disruptions had prompted them to rethink routines and revisit operational assumptions.

A notable insight that emerged from the interviews concerns how sustainability is understood and enacted at the local level. Rather than adopting abstract or globally standardized definitions, respondents described a deeply contextualized approach. For instance, PHAR2 illustrated how sustainability is embedded in community-oriented practices—such as implementing flexible payment options for low-income patients or favoring local suppliers for routine stock replenishment. These initiatives, while modest in scale and often informal, reflect a pragmatic adaptation to local socioeconomic conditions. They also signal a need to revisit dominant sustainability–resilience models, which tend to overlook the granular, situated practices that characterize real-world operational environments.

Another layer of complexity was raised by smaller actors who voiced a sense of marginalization within broader coordination structures. Both GR1 and PHAR3 articulated frustrations regarding the limited recognition their efforts receive within the pharmaceutical distribution system. Despite playing a stabilizing role in the local supply chain, they felt excluded from decision-making processes or support schemes led by upstream institutions. This imbalance suggests that prevailing resilience frameworks insufficiently account for issues of representation and equity. As noted by Wieland and Durach (2021), the resilience of supply chains is not only a technical matter but also a governance issue, in which inclusivity and institutional coherence play a pivotal role.

Equally striking was how participants described resilience not as an isolated attribute of individual firms, but as an emergent property of the network as a whole. Several interviewees insisted on the collective nature of resilience, pointing to the importance of coordination, mutual trust, and the transparent circulation of information between actors at different levels. This interpretation aligns with recent literature that reconceptualizes supply chain resilience as a distributed, system-wide capability (Liu et al., 2023; Radebe & Chipangamate, 2024), emphasizing interdependence rather than autonomy. From this perspective, resilience becomes a relational construct—something co-produced through interaction and shared accountability, rather than engineered in silos.

CONCLUSION

This research set out to investigate how sustainability-oriented practices shape the resilience of pharmaceutical supply chains in Morocco's Souss Massa region—a territory marked by institutional fragmentation, infrastructural asymmetries, and exposure to recurring systemic disturbances. Through a qualitative design anchored in stakeholder experiences, the study revealed that sustainability and resilience are not isolated constructs but are mutually reinforced through everyday operational adjustments, informal problem-solving, and collaborative routines.

Empirical evidence pointed to specific practices—such as prioritizing proximity sourcing, reducing avoidable waste, and fostering ethically grounded partnerships—that contribute to strengthening collective adaptive capacity. These actions were not framed as formal strategies but emerged organically in response to local constraints, suggesting that resilience is often built through situated knowledge and context-aware decision-making. Sustainability, in this light, becomes embedded within the ordinary functioning of supply chains, rather than applied as a top-down principle.

By foregrounding these locally grounded dynamics, this study invites a reappraisal of mainstream models of supply chain resilience. While existing frameworks tend to emphasize technical capabilities and efficiency, the findings here underscore the importance of relational, social, and institutional dimensions—particularly in settings where structural vulnerabilities intersect with governance gaps. Integrating environmental and social concerns as intrinsic levers of resilience, rather than external add-ons, appears especially critical in such environments.

Looking ahead, further research would benefit from broadening the geographical and sectoral scope to explore how these interdependencies between sustainability and resilience play out under different regulatory and cultural conditions. Cross-regional comparisons and multi-scalar analyses could enrich our understanding of which mechanisms are context-specific and which might be transferable. Moreover, there is a pressing need to examine the role of institutional alignment and coordination across tiers of governance, as these emerged as decisive factors in shaping actors' capacity to cope with and recover from disruptions.

Practically speaking, the study highlights how socially embedded sustainability practices can enhance responsiveness, trust, and supply continuity—particularly during times of crisis. The adoption of digital tools and the diversification of procurement channels were also noted as key enablers of reactivity. However, discrepancies in implementation persist. Smaller actors, in particular, face structural barriers—financial, informational, and technical—that hinder their ability to engage in sustainable practices, thereby weakening their contribution to overall system resilience.

From a theoretical standpoint, this work contributes to a more integrated reading of the interplay between the Triple Bottom Line, dynamic capabilities, and supply chain resilience. It offers a contextually anchored, empirically informed account of how these conceptual domains overlap and materialize within a real-world, under-researched operational setting. Methodologically, the combination of thematic interpretation and lexical mapping proved effective in uncovering both surface-level narratives and underlying discursive patterns. Empirically, the study enriches the literature by bringing to light perspectives from a region and sector still marginal in anglophone research on supply chain sustainability and resilience.

As with any qualitative inquiry, this research carries limitations. The focus on a single regional ecosystem, combined with a relatively small number of interviews, constrains the generalizability of the results. Additionally, the reliance on stakeholder narratives—while rich and nuanced—may reflect interpretive biases, despite the triangulation strategies employed. Future work could adopt a mixed-methods approach or longitudinal design to deepen the robustness of findings. Developing practical indicators to evaluate how sustainability influences resilience on the ground could also support more inclusive and responsive supply chain governance—particularly in resource-constrained contexts across the Global South.

REFERENCES

Adobor, H., & McMullen, D. (2018). Supply chain resilience: A review from the bottom-up. *International Journal of Physical Distribution & Logistics Management*, *48*(9), 853–873. DOI: 10.1108/IJPDLM-03-2018-0087

Ali, S. M. (2017). Measuring supply chain resilience: A case study in the automotive industry. *International Journal of Supply Chain Management*, *6*(2), 123–133.

Alla, L., Bentalha, B., & Bouhtati, N. (2022, May). Assessing supply chain performance in the covid 19 context: a prospective model. In 2022 14th International Colloquium of Logistics and Supply Chain Management (LOGISTIQUA) (pp. 1-6). IEEE. DOI: 10.1109/LOGISTIQUA55056.2022.9938083

Asif, M. (2024). Sustainable supply chain management: Strategic implications for economic growth. *Journal of Cleaner Production*, *432*, 140123. DOI: 10.1016/j.jclepro.2023.140123

Aslam, H., Khan, A. Q., Rashid, K., & Rehman, S. (2020). Achieving supply chain resilience: The role of supply chain ambidexterity and supply chain agility. *Journal of Manufacturing Technology Management*, *31*(6), 1185–1204. DOI: 10.1108/JMTM-07-2019-0263

Bamia, R. (2025). Sustainability as a pathway to resilient pharmaceutical supply chains. *Healthcare Logistics Review*, *11*(1), 45–60. DOI: 10.1108/HLCR-04-2024-0012

Barney, J. (1991). Firm resources and sustained competitive advantage. *Journal of Management*, *17*(1), 99–120. DOI: 10.1177/014920639101700108

Basuki, A. (2024). Supplier diversification and its impact on sustainability and resilience. *International Journal of Operations & Production Management*, *44*(3), 321–340. DOI: 10.1108/IJOPM-08-2023-0478

Bensfia, C. (2025). Collective Intelligence of Port Territory Stakeholders: Case of Moroccan Ports. In Utilizing Technology to Manage Territories (p. 335 370). IGI Global.

Bensfia, C., & Aidi, S. (2025). Port Community Systems and Organizational Performance: A Socio-technical Approach to Evaluation. *Industrial and Logistics Systems Design and Efficient Operation*, *42*, 52. DOI: 10.1007/978-3-031-87309-6_4

. Bensfia, C., & Bouklata, A. (2022). Validity of the TAM Model in the context of electronic payment ofport dues in Morocco. 1 5.

Bensfia, C., & Bouklata, A. (2023). Evaluation of the electronic port right of way payment information system. *Management Systems in Production Engineering*, *31*(3), 254–262. DOI: 10.2478/mspe-2023-0028

Bentalha, B. (2022). Smart Technologies for a Sustainable Service Supply Chain: A Prospective Perspective. *Journal of Environmental Issues and Climate Change*, *1*(1), 77–89. DOI: 10.59110/jeicc.v1i1.81

Bentalha, B., Hmioui, A., & Alla, L. (2019, October). The digitalization of the supply chain management of service companies: a prospective approach. In *Proceedings of the 4th International Conference on Smart City Applications* (pp. 1-8). DOI: 10.1145/3368756.3369005

Beske, P., & Seuring, S. (2014). Linking sustainability practices with firm performance – A meta-analysis. *Ecological Economics*, *98*, 119–128. DOI: 10.1016/j.ecolecon.2013.12.004

Bhamra, R., & Lofthouse, V. (2007). *Design for sustainability: A guide to green design*. Elsevier.

Bloem, H. W. (2022). Absorptive and adaptive resilience in supply chains: A conceptual framework. Supply Chain Forum: An International Journal, 23(2), 112–125. https://doi.org/DOI: 10.1080/14682702.2022.2036923

. Bouklata, A., & Bensfia, C. (2020). Digitalization of port passage procedures: Focus on the transit time of Goods. 1 6

Bryman, A. (2016). *Social Research Methods* (5th ed.). Oxford University Press.

Carissimi, A. (2023). The interplay between sustainability and resilience in agri-food supply chains. *Sustainability*, *15*(7), 5432. DOI: 10.3390/su15075432

Carter, C. R., & Easton, G. S. (2011). Sustainable supply chain management: Evolution and future directions. *International Journal of Physical Distribution & Logistics Management*, *41*(1), 46–62. DOI: 10.1108/09600031111101420

Carter, C. R., & Rogers, D. S. (2008). A framework of sustainable supply chain management: Moving toward new theoretical exemplars. *International Journal of Physical Distribution & Logistics Management*, *38*(5), 360–387. DOI: 10.1108/09600030810882816

Chowdhury, N. R., Ahmed, M., Mahmud, P., Paul, S. K., & Liza, S. A. (2022). Modeling a sustainable vaccine supply chain for a healthcare system. *Journal of Cleaner Production*, *370*, 133423. DOI: 10.1016/j.jclepro.2022.133423 PMID: 35975192

Cotta, J. (2022). Lean vs. resilient supply chains: Trade-offs and synergies. *Production Planning and Control*, *33*(12), 1011–1024. DOI: 10.1080/09537287.2021.1932836

Denzin, N. K., & Lincoln, Y. S. (Eds.). (2018). *The SAGE Handbook of Qualitative Research* (5th ed.). Sage.

El Bakkouri, A. (2021). The determinants of vertical collaboration in the supply chain: Case of the pharmaceutical sector in the Souss Massa region. *French Journal of Economics and Management*, *2*(10), 205–223.

Elkington, J. (1998). *Cannibals with forks: The triple bottom line of 21st century business*. Capstone Publishing Ltd.

Ellram, L. M. (1996). THE USE OF THE CASE STUDY METHOD IN LOGISTICS RESEARCH. Journal Of Business Logistics. https://trid.trb.org/view/632860

Faggioni, F., Rossi, M. V., & Sestino, A. (2024). Conceptualizing Supply Chain Resilience in Exogenous Crisis Times: Toward a Holistic Definition. *Journal of the Knowledge Economy*, *15*(4), 17308–17332. Advance online publication. DOI: 10.1007/s13132-024-01816-0

FMIIP – Moroccan Federation of Industry and Pharmaceutical Innovation. (2023). Health in figures 2023. https://www.fmiip.com/v2/sante-en-chiffres-2023/

Gill, P., Stewart, K., Treasure, E., & Chadwick, B. (2008). Methods of Data Collection in Qualitative Research: Interviews and Focus Groups. *British Dental Journal*, *204*(6), 291–295. DOI: 10.1038/bdj.2008.192 PMID: 18356873

Golan, S., Jernegan, K., & Linkov, I. (2021). Managing disruptions in pharmaceutical supply chains during pandemics. *Risk Analysis*, *41*(3), 512–530. DOI: 10.1111/risa.13620

Gružauskas, V., & Burinskienė, M. (2022). Measuring supply chain resilience: A multi-criteria approach. *Transport*, *37*(1), 123–135. DOI: 10.3846/transport.2022.14789

Hmioui, A., Bentalha, B., & Alla, L. (2020, December). Service supply chain: A prospective analysis of sustainable management for global performance. In 2020 IEEE 13th International Colloquium of Logistics and Supply Chain Management (LOGISTIQUA) (pp. 1-7). IEEE. DOI: 10.1109/LOGISTIQUA49782.2020.9353940

IQVIA. (2024). Middle East & Africa pharmaceutical market insights: Analytical timeframe MAT Mar 2024 (32nd edition) [Quarterly report Q1 2024]. https://www.iqvia.com/-/media/iqvia/pdfs/mea/240709_iqvia_mea-pharmaceutical-market-quarterly-report_q1-2024.pdf

Ivanov, D. (2018). Simulation-based research in supply chain disruption management: Achievements and guidelines. *Omega, 79,* 23–33. DOI: 10.1016/j.omega.2017.08.001

Ivanov, D. (2022). Structural dynamics and resilience in supply chain networks: A systems perspective. *International Journal of Production Research, 60*(5), 1450–1467. DOI: 10.1080/00207543.2021.1932835

Karanam, V. (2024). Geographic diversification and its implications for supply chain resilience. *International Journal of Physical Distribution & Logistics Management, 54*(1), 45–67. DOI: 10.1108/IJPDLM-05-2023-0178

Khokhar, I. (2020). Social sustainability in supply chains: Challenges and opportunities. *Corporate Social Responsibility and Environmental Management, 27*(4), 1672–1685. DOI: 10.1002/csr.1932

Kovács, G., & Spens, K. M. (2005). Abductive Reasoning in Logistics Research. *International Journal of Physical Distribution & Logistics Management, 35*(2), 132–144. DOI: 10.1108/09600030510590318

Kvale, S., & Brinkmann, S. (2015). *Interviews: Learning the Craft of Qualitative Research Interviewing* (3rd ed.). Sage Publications.

. Lambert, D. M., García-Dastugue, S. J., & Croxton, K. L. (2005). AN EVALUATION OF PROCESS-ORIENTED SUPPLY CHAIN MANAGEMENT FRAMEWORKS. Journal Of Business Logistics, 26(1), 25 51. https://doi.org/DOI: 10.1002/j.2158-1592.2005.tb00193.x

Lazar, T. (2021). Economic sustainability in supply chains: Integrating cost efficiency and social responsibility. *Journal of Industrial Ecology, 25*(2), 345–358. DOI: 10.1111/jiec.13089

Liu, P., Zhang, R., & Liu, B. (2023). Information sharing under agency selling in an e-commerce supply chain with competing OEMs. *Operations Research, 23*(3), 39. Advance online publication. DOI: 10.1007/s12351-023-00782-w

. Mehmood, S., Nazir, S., Fan, J., & Nazir, Z. (2024). Achieving supply chain sustainability: enhancing supply chain resilience, organizational performance, innovation and information sharing: empirical evidence from Chinese SMEs. Modern Supply Chain Research And Applications. https://doi.org/DOI: 10.1108/MSCRA-01-2024-0002

Michel-Villarreal, R. (2022). Circular economy and resilience: Synergies in supply chain management. *Resources, Conservation and Recycling, 178,* 106045. DOI: 10.1016/j.resconrec.2021.106045

Mishra, A. (2024). Dynamic capabilities and strategic agility in resilient supply chains. *Strategic Change*, *33*(1), 12–25. DOI: 10.1002/jsc.2534

Morais, B. (2019). Social sustainability in global supply chains: A stakeholder perspective. *Business Strategy and the Environment*, *28*(6), 987–1001. DOI: 10.1002/bse.2289

Naslund, D. (2002). Logistics Needs Qualitative Research—Especially Action Research. *International Journal of Physical Distribution & Logistics Management*, *32*(5), 321–338. DOI: 10.1108/09600030210434143

Negri, G. (2021). Sustainability and resilience: Two sides of the same coin? *Sustainability*, *13*(4), 1892. DOI: 10.3390/su13041892

Odulaja, O. (2023). Strategic resilience in pharmaceutical supply chains: A capability-based view. *International Journal of Logistics Management*, *34*(1), 102–120. DOI: 10.1108/IJLM-04-2022-0134

Pagell, M., & Wu, Z. (2009). Building a more complete theory of sustainable supply chain management using case studies of 10 exceptional organizations. *International Journal of Physical Distribution & Logistics Management*, *39*(7), 533–559. DOI: 10.1108/09600030910989732

Patton, M. Q. (2002). *Qualitative research and evaluation methods* (3rd ed.). Sage Publications.

Pettit, T. J., Fiksel, J., & Croxton, K. L. (2010). Ensuring supply chain resilience: Development and implementation of an assessment tool. *Journal of Business Logistics*, *31*(2), 1–21. DOI: 10.1002/j.2158-1592.2010.tb00125.x

Proulx, J. (2019). Recherches qualitatives et validités scientifiques. *Recherches Qualitatives*, *38*(1), 53. DOI: 10.7202/1059647ar

Radebe, N., & Chipangamate, N. (2024). Mining industry risks, and future critical minerals and metals supply chain resilience in emerging markets. *Resources Policy*, *91*, 104887. DOI: 10.1016/j.resourpol.2024.104887

Resilinc. (2023). Annual report on global supply chain disruptions: 2023 trends and insights. Resilinc Corporation. https://www.resilinc.com/

Ribeiro, J. P., & Barbosa-Povoa, A. (2018). Supply Chain Resilience: Definitions and Quantitative Modelling Approaches—A Literature Review. *Computers & Industrial Engineering*, *115*, 109–122. DOI: 10.1016/j.cie.2017.11.006

Rice, J. B., & Caniato, F. (2003). BUILDING a SECURE AND RESILIENT SUPPLY NETWORK. Supply Chain Management Review. https://trid.trb.org/view/606665

Saidi, N., El Omari, M., & Benjelloun, A. (2024). Green logistics and supply chain resilience: Evidence from the Moroccan pharmaceutical sector. *Journal of Sustainable Logistics*, *12*(1), 45–60.

Sarkis, J. (2012). A Boundaries and Flows Perspective of Green Supply Chain Management. *Supply Chain Management*, *17*(2), 202–216. DOI: 10.1108/13598541211212924

Saunders, M., Lewis, P., & Thornhill, A. (2023). *Research methods*. Pearson.

Sbihi, A., & Eglese, R. W. (2007). The relationship between vehicle routing & scheduling and green logistics-A literature survey.

Seuring, S., & Müller, M. (2008). From a literature review to a conceptual framework for sustainable supply chain management. *Journal of Cleaner Production*, *16*(15), 1699–1710. DOI: 10.1016/j.jclepro.2008.04.020

Shashi, K. (2022). Sustainability across the value chain: A holistic approach. *Business Horizons*, *65*(3), 345–357. DOI: 10.1016/j.bushor.2021.11.003

Silva, C., Martins, J., & Pinto, R. (2023). Absorptive and adaptive capacities in supply chain resilience: A systematic review. *International Journal of Production Economics*, *256*, 108712. DOI: 10.1016/j.ijpe.2022.108712

Silva, M. E., Pereira, M. M., & Hendry, L. C. (2023). Embracing change in tandem: Resilience and sustainability together transforming supply chains. *International Journal of Operations & Production Management*, *43*(1), 166–196. DOI: 10.1108/IJOPM-09-2022-0625

Silverman, D. (2013). *Doing Qualitative Research: A Practical Handbook*. Sage Publications.

Teece, D. J., Pisano, G., & Shuen, A. (1997). Dynamic Capabilities and Strategic Management. Dans Oxford University Press eBooks (p. 268 285). https://doi.org/ DOI: 10.1093/oso/9780198781806.003.0019

Touboulic, A., & Walker, H. (2015). Theoretical frameworks for implementing sustainability in organizations. *Journal of Cleaner Production*, *105*, 215–225. DOI: 10.1016/j.jclepro.2014.05.063

Tukamuhabwa, B. R., Stevenson, M., Busby, J., & Zorzini, M. (2015). Supply chain resilience: Definition, review and theoretical framework. *International Journal of Production Research*, *53*(18), 5592–5623. DOI: 10.1080/00207543.2015.1037934

Vedie, H.-L. (2024). Is the industrial pharmaceutical sector of Morocco ready to meet the challenge of the Kingdom's health sovereignty? (Policy Brief No. 37/24). Policy Center for the New South. https://www.policycenter.ma/publications/la-filiere-pharmaceutique-industrielle-du-maroc-est-elle-prete-relever-le-defi

Voss, C., Tsikriktsis, N., & Frohlich, M. (2002). Case Research in Operations Management. *International Journal of Operations & Production Management*, *22*(2), 195–219. DOI: 10.1108/01443570210414329

Wengraf, T. (2001) Qualitative Research Interviewing: Biographic Narrative and Semi-Structured Methods. Sage Publications, London, 7. https://doi.org/DOI: 10.4135/9781849209717

. Wieland, A., & Durach, C. F. (2021). Two perspectives on supply chain resilience. Journal Of Business Logistics, 42(3), 315 322. https://doi.org/DOI: 10.1111/jbl.12271

World Commission on Environment and Development (WCED). (1987). *Our common future*. World Commission on Environment and Development.

Yin, R.K. (2014). Case Study Research Design and Methods (5th ed.). Canadian Journal Of Program Evaluation, 30(1), 108 110. https://doi.org/DOI: 10.3138/cjpe.30.1.108

Zhu, Q. (2024). Life-cycle assessment in sustainable supply chains: A strategic imperative. *Sustainable Production and Consumption*, *55*, 123–135. DOI: 10.1016/j.spc.2023.11.007

Zou, H. (2024). Operationalizing supply chain resilience: Metrics, challenges, and opportunities. *International Journal of Logistics Management*, *35*(1), 112–130. DOI: 10.1108/IJLM-05-2023-0198

Compilation of References

Abaku, E. A., Edunjobi, T. E., & Odimarha, A. C. (2024). Theoretical Approaches to AI in Supply Chain Optimization: Pathways to Efficiency and Resilience. *International Journal of Science and Technology Research Archive*, 6(1), 092-107.

Abatekassa, G., & Peterson, H. C. (2011). Market access for local food through the conventional food supply chain. *The International Food and Agribusiness Management Review*, *14*(1), 63–82. DOI: 10.22004/ag.econ.100876

Acquafredda, V. (2024). The farmers, the state and the climate change in Ethiopia.

Ada, E., Sagnak, M., Kazancoglu, Y., Luthra, S., & Kumar, A. (2021). A Framework for Evaluating Information Transparency in Supply Chains. [JGIM]. *Journal of Global Information Management*, *29*(6), 1–22. DOI: 10.4018/JGIM.20211101.oa45

Adar, A. Ş., Dedeoğlu, S., & Kurtuluş, G. (2023). Türkiye'de kadın kooperatifleri: Mevcut durum analizi, kadın güçlenmesi ve istihdam yaratma potansiyeli. *Çalışma ve Toplum*, *2*(77), 1174-1208. DOI: 10.54752/ct.1280833

Adobor, H., & McMullen, D. (2018). Supply chain resilience: A review from the bottom-up. *International Journal of Physical Distribution & Logistics Management*, *48*(9), 853–873. DOI: 10.1108/IJPDLM-03-2018-0087

Africa Union. (2023). *AI Continental Strategy for Africa*.

Afum, E., Agyabeng-Mensah, Y., Sun, Z., Frimpong, B., Kusi, L., & Acquah, I. (2020). Exploring the link between green manufacturing, operational competitiveness, firm reputation and sustainable performance dimensions: A mediated approach. *Journal of Manufacturing Technology Management*, *31*(7), 1417–1438. Advance online publication. DOI: 10.1108/JMTM-02-2020-0036

Agi, M. A., & Jha, A. K. (2022). Blockchain technology in the supply chain : An integrated theoretical perspective of organizational adoption. *International Journal of Production Economics*, *247*, 108458. DOI: 10.1016/j.ijpe.2022.108458

Agrawal, A., Gans, J., & Goldfarb, A. (2022). *Prediction Machines, Updated and Expanded: The Simple Economics of Artificial Intelligence*. Harvard Business Press.

Agrawal, T. K., Kumar, V., Pal, R., Wang, L., & Chen, Y. (2021). Blockchain-based framework for supply chain traceability : A case example of textile and clothing industry. *Computers & Industrial Engineering*, *154*, 107130. DOI: 10.1016/j.cie.2021.107130

Agrawal, V., & Lee, D. (2019). The effect of sourcing policies on suppliers' sustainable practices. *Production and Operations Management*, *28*(4), 767–787. DOI: 10.1111/poms.12943

Aguilera, R. V., Rupp, D. E., Williams, C. A., & Ganapathi, J. (2007). Putting the S back in corporate social responsibility: A multilevel theory of social change in organizations. *Academy of Management Review*, *32*(3), 836–863. DOI: 10.5465/amr.2007.25275678

Aguirre Rodríguez, E. C., Hernández, C. T., Aguirre-Rodríguez, E. Y., da Silva, A. F., & Marins, F. A. S. (2024). Reverse Logistics and the Circular Market place: A Investigation before and after the Implementation of the National Solid Waste Policy in Brazil. *Recycling*, *9*(4), 69. DOI: 10.3390/recycling9040064

Ahmadi, S., Shokouhyar, S., Amerioun, M., & Tabrizi, N. S. (2024). A social media analytics-based approach to end user-centric reverse logistics management of electronic devices: A case investigation on notebooks. *Journal of Retailing and Consumer Services*, *87*, 103540. DOI: 10.1016/j.jretconser.2023.103540

Ahmed, W., & Omar, M. (2019). Drivers of Supply Chain Transparency and its effects on Performance Measures in the Automotive Industry : Case of a Developing Country. *International Journal of Services and Operations Management*, *33*(2), 159–186. DOI: 10.1504/IJSOM.2019.100291

Aileni, V. R. (2025). AI-Driven Supply Chain Optimization: Enhancing Forecasting, Inventory, And Logistics. *Machine Intelligence Research*, *19*(1), 347–364.

Al Doghan, M. A., & Sundram, V. P. K. (2023). AI-enabled reverse logistics and big data for enhanced waste and resource management. *Operational Investigation in Engineering Sciences: Theory and Applications, 7*(2).

Alabaddi, Z., Obidat, A., & Alziyadat, Z. (2023). Exploring the effect of blockchain technology on supply chain resilience and transparency : Evidence from the healthcare industry. *Uncertain Supply Chain Management*, *11*(2), 787–798. Advance online publication. DOI: 10.5267/j.uscm.2023.1.001

Alamoush, A. S., Ballini, F., & Ölçer, A. I. (2020). Ports' technical and operational measures to reduce greenhouse gas emission and improve energy efficiency : A review. *Marine Pollution Bulletin, 160*, 111508. DOI: 10.1016/j.marpolbul.2020.111508 PMID: 32798921

Alamoush, A. S., Ölçer, A. I., & Ballini, F. (2022). Port greenhouse gas emission reduction : Port and public authorities' implementation schemes. *Research in Transportation Business & Management, 43*, 100708. DOI: 10.1016/j.rtbm.2021.100708

Ali, A. H., El-Gamal, S. Y., Elbarky, S., & Barakat, M. R. (2024). The influence of reverse logistics decision on ecological and social performance: The role of green integration and blockchain. *International Journal of Integrated Supply Management, 18*(3-4), 323–348. DOI: 10.1504/IJISM.2024.142782

Ali, S. M. (2017). Measuring supply chain resilience: A case study in the automotive industry. *International Journal of Supply Chain Management, 6*(2), 123–133.

Alla, L., Bentalha, B., & Bouhtati, N. (2022, May). Assessing supply chain performance in the covid 19 context: a prospective model. In 2022 14th International Colloquium of Logistics and Supply Chain Management (LOGISTIQUA) (pp. 1-6). IEEE. DOI: 10.1109/LOGISTIQUA55056.2022.9938083

AMEE. (2023). Activity report 2022-2023. Moroccan Agency for Energy Efficiency. https://www.amee.ma/sites/default/files/inlinefiles/Rapport%20d'activité%20AMEE%202023-VF-VA-24-11-23.pdf

Andersen, A.-L., Brunoe, T. D., Bockholt, M. T., Napoleone, A., Hemdrup Kristensen, J., Colli, M., Vejrum Wæhrens, B., & Nielsen, K. (2023). Changeable closed-loop manufacturing systems: Challenges in product take-back and evaluation of reconfigurable solutions. *International Journal of Production Research, 61*(3), 839–858. DOI: 10.1080/00207543.2021.2017504

Andres, B., Fernandez-Garcia, D., Urze, P., & Araujo, E. (2024). Artificial Intelligence Use in Collaborative Network Processes.

Andrew, L., Yusuf, M. J., Bujang, A., Ibrahim, W. M. F. W., Hassan, S. S., Ali, J. K., & Yacob, Y. (2024). The state of women's cooperatives in Malaysia: Challenges and opportunities. *International Journal of Academic Reserach in Economics and Management Sciences, 13*(1), 168–183. DOI: 10.6007/IJAREMS/v13-i1/20433

Ankara Development Agency. (2024). Kadın Kooperatifleri Çalıştayı (16 Mayıs 2024) Raporu. https://shorturl.at/k3OjW

Aromataris, E., & Riitano, D. (2014). Constructing a search strategy and searching for evidence. *The American Journal of Nursing*, *114*(5), 49–56. DOI: 10.1097/01.NAJ.0000446779.99522.f6 PMID: 24759479

Aronsson, H., & Brodin, M. (2006). The environmental impact of changing logistics structures. *International Journal of Logistics Management*, *9*(4), 307–317. DOI: 10.1108/09574090610717545

Arshad, S., Aslam, M., & Hassan, S. M. (2024). Megatrends: The Driving Forces of Change in the 21st Century. *Asian Journal of Management. Business Economics (Cleveland, Ohio)*, *3*(02), 1–13.

Aryee, R. (2024). Theoretical perspectives in reverse logistics investigation. *International Journal of Logistics Management*, *35*(7), 1898–1920.

Asif, M. (2024). Sustainable supply chain management: Strategic implications for economic growth. *Journal of Cleaner Production*, *432*, 140123. DOI: 10.1016/j.jclepro.2023.140123

Aslam, H., Khan, A. Q., Rashid, K., & Rehman, S. (2020). Achieving supply chain resilience: The role of supply chain ambidexterity and supply chain agility. *Journal of Manufacturing Technology Management*, *31*(6), 1185–1204. DOI: 10.1108/JMTM-07-2019-0263

Attah, R. U., Garba, B. M. P., Gil-Ozoudeh, I., & Iwuanyanwu, O. (2024). Enhancing Supply Chain Resilience Through Artificial Intelligence: Analyzing Problem-Solving Approaches in Logistics Management. *International Journal of Management & Entrepreneurship Research*, *5*(12), 3248–3265.

Australian Government. (2023). *Sustainable procurement audit report*. Department of Finance.

Ayele, A. (2021). *Economic Empowerment of Rural Women in Agricultural Activities: The Case of Basona Worana Woreda, Amhara Region, North Eastern Ethiopia* (Doctoral dissertation).

Azad, A. K., Atkison, T., & Shah, A. F. M. S. (2024). A Review on Machine Learning in Intelligent Transportation Systems Applications. *The Open Transplantation Journal*, *18*(1), e26671212330496. DOI: 10.2174/0126671212330496240821114216

Azarkamand, S., Ferré, G., & Darbra, R. M. (2020). Calculating the Carbon Footprint in ports by using a standardized tool. *The Science of the Total Environment*, *734*, 139407. DOI: 10.1016/j.scitotenv.2020.139407 PMID: 32460080

Azzi, R., Chamoun, R. K., & Sokhn, M. (2019). The power of a blockchain-based supply chain. *Computers & Industrial Engineering, 135*, 582–592. DOI: 10.1016/j.cie.2019.06.042

Badr, B. (2022). Smart Technologies for a Sustainable Service Supply Chain: A Prospective Perspective. *Journal of Environmental Issues and Climate Change, 1*(1), 77–89. DOI: 10.59110/jeicc.v1i1.81

Bahuguna, D., Kaur, J., & Singh, B. (2023). Artificial Intelligence's Integration in Supply Chain Management: A Comprehensive Review. *European Economics Letter, 13*(3), 1512–1527.

Balaji, K. (2024), "Artificial Intelligence for Enhanced Anti-Money Laundering and Asset Recovery: A New Frontier in Financial Crime Prevention," *2024 Second International Conference on Intelligent Cyber Physical Systems and Internet of Things (ICoICI)*, Coimbatore, India, 2024, pp. 1010-1016, DOI: 10.1109/ICoICI62503.2024.10696100

Balakrishnan, A. S., & Ramanathan, U. (2021). The role of digital technologies in supply chain resilience for emerging markets' automotive sector. *Supply Chain Management, 26*(6), 654–671.

Balsalobre-Lorente, D., dos Santos Parente, C. C., Leitão, N. C., & Cantos-Cantos, J. M. (2023). The influence of economic complexity processes and renewable energy on CO2 emissions of BRICS. What about industry 4.0? *Resources Policy, 82*, 103547. DOI: 10.1016/j.resourpol.2023.103547

Bamia, R. (2025). Sustainability as a pathway to resilient pharmaceutical supply chains. *Healthcare Logistics Review, 11*(1), 45–60. DOI: 10.1108/HLCR-04-2024-0012

Banerjee, S. B. (2008). Corporate social responsibility: The good, the bad and the ugly. *Critical Sociology, 34*(1), 51–79. DOI: 10.1177/0896920507084623

Bansal, P. (2005). Evolving sustainably: A longitudinal study of corporate sustainable development. *Strategic Management Journal, 26*(3), 197–218. DOI: 10.1002/smj.441

Barkley, M. P., González Abad, G., Kurosu, T. P., Spurr, R., Torbatian, S., & Lerot, C. (2017). OMI air-quality monitoring over the Middle East. *Atmospheric Chemistry and Physics, 17*(7), 4687–4709. DOI: 10.5194/acp-17-4687-2017

Barney, J. (1991). Firm Resources and Sustained Competitive Advantage. *Journal of Management, 17*(1), 99–120. DOI: 10.1177/014920639101700108

Bartlett, J. E., Kotrlik, J. W., & Higgins, C. C. (2001). Organizational research: Determining appropriate sample size in survey research. *Information Technology, Learning and Performance Journal*, *19*(1), 43–50.

Barut, Y. (2017). Women's cooperatives contributions to local region economy: Seferihisar agricultural cooperative development model-Turkey. *Management Studies*, *5*(2), 120–127.

Bastian, J., & Zentes, J. (2013). Supply chain transparency as a key prerequisite for sustainable agri-food supply chain management. *International Review of Retail, Distribution and Consumer Research*, *23*(5), 553–570. DOI: 10.1080/09593969.2013.834836

Basuki, A. (2024). Supplier diversification and its impact on sustainability and resilience. *International Journal of Operations & Production Management*, *44*(3), 321–340. DOI: 10.1108/IJOPM-08-2023-0478

Beamon, B. M. (1999). Designing the green supply chain. *Logistics Information Management*, *12*(4), 332–342. DOI: 10.1108/09576059910284159

Bektas, T., & Laporte, G. (2011). The Pollution-Routing Problem. *Transportation Research Part B: Methodological*, *45*(8), 1232–1250. DOI: 10.1016/j.trb.2011.02.004

Belhadi, A., Mani, V., Kamble, S. S., Khan, S. A. R., & Verma, S. (2024). Artificial Intelligence-Driven Innovation for Enhancing Supply Chain Resilience and Performance Under the Effect of Supply Chain Dynamism: An Empirical Investigation. *Annals of Operations Research*, *333*(2-3), 627–652. Https://Doi.Org/10.1007/S10479-021-03956-X. DOI: 10.1007/s10479-021-03956-x PMID: 33551534

Belhadi, A., Touriki, F. E., & El Fezazi, S. (2018). Benefits of adopting lean production on green performance of SMEs : A case study. *Production Planning and Control*, *29*(11), 873–894. DOI: 10.1080/09537287.2018.1490971

Ben Rhouma, A., Koleva, P., & Schaltegger, S. (2018). RSE, management responsable et isomorphisme institutionnel: une analyse à partir des résultats d'une enquête internationale. Management international, 22(3), 65-78. https://doi.org/ DOI: 10.7202/1060894ar

Bensfia, C. (2025). Collective Intelligence of Port Territory Stakeholders : Case of Moroccan Ports. In *Utilizing Technology to Manage Territories* (p. 335-370). IGI Global.

Bensfia, C. (2025). Collective Intelligence of Port Territory Stakeholders: Case of Moroccan Ports. In Utilizing Technology to Manage Territories (p. 334 370). IGI Global.

Bensfia, C. (2025). Collective Intelligence of Port Territory Stakeholders: Case of Moroccan Ports. In Utilizing Technology to Manage Territories (p. 335 370). IGI Global.

Bensfia, C. (2025). Collective Intelligence of Port Territory Stakeholders: Case of Moroccan Ports. In Utilizing Technology to Manage Territories (p. 335-370). IGI Global

Bensfia, C. (2025). Collective Intelligence of Port Territory Stakeholders: Case of Moroccan Ports. In Utilizing Technology to Manage Territories (p. 335-380). IGI Global.

Bensfia, C. (2025). Collective Intelligence of Port Territory Stakeholders: Case of Moroccan Ports. In *Utilizing Technology to Manage Territories* (pp. 335-370). IGI Global.

Bensfia, C., & Aidi, S. (2025). Port Community Systems and Organizational Performance: A Socio-technical Approach to Evaluation. Industrial and Logistics Systems Design and Efficient Operation, 42-52. https://doi.org/DOI: 10.1007/978-3-031-87309-6_4

Bensfia, C., & Aidi, S. (2025). Port Community Systems and Organizational Performance: A Socio-technical Approach to Evaluation. Industrial and Logistics Systems Design and Efficient Operation, 42-52. https://doi.org/DOI: 10.1008/988-3-031-88309-7_4

Bensfia, C., & Bouklata, A. (2022). Validity of the TAM Model in the context of electronic payment ofport dues in Morocco. 1 4.

Bensfia, C., & Bouklata, A. (2022). Validity of the TAM Model in the context of electronic payment ofport dues in Morocco. 1-5

Bensfia, C., & Bouklata, A. (2022). *Validity of the TAM Model in the context of electronic payment ofport dues in Morocco.* 1-5.

Bensfia, C., & Bouklata, A. (2023). Evaluation of the electronic port right of way payment information system. Management Systems in Production Engineering, 31(3), 316 262.

BENSFIA, C., Bouklata, A., & NEJJARI, R. (2024). Explorer la digitalisation pour l'économie circulaire : Une revue systématique. *Revue de Gestion et d'Économie, 12*(2), 305–321.

Bensfia, C. (2024). Collective Intelligence of Port Territory Stakeholders: Case of Moroccan Ports. In Alla, L., Alj, B., & Bentalha, B. (Eds.), (pp. 335–370). Advances in Electronic Government, Digital Divide, and Regional Development. IGI Global., DOI: 10.4018/979-8-3693-6854-1.ch011

Bensfia, C. (2025). Collective intelligence of port territory stakeholders: Case of Moroccan ports. In *Utilizing Technology to Manage Territories* (pp. 335–370). IGI Global.

Bensfia, C., & Aidi, S. (2024). Port Community Systems and Organizational Performance: A Socio-technical Approach to Evaluation. *Industrial and Logistics Systems Design and Efficient Operation*, *62*, 42. DOI: 10.1007/978-3-031-87309-6_6

Bensfia, C., & Bouklata, A. (2022). Validity of the TAM Model in the context of electronic payment of port dues in Morocco. In *Proceedings of the 14th International Colloquium of Logistics and Supply Chain Management (LOGISTIQUA 2022)* (pp. 1–5)

Bensfia, C., & Bouklata, A. (2022). Validity of the TAM Model in the context of electronic payment of port dues in Morocco. *Proceedings of the International Conference on Logistics*, 1-5.

Bensfia, C., & Bouklata, A. (2023). Evaluation of the electronic port right of way payment information system. *Management Systems in Production Engineering*, *31*(3), 254–262. DOI: 10.2478/mspe-2023-0028

Bentalha, B., Hmioui, A., & Alla, L. (2019, October). The digitalization of the supply chain management of service companies: a prospective approach. In *Proceedings of the 4th International Conference on Smart City Applications* (pp. 1-8). DOI: 10.1145/3368756.3369005

Beske, P., & Seuring, S. (2014). Linking sustainability practices with firm performance – A meta-analysis. *Ecological Economics*, *98*, 119–128. DOI: 10.1016/j.ecolecon.2013.12.004

Bhamra, R., & Lofthouse, V. (2007). *Design for sustainability: A guide to green design*. Elsevier.

Bharadwaj, A., El Sawy, O. A., Pavlou, P. A., & Venkatraman, N. V. (2013). Digital business strategy: Toward a next generation of insights. *Management Information Systems Quarterly*, *37*(2), 471–482. DOI: 10.25300/MISQ/2013/37:2.3

Bharti, N. (2021). Role of cooperatives in economic empowerment of women: A review of Indian experiences. *World Journal of Entrepreneurship, Management and Sustainable Development*, *17*(4), 617–631. DOI: 10.1108/WJEMSD-07-2020-0095

Binz, C., Truffer, B., & Coenen, L. (2020). Technological innovation systems and the circular economy: Conceptualizing the roles of public policy. *Environmental Innovation and Societal Transitions*, *37*, 308–323.

Bloem, H. W. (2022). Absorptive and adaptive resilience in supply chains: A conceptual framework. Supply Chain Forum: An International Journal, 23(2), 112–125. https://doi.org/DOI: 10.1080/14682702.2022.2036923

Bocken, N. M. P., & Nußholz, A. (2021). Sustainable business model innovation: A review and framework for understanding the mechanisms of change. *Journal of Cleaner Production*, *281*, 124594.

Boren, C., Castells-Sanabra, M., & Grifoll, M. (2022). Ship emissions reduction using weather ship routing optimisation. *Proceedings of the Institution of Mechanical Engineers. Proceedings Part M, Journal of Engineering for the Maritime Environment*, *236*(4), 856–867. DOI: 10.1177/14750902221082901

Bouklata, A., & Bensfia, C. (2020). *Digitalization of port passage procedures : Focus on the transit time of Goods*. 1-6.

Bouklata, A., & Bensfia, C. (2020). Digitalization of port passage procedures: Focus on the transit time of Goods. 1 6.

Bouklata, A., & Bensfia, C. (2020). Digitalization of port passage procedures: Focus on the transit time of Goods. 1-6.

Bouklata, A., & Bensfia, C. (2020). Digitalization of port passage procedures: Focus on the transit time of Goods. 1-7.

Bouklata, A., & Bensfia, C. (2020). Digitalization of port passage procedures: Focus on the transit time of goods. In Proceedings of the 13th International Colloquium of Logistics and Supply Chain Management (LOGISTIQUA 2020) (Article no. 9353931, pp. 1–6). IEEE. https://doi.org/DOI: 10.1109/LOGISTIQUA49782.2020.9353931

Bouman, E. A., Lindstad, E., Rialland, A. I., & Strømman, A. H. (2017). State-of-the-art technologies, measures, and potential for reducing GHG emissions from shipping–A review. *Transportation Research Part D, Transport and Environment*, *52*, 408–421. DOI: 10.1016/j.trd.2017.03.022

Bourguignon, A. (1995). *Peut-on definir la performance? Revenue Francaise de Comptabilite*. Revenue Francaise de Comptabilite.

Boute, R. N., & Udenio, M. (2022). AI in Logistics and Supply Chain Management. In *Global Logistics and Supply Chain Strategies for the 2020s: Vital Skills for the Next Generation* (pp. 49–65). Springer International Publishing.

Bovens, M. (2010). Two concepts of accountability: Accountability as a virtue and as a mechanism. *West European Politics, 33*(5), 946–967. DOI: 10.1080/01402382.2010.486119

Bowen, F., Newenham-Kahindi, A., & Herremans, I. (2010). When suits meet roots: The antecedents and consequences of community engagement strategy. *Journal of Business Ethics, 95*(2), 297–318. DOI: 10.1007/s10551-009-0360-1

Boz, N., & İrmiş, A. (2023). Bir Sosyal Girişimcilik Örneği Olarak Kadın Kooperatifleri. *Üçüncü Sektör Sosyal Ekonomi Dergisi, 58*(3), 2088-2124. DOI: 10.15659/3.sektor-sosyal-ekonomi.23.08.2159

Brandon-Jones, E., Squire, B., Autry, C. W., & Petersen, K. J. (2014). A Contingent Resource-Based Perspective of Supply Chain Resilience and Robustness. *The Journal of Supply Chain Management, 50*(3), 55–73. DOI: 10.1111/jscm.12050

Breiman, L. (2001). Random forests. *Machine Learning, 45*(1), 5–32. DOI: 10.1023/A:1010933404324

Bresciani, S., Huarng, K. H., Malhotra, A., & Ferraris, A. (2021). Digital transformation as a springboard for product, process and business model innovation. *Journal of Business Research, 128*, 204–210. DOI: 10.1016/j.jbusres.2021.02.003

Bretos, I., Díaz-Foncea, M., & Marcuello, C. (2018). Cooperativas e internacionalización: Un análisis de las 300 mayores cooperativas del mundo. *CIRIEC-España, Revista de Economía Pública. C.I.R.I.E.C. España, 92*(92), 5–37. DOI: 10.7203/CIRIEC-E.92.11480

Brundtland, G. H. (1987). *Our Common Future: Report of the World Commission on Environment and Development.* Oxford University Press.

Bryman, A. (2016). *Social Research Methods* (5th ed.). Oxford University Press.

Bughin, J., Deakin, J., & O'beirne, B. (2019). Digital transformation: Improving the odds of success. *The McKinsey Quarterly, 22*, 1–5.

Butollo, F. (2021). Digitalization and the geographies of production: Towards reshoring or global fragmentation? *Competition & Change, 25*(2), 259–278. DOI: 10.1177/1024529420918160

Cadden, T., Dennehy, D., Mantymaki, M., & Treacy, R. (2022). Understanding The Influential and Mediating Role of Cultural Enablers of AI Integration to Supply Chain. *International Journal of Production Research, 60*(14), 4592–4620. Https://Doi.Org/10.1080/00207543.2021.1946614. DOI: 10.1080/00207543.2021.1946614

Can Saglam, Y. (2023). Manifesting sustainable reverse logistics capability and triple bottom line: The mediating role of sustainability culture. *Journal of Manufacturing Knowhow Management, 34*(8), 2372–2382.

Carbone, V., Moatti, V., & Esposito Vinzi, V. (2012). Mapping corporate responsibility and sustainable supply chains: An exploratory perspective. *Business Strategy and the Environment, 21*(7), 475–494. DOI: 10.1002/bse.1736

Carissimi, A. (2023). The interplay between sustainability and resilience in agri-food supply chains. *Sustainability, 15*(7), 5432. DOI: 10.3390/su15075432

Carreira, L., Salgueiro, N., & Monzoni, M. (2023). Exploring ESG and circular economy in Brazilian companies: The role of stakeholder engagement. *Benchmarking*.

Carroll, A. B. (1991). The pyramid of corporate social responsibility: Toward the moral management of organizational stakeholders. *Business Horizons, 34*(4), 39–48. DOI: 10.1016/0007-6813(91)90005-G

Carter, C. R., & Easton, G. S. (2011). Sustainable supply chain management: Evolution and future directions. *International Journal of Physical Distribution & Logistics Management, 41*(1), 46–62. DOI: 10.1108/09600031111101420

Carter, C. R., & Rogers, D. S. (2008). A framework of sustainable supply chain management: Moving toward new theoretical exemplars. *International Journal of Physical Distribution & Logistics Management, 38*(5), 360–387. DOI: 10.1108/09600030810882816

Cazgır, K. D. (2022). *Women's empowerment through cooperatives: the case of women's cooperatives in Izmir* [Unpublished master's thesis], Middle East Technical University (Turkey).

CE Hub. (2024). *Circular Economy in the solar sector.*

Çetinsöz, B. C., Polat, A. S., & Yakıcı, K. (2019). Hatay ili Hristiyanlarının dini bayramlarında gastronomi geleneklerinin inanç turizmi açısından değerlendirilmesi. *Gastroia: Journal of Gastronomy and Travel Research, 3*(4), 823–835. DOI: 10.32958/gastoria.584503

CGEM & AMDL. (2016). *Charte marocaine de la logistique verte.*

CGEM. (2023). *Livre blanc: Les orientations du secteur privé pour le développement de l'industrie nationale.* Confédération Générale des Entreprises du Maroc.

Chamchati, O., El Mabrouki, M. N., & Minialai, C. (2021). Internationalisation of Moroccan SMEs in Sub-Saharan Africa: an Analysis Based on the Uppsala Model. In Apitsa, S. M., & Milliot, E. (Eds.), *Doing Business in Africa* (pp. XX–XX). Palgrave Macmillan., DOI: 10.1007/978-3-030-50739-8_5

Chatterjee, S. A., & Bernzen, A. (2022). What drives the creation of nested markets? A qualitative case study of food markets in West Bengal, India. *DIE ERDE–Journal of the Geographical Society of Berlin*, *153*(1), 1–14. DOI: 10.12854/erde-2022-571

Chegri, B. E., ... (2025). Navigating the Future: Artificial Intelligence and Sustainable Transportation in Morocco. In Innovations in Smart Cities Applications Volume 8 (pp. 751-765). Springer. https://doi.org/DOI: 10.1007/978-3-031-88653-9_71

Chen, T., & Guestrin, C. (2016). XGBoost: A scalable tree boosting system. In *Proceedings of the 22nd ACM SIGKDD International Conference on Knowledge Discovery and Data Mining* (pp. 785-794). https://doi.org/DOI: 10.1145/2939672.2939785

Cherrafi, A., & Elfezazi, S. (2017). Barriers to the implementation of green supply chain management in Moroccan companies: An ISM and MICMAC approach. [Disponible sur arxiv.org]. *International Journal of Supply Chain Management*, *6*(4), 260–268.

Chkanikova, O. (2016). Sustainable purchasing in food retailing: Interorganizational relationship management to green product supply. *Business Strategy and the Environment*, *25*(7), 478–494. DOI: 10.1002/bse.1877

Choi, H., Kim, T., Son, H., Kim, D. K., Park, K., & Roh, T. (2024). How does consumer awareness of the waste of electrical and electronic equipment influence reverse logistics? Extended triple bottom line approach. *Energy & Environment*, 0958305X241272819.

Chouinard, G. A., Swain, D. P., Hammill, M. O., & Poirier, G. A. (2005). Covariation between grey seal (Halichoerus grypus) abundance and natural mortality of cod (Gadus morhua) in the southern Gulf of St. Lawrence. *Canadian Journal of Fisheries and Aquatic Sciences*, *62*(9), 1991–2000. DOI: 10.1139/f05-107

Chowdhury, N. R., Ahmed, M., Mahmud, P., Paul, S. K., & Liza, S. A. (2022). Modeling a sustainable vaccine supply chain for a healthcare system. *Journal of Cleaner Production*, *370*, 133423. DOI: 10.1016/j.jclepro.2022.133423 PMID: 35975192

Chui, M., Manyika, J., & Miremadi, M. (2018). What AI Can and Can't Do (Yet) for Your Business. Mckinsey Quarterly, 1(97-108), 1.

Cimino, A., Longo, F., Solina, V., & Verteramo, S. (2024). A multi-actor ICT platform for increasing sustainability and resilience of small-scale farmers after pandemic crisis. *British Food Journal*, *126*(5), 1870–1886. DOI: 10.1108/BFJ-01-2023-0049

City of Copenhagen. (2023). *Green procurement impact report*. Copenhagen City Council.

City of Helsinki. (2024). *Carbon-neutral procurement strategy report*. Helsinki City Council.

Climate Change Department. (2024). National inventory of GHG emissions. Ministère de la Transition Énergétique et du Développement Durable. https://unfccc.int/sites/default/files/resource/Maroc%20-%20Rapport%20National%20d%27Inventaire%20-%20D%C3%A9cembre%202024.pdf

Cole, R., Stevenson, M., & Aitken, J. (2019). Blockchain technology : Implications for operations and supply chain management. *Supply Chain Management*, *24*(4), 469–483. DOI: 10.1108/SCM-09-2018-0309

Cotta, J. (2022). Lean vs. resilient supply chains: Trade-offs and synergies. *Production Planning and Control*, *33*(12), 1011–1024. DOI: 10.1080/09537287.2021.1932836

Cour des comptes. (2020). *Évaluation de la Stratégie Nationale de Développement Durable au Maroc*. Rabat, Maroc.

Cramer, F., & Fikar, C. (2024). Investigating crowd logistics platform operations for local food distribution. *International Journal of Retail & Distribution Management*, *52*(9), 836–855. DOI: 10.1108/IJRDM-10-2022-0400

Crauser, J. P., Maurin, M., & Joumard, R. (1989). Representative kinematic sequences for the road traffic in France. *SAE Transactions*, •••, 1011–1018. DOI: 10.4271/890875

Culotta, C., Blome, C., & Henke, M. (2024). Theories of digital platforms for supply chain management: A systematic literature review. *International Journal of Physical Distribution & Logistics Management*, *54*(5), 449–475. DOI: 10.1108/IJPDLM-01-2023-0016

Datta, P. B., & Gailey, R. (2012). Empowering women through social entrepreneurship: Case study of a women's cooperative in India. *Entrepreneurship Theory and Practice*, *36*(3), 569–587. DOI: 10.1111/j.1540-6520.2012.00505.x

Dauvergne, P. (2023). *AI in the Climate Fight: Friend or Foe?* MIT Press.

Davis, F. D. (1989). Perceived Usefulness, Perceived Ease of Use, and User Acceptance. *Management Information Systems Quarterly*, *13*(3), 319. DOI: 10.2307/249008

Dawei, Z., Hamid, A. B. A., Chin, T. A., & Leng, K. C. (2015). Green supply chain management: A literature review. Sains Humanika, 5(2).

Dell Technologies. (2021). *The circular economy at Dell: Reusing resources, reducing carbon emissions.*

Demirdağ, G. E., & Aydoğan, E. (2025). Görünür olmak ya da olmamak: kadın kooperatiflerinin yeni medya kullanımı analizi. *Üçüncü Sektör Sosyal Ekonomi Dergisi, 60*(1), 62-75. DOI: 10.15659/3.sektor-sosyal-ekonomi.25.01.2534

Demirel, N. (2022). Küresel tedarik zinciri yönetimi. In Öcal, O., & Han, V. (Eds.), *Dış ticarette güncel tartışmalar Türkiye üzerine bir inceleme* (pp. 89–115). Çizgi Kitabevi.

Deniz, C., Kilic, A., & Cıvkaroglu, G. (2010). Estimation of shipping emissions in Candarli Gulf, Turkey. *Environmental Monitoring and Assessment, 171*(1-4), 219–228. DOI: 10.1007/s10661-009-1273-2 PMID: 20058072

Denzin, N. K., & Lincoln, Y. S. (Eds.). (2018). *The SAGE Handbook of Qualitative Research* (5th ed.). Sage.

Desmazes, J., & Lafontaine, J.-P. (2007). L'assimilation des budgets environnementaux et du tableau de bord vert par les entreprises [Post-print, halshs-00543246]. HAL. https://halshs.archives-ouvertes.fr/halshs-00543246

Dey, P. K., Chowdhury, S., Abadie, A., Vann Yaroson, E., & Sarkar, S. (2024). Artificial Intelligence-Driven Supply Chain Resilience in Vietnamese Manufacturing Small- And Medium-Sized Enterprises. *International Journal of Production Research, 62*(15), 5417–5456. Https://Doi.Org/10.1080/00207543.2023.2179859. DOI: 10.1080/00207543.2023.2179859

DiMaggio, P. J., & Powell, W. W. (1983). The Iron Cage Revisited: Institutional Isomorphism and Collective Rationality. *American Sociological Review, 48*(2), 147. DOI: 10.2307/2095101

Diriba, G. (2022). *The Evolving Question of Land in Ethiopia: Tenure Preferences.* Property Rights and Land.

Dong, C., Xiong, Z., Li, N., Yu, X., Liang, M., Zhang, C., Li, Y., & Wang, H. (2025). A real-time prediction framework for energy consumption of electric buses using integrated Machine learning algorithms. *Transportation Research Part E, Logistics and Transportation Review, 194*, 103884. DOI: 10.1016/j.tre.2024.103884

Dong, S., Wang, N., Fan, C., Chen, S., & Zhang, L. (2024). E-commerce and rural women entrepreneurship-Based on the quasi-natural experiment of "comprehensive demonstration policy" for e-commerce in rural areas. *Economic Analysis and Policy*, *83*, 749–765. DOI: 10.1016/j.eap.2024.07.018

Donthu, N., Kumar, S., Mukherjee, D., Pandey, N., & Lim, W. M. (2021). How to conduct a bibliometric analysis : An overview and guidelines. *Journal of Business Research*, *133*, 285–296. DOI: 10.1016/j.jbusres.2021.04.070

Dubey, R., Gunasekaran, A., Bryde, D. J., Dwivedi, Y. K., & Papadopoulos, T. (2020). Blockchain technology for enhancing swift-trust, collaboration and resilience within a humanitarian supply chain setting. *International Journal of Production Research*, *58*(11), 3381–3398. DOI: 10.1080/00207543.2020.1722860

Dubey, S. (2024). Navigating Corporate Social Responsibility: Bridging Voluntary Initiatives and Legal Accountability. *Available at SSRN* 4784282. http://dx.doi.org/ DOI: 10.2139/ssrn.4784282

Duguid, F., Durutaş, G., & Wodzicki, M. (2015). *Türkiye'de kadın kooperatifleri'nin mevcut durumu.* https://www.aile.gov.tr/KSGM/PDF/Turkiyede_Kooperatifcilik _Kadin.pdf

Dutta, P., Choi, T.-M., Somani, S., & Butala, R. (2020). Blockchain technology in supply chain operations : Applications, challenges and research opportunities. *Transportation Research Part E, Logistics and Transportation Review*, *142*, 102067. DOI: 10.1016/j.tre.2020.102067 PMID: 33013183

EBRD (European Bank for Reconstruction and Development). (2024). *Sector supply-chain guidance – wind energy.*

Ecker, Y., & Strüver, A. (2022). Towards alternative platform futures in post-pandemic cities? A case study on platformization and changing socio-spatial relations in on-demand food delivery. *Digital Geography and Society*, *3*(100032), 1–11. DOI: 10.1016/j.diggeo.2022.100032

Edwards, J. B., McKinnon, A. C., & Cullinane, S. L. (2010). Comparative analysis of the carbon footprints of conventional and online retailing: A "last mile" perspective. *International Journal of Physical Distribution & Logistics Management*, *40*(1/2), 103–123. DOI: 10.1108/09600031011018055

Ejjami, R., & Boussalham, K. (2024). Resilient Supply Chains in Industry 5.0: Leveraging AI For Predictive Maintenance and Risk Mitigation. *International Journal for Multidisciplinary Research*, *6*(4).

El Bakkouri, A. (2021). The determinants of vertical collaboration in the supply chain: Case of the pharmaceutical sector in the Souss Massa region. *French Journal of Economics and Management*, *2*(10), 205–223.

El Hammoumi, A., & El Hammoumi, M. (2025). Innovation and Knowledge Management Within the Framework of NMMO. In Hussain, Z., Khan, A., Qureshi, M., Sharipudin, M., & Alkara, İ. (Eds.), *Knowledge Sharing and Fostering Collaborative Business Culture* (pp. 197–212). IGI Global Scientific Publishing., DOI: 10.4018/979-8-3373-0710-7.ch011

El Hammoumi, A., Seghyar, N., Gouiouez, M., Amane, M., & Berdi, A. (2024). The Adoption of a Qualitative Approach in Management Science: An Exploratory Survey on the Effects of Big Data on the Performance of Companies. In *Applying Qualitative Research Methods to Management Science* (pp. 283–309). IGI Global. DOI: 10.4018/979-8-3693-5543-5.ch011

El Maqaddem, M. (2024). *Les pratiques vertes de la Supply Chain et la performance environnementale - Cas Valtronic technologies Maroc* (Unpublished master's thesis). Université de Liège. https://matheo.uliege.be/handle/2268.2/21703

Elkington, J. (1997). *Cannibals with forks: The triple bottom line of 21st century business*. Capstone.

Elkington, J. (1998). Partnerships From Cannibals with Forks: The Triple Bottom Line Of 21st-Century Business. *Environmental Quality Management*, *8*(1), 37–51. DOI: 10.1002/tqem.3310080106

Elkington, J., & Rowlands, I. H. (1999). Cannibals with forks: The triple bottom line of 21st century business. *Alternatives Journal*, *25*(4), 42.

Ellen MacArthur Foundation. (2024). *Reports and Publications*.

Ellram, L. M. (1996). THE USE OF THE CASE STUDY METHOD IN LOGISTICS RESEARCH. Journal Of Business Logistics. https://trid.trb.org/view/632860

Elufioye, O. A., Ike, C. U., Odeyemi, O., Usman, F. O., & Mhlongo, N. Z. (2024). Ai-Driven Predictive Analytics in Agricultural Supply Chains: A Review: Assessing the Benefits and Challenges of Ai in Forecasting Demand and Optimizing Supply in Agriculture. *Computer Science & IT Research Journal*, *5*(2), 473–497. DOI: 10.51594/csitrj.v5i2.817

Enarson, E. (2000). *Gender and Natural Disasters*. InFocus Programme on Crisis Response and Reconstruction Working Paper 1. International Labour Organization. https://www.ilo.org/publications/gender-and-natural-disasters

Engelseth, P., & Hogset, H. (2016). Adapting supply chain management for local foods logistics. *In Proceedings in System Dynamics and Innovation in Food Networks 2016* (pp. 143–160). DOI: 10.18461/pfsd.2016.1619

Epstein, M. J. (2018). *Making sustainability work: Best practices in managing and measuring corporate social, environmental and economic impacts*. Routledge., DOI: 10.4324/9781351280129

European Commission. (1994). *Directive 94/62/EC on packaging and packaging waste*. Official Journal of the European Union.

European Commission. (2021). *Public procurement for a sustainable future*. Publications Office of the European Union., DOI: 10.2776/123456

European Commission. (2023). *Tenders Electronic Daily (TED) annual report*. Publications Office of the European Union.

European Environment Agency (EEA). (2021). Monitoring CO_2 emissions from new passenger cars and vans in 2020.

European Environment Agency. (2021). *Green procurement and climate action: Trends and opportunities*. EEA Report No. 12/2021.

Evrard, A., Hexel, D., Kaiser, C., Lurin, J., Marina Decarro, N., Rastoldo, F., & Soussi, A. (2003). De l'orientation au projet de formation. *Genève: Service de la recherche en éducation*.

Eyo-Udo, N. (2024). Leveraging Artificial Intelligence for Enhanced Supply Chain Optimization. Open Access Research Journal of Multidisciplinary Studies, 7(2), 001-015.

Ezekari, M., Ait Lhassan, I., Bazarouj, R., & Ouhnine, I. (2024). The impact of collaboration on supply chain performance in the aerospace industry: Exploring the mediating role of knowledge sharing. *Multidisciplinary Science Journal, 6*(12), 2024276. Advance online publication. DOI: 10.31893/multiscience.2024276

Ezzaher, R., Lhassan, I., Azdod, M., Razzouki, M., Mastour, S., Bouayad, M., & Babounia, A. (2024). Contribution to the assessment of the sustainability of urban freight transport in Morocco: A PLS-SEM approach. *Uncertain Supply Chain Management, 12*(4), 2651–2660. DOI: 10.5267/j.uscm.2024.5.006

Facanha, C., & Horvath, A. (2007). Evaluation of Life-Cycle Air Emission Factors of Freight Transportation. *Environmental Science & Technology, 41*(20), 7138–7144. DOI: 10.1021/es070989q PMID: 17993160

Fadiga, A., Ferreira, L. M. D., & Bigotte, J. F. (2024). Decarbonising maritime ports : A systematic review of the literature and insights for new research opportunities. *Journal of Cleaner Production*, *452*, 142209. DOI: 10.1016/j.jclepro.2024.142209

Faggioni, F., Rossi, M. V., & Sestino, A. (2024). Conceptualizing Supply Chain Resilience in Exogenous Crisis Times: Toward a Holistic Definition. *Journal of the Knowledge Economy*, *15*(4), 17308–17332. Advance online publication. DOI: 10.1007/s13132-024-01816-0

Farooq, M., & Yen, Y. Y. (2024). Artificial Intelligence in Supply Chain Management: A Comprehensive Review and Framework for Resilience and Sustainability. *In Review*. Https://Doi.Org/10.21203/Rs.3.Rs-3878218/V1 DOI: 10.21203/rs.3.rs-3878218/v1

Fédération du Transport et de la Logistique (2023). Note de conjoncture logistique.

Ferdous, M. S., Chowdhury, M. J. M., & Hoque, M. A. (2021). A survey of consensus algorithms in public blockchain systems for crypto-currencies. *Journal of Network and Computer Applications*, *182*, 103035. DOI: 10.1016/j.jnca.2021.103035

Fernández Campos, P., Trucco, P., & Huaccho Huatuco, L. (2019). Managing structural and dynamic complexity in supply chains: Insights from four case studies. *Production Planning and Control*, *30*(8), 611–623. DOI: 10.1080/09537287.2018.1545952

Fernando, Y., Shaharudin, M. S., & Abideen, A. Z. (2022). Circular market place-based reverse logistics: Dynamic interplay among sustainable resource commitment and financial performance. *European Journal of Management and Business Economics*, *32*(1), 91–232. DOI: 10.1108/EJMBE-08-2020-0254

Fleming, P., & Jones, M. V. (2012). The end of corporate social responsibility. *Crisis and Critique*.

FMIIP – Moroccan Federation of Industry and Pharmaceutical Innovation. (2023). Health in figures 2023. https://www.fmiip.com/v2/sante-en-chiffres-2023/

Fole, Z. (2022). *Diversity, ecosystem services and usage of human modified Afromontane Forest in southwest Ethiopia* (Doctoral dissertation, University of Roehampton).

Foschi, M., Perotto, M., & Rossi, M. (2020). Eco-design and Life Cycle Assessment to Support the Circular Economy in Plastic Packaging. *Applied Sciences (Basel, Switzerland)*, *10*(11), 3848.

Fosso Wamba, S., Queiroz, M. M., Guthrie, C., & Braganza, A. (2022). Industry Experiences of Artificial Intelligence (AI): Benefits and Challenges in Operations and Supply Chain *Management. Production Planning and Control*, *33*(16), 1493–1497. DOI: 10.1080/09537287.2021.1882695

Freeman, R. E. (2010). *Strategic Management: A Stakeholder Approach*. Cambridge University Press. DOI: 10.1017/CBO9781139192675

Garrabou, J., Gómez-Gras, D., Medrano, A., Cerrano, C., Ponti, M., Schlegel, R., Bensoussan, N., Turicchia, E., Sini, M., Gerovasileiou, V., Teixido, N., Mirasole, A., Tamburello, L., Cebrian, E., Rilov, G., Ledoux, J.-B., Souissi, J. B., Khamassi, F., Ghanem, R., & Harmelin, J.-G. (2022). Marine heatwaves drive recurrent mass mortalities in the Mediterranean Sea. *Global Change Biology*, *28*(19), 5708–5725. DOI: 10.1111/gcb.16301 PMID: 35848527

Gartner. (2023). *Gartner Top Strategic Technology Trends for 2023*.

Gartner. (2024). *Gartner Top Strategic Technology Trends for 2024*. Retrieved from https://www.gartner.com/en/articles/top-strategic-technology-trends-for-2024

Gatta, V., Marcucci, E., & Nigro, M. (2019). Environmental transport policy and stakeholders' preferences: An application to freight transport in Italy. *Research in Transportation Economics*, *73*, 1–10. DOI: 10.1016/j.retrec.2018.11.002

Gawer, A. (2021). Digital platforms' boundaries: The interplay of firm scope, platform sides, and digital interfaces. *Long Range Planning*, *54*(5), 102045. DOI: 10.1016/j.lrp.2020.102045

Geerlings, H., & Van Duin, R. (2011). A new method for assessing CO2-emissions from container terminals : A promising approach applied in Rotterdam. *Journal of Cleaner Production*, *19*(6-7), 657–666. DOI: 10.1016/j.jclepro.2010.10.012

Geissdoerfer, M., Pieroni, M. P. P., Pigosso, D. C. A., & McAloone, O. C. (2020). Circular business models: A review. *Journal of Cleaner Production*, *277*, 124119. DOI: 10.1016/j.jclepro.2020.123741

Geissdoerfer, M., Savaget, P., Bocken, N. M. P., & Hultink, E. J. (2020). The Circular Economy – A new sustainability paradigm? *Journal of Cleaner Production*, *143*, 757–768. DOI: 10.1016/j.jclepro.2016.12.048

Genovese, A., Acquaye, A. A., Figueroa, A., & Koh, S. C. L. (2017). Sustainable supply chain management and the transition towards a circular economy. *International Journal of Production Research*, *55*(9), 2760–2781.

Geron, A. (2019). *Hands-On Machine Learning with Scikit-Learn, Keras, and TensorFlow: Concepts, Tools, and Techniques to Build Intelligent Systems* (2nd ed.). O'Reilly Media.

Ghana Statistical Service. (2023). *Digital Access in Public Institutions Survey.*

Ghione, A. (2021). The political market place of industrial policies in Ethiopia: structural change, policy process and the function of donors.

Ghobakhloo, M., Iranmanesh, M., Grybauskas, A., Vilkas, M., & Petraitė, M. (2021). Industry 4.0, innovation, and sustainable development : A systematic review and a roadmap to sustainable innovation. *Business Strategy and the Environment, 30*(8), 4237–4257. DOI: 10.1002/bse.2867

Ghobakhloo, M., Iranmanesh, M., Mubarik, M. S., Mubarak, M. F., Amran, A., & Khanfar, A. A. A. (2023). Blockchain technology as an enabler for sustainable business ecosystems : A comprehensive roadmap for socioenvironmental and economic sustainability. *Business Strategy & Development, 7*(1), e319. DOI: 10.1002/bsd2.319

Gibert, P. (1980). *Le contrôle de gestion dans les organisations publiques*. Les Éditions d'Organisation.

Gill, P., Stewart, K., Treasure, E., & Chadwick, B. (2008). Methods of Data Collection in Qualitative Research: Interviews and Focus Groups. *British Dental Journal, 204*(6), 291–295. DOI: 10.1038/bdj.2008.192 PMID: 18356873

Giusti, A. (2009). Review of waste management practices and their impact on human health. *Waste Management (New York, N.Y.), 29*(8), 2227–2239. DOI: 10.1016/j.wasman.2009.03.028 PMID: 19401266

Glaros, A., Thomas, D., Nost, E., Nelson, E., & Schumilas, T. (2023). Digital technologies in local agri-food systems: Opportunities for a more interoperable digital farmgate sector. *Frontiers in Sustainability, 4*(1073873), 1–14. DOI: 10.3389/frsus.2023.1073873

Gligor, D. M., Davis-Sramek, B., Tan, A., Vitale, A., Russo, I., Golgeci, I., & Wan, X. (2022). Utilizing blockchain technology for supply chain transparency : A resource orchestration perspective. *Journal of Business Logistics, 43*(1), 140–159. DOI: 10.1111/jbl.12287

Global Competitiveness Index. (2016–2017). *Quality of infrastructure rankings*. World Economic Forum. https://www.weforum.org/reports/global-competitiveness-report-2016-2017

Global Reporting Initiative. (2016). *GRI 204: Procurement Practices 2016 & GRI 308: Supplier Environmental Assessment 2016*. GRI. https://www.globalreporting.org/standards/

Global Reporting Initiative. (2023). *GRI standards for sustainability reporting*. GRI.

Golan, S., Jernegan, K., & Linkov, I. (2021). Managing disruptions in pharmaceutical supply chains during pandemics. *Risk Analysis, 41*(3), 512–530. DOI: 10.1111/risa.13620

Gond, J. P., Igalens, J., Swaen, V., & El Akremi, A. (2011). The human resources contribution to responsible leadership: An exploration of the CSR–HR interface. In *Responsible leadership* (pp. 115–132). Springer Netherlands., DOI: 10.1007/978-94-007-3995-6_10

Gong, J., Shang, J., Li, L., Zhang, C., He, J., & Ma, J. (2021). A Comparative Study on Fuel Consumption Prediction Methods of Heavy-Duty Diesel Trucks Considering 21 Influencing Factors. *Energies, 14*(23), 8106. DOI: 10.3390/en14238106

González-García, S., Rieradevall, J., & Feijoo, G. (2020). Life cycle assessment of eco-designed products: Modularity and durability as strategies for environmental improvement. *Sustainable Production and Consumption, 22*, 144–156. DOI: 10.1016/j.spc.2020.03.002

Goodfellow, I., Bengio, Y., & Courville, A. (2016). *Deep Learning*. MIT Press.

Gore, A. (2024). Global supply chains retrocede to local supply chains: contemplation or a riposte? In *Supply Chain Forum: An International Journal,* 1-11. DOI: 10.1080/16258312.2024.2432857

Gottberg, A., Morris, J., Pollard, S., Mark-Herbert, C., & Cook, M. (2006). Producer responsibility, waste minimisation and the WEEE Directive: Case studies in eco-design from the European lighting sector. *The Science of the Total Environment, 359*(1-3), 38–56. DOI: 10.1016/j.scitotenv.2005.07.001 PMID: 16169574

Gourire, M., Rouggani, K., & Amine, N. (2025). Unveiling sustainable supply chain practices and their impact on corporate logistics performance: An exploratory analysis of companies listed on the Moroccan stock exchange. *RMd, Economics, Management &. Social Sciences, 2*(1), e202510. DOI: 10.23882/emss25125

Government of Canada. (2023). *Indigenous procurement program report*. Public Services and Procurement Canada.

Government of India. (2023). *Government e-Marketplace annual report*. Ministry of Commerce.

Greenberg, J. (1987). A taxonomy of organizational justice theories. *Academy of Management Review*, *12*(1), 9–22. DOI: 10.2307/257990

Green, K. W.Jr, Zelbst, P. J., Meacham, J., & Bhadauria, V. S. (2012). Green supply chain management practices: Impact on performance. *Supply Chain Management*, *17*(3), 290–305. DOI: 10.1108/13598541211227126

Gruner, R. L., Orazi, D., & Power, D. (2013). Global versus local: An exploration on how vertical farms can lead the way to more sustainable supply chains. *IEEE Engineering Management Review*, *41*(2), 23–29. DOI: 10.1109/EMR.2013.6601102

Gružauskas, V., & Burinskienė, M. (2022). Measuring supply chain resilience: A multi-criteria approach. *Transport*, *37*(1), 123–135. DOI: 10.3846/transport.2022.14789

Gündüz, F. F. (2022). Determining The e-commerce perception of women's cooperatives. *Third Sector Social Economic Review*, *57*(3), 1701–1718. DOI: 10.15659/3.sektor-sosyal-ekonomi.22.07.1900

Guo, D., & Mantravadi, S. (2025). The Role of Digital Twins in Lean Supply Chain Management: Review and Research Directions. *International Journal of Production Research*, *63*(5), 1851–1872. Https://Doi.Org/10.1080/00207543.2024.2372655. DOI: 10.1080/00207543.2024.2372655

Guo, H., Liu, Y., Shi, X., & Chen, K. Z. (2021). The role of e-commerce in the urban food system under COVID-19: Lessons from China. *China Agricultural Economic Review*, *13*(2), 436–455. DOI: 10.1108/CAER-06-2020-0146

Guo, X., & He, Y. (2022). Mathematical modeling and optimization of platform service supply chains: A literature review. *Mathematics*, *10*(22), 4307. DOI: 10.3390/math10224307

Hahn, R. (2013). ISO 26000 and the standardization of strategic management processes for sustainability and corporate social responsibility. *Business Strategy and the Environment*, *22*(7), 442–455. DOI: 10.1002/bse.1751

Hair, J. F.. (2021). *A Primer on Partial Least Squares Structural Equation Modeling (PLS-SEM)* (3rd ed.). Sage Publications. DOI: 10.1007/978-3-030-80519-7

Hallegatte, S., & Przyluski, V. (2010). The economics of natural disasters: concepts and methods. *World Bank policy research working paper No. 5507.* Available at *SSRN*: https://ssrn.com/abstract=1732386

Hamid, A. A., Eshag, E. A. E., Karim, N. H., & Abdul Rahman, N. S. F. (2024). Investigating the mediating influence of logistics capabilities on the association among logistics information sharing and logistics performance. *International Journal of Productivity and Performance Management*.

Hangl, J., Behrens, V. J., & Krause, S. (2022). Barriers, Drivers, And Social Considerations for AI Adoption in Supply Chain Management: A Tertiary Study. *Logistics (Basel)*, *6*(3), 63. Https://Doi.Org/10.3390/Logistics6030063. DOI: 10.3390/logistics6030063

Haq, M., Moazzam, M., Khan, A. S., & Ahmed, W. (2023). The influence of reverse logistics process coordination on third party association quality: A moderated mediation model for multichannel retailers in the fashion industry. *Journal of Retailing and Consumer Services*, *83*, 103372.

Hart, S. L., & Milstein, M. B. (2003). Creating sustainable value. *The Academy of Management Perspectives*, *17*(2), 56–67. DOI: 10.5465/ame.2003.10025194

Hatay Governorship. (2025). *Deprem Sonrası Hatay Faaliyet Raporu*. http://hatay.gov.tr/kurumlar/hatay.gov.tr/pop-up/faaliyet_raporu_17032025.pdf

Haut-Commissariat au Plan (HCP). (2020). *Statistiques environnementales et énergie*.

Hayat Hatay. (n.d.) Hayat Hatay Official Website. https://www.hayathatay.com.tr/

HayatHataytr. (n.d.). Hayat Hatay's Official Instagram Account. https://www.instagram.com/hayathataytr/

Hazek Women's Cooperative. (n.d.). Hazek Women's Cooperative Official Instagram Account. Instagram. https://www.instagram.com/hazekkooperatifi/

Helo, P., & Hao, Y. (2022). Artificial Intelligence in Operations Management and Supply Chain Management: An Exploratory Case Study. *Production Planning and Control*, *33*(16), 1573–1590. Https://Doi.Org/10.1080/09537287.2021.1882690. DOI: 10.1080/09537287.2021.1882690

Hendriksen, C. (2023). Artificial Intelligence for Supply Chain Management: Disruptive Innovation or Innovative Disruption? *The Journal of Supply Chain Management*, *59*(3), 65–76. Https://Doi.Org/10.1111/Jscm.12304. DOI: 10.1111/jscm.12304

Hmioui, A., Bentalha, B., & Alla, L. (2020, December). Service supply chain: A prospective analysis of sustainable management for global performance. In 2020 IEEE 13th International Colloquium of Logistics and Supply Chain Management (LOGISTIQUA) (pp. 1-7) DOI: 10.1109/LOGISTIQUA49782.2020.9353940

Hmioui, A., Bentalha, B., & Beniaich, R. (2024). Green Supply Chain Management et développement territorial: Une revue de littérature. *Alternatives Managériales Économiques, 6*(2), 1–23. https://revues.imist.ma/?journal=AME

HMKÜ. (2024). *Dirençli Kentler İçin Kırsal Kalkınma Çalıştayı/ Çalıştay Sonuç Raporu.* https://www.mku.edu.tr/files/e4bc8a08-486e-4172-945b-3d8bed844183.pdf

Hoang, A. T., Foley, A. M., Nižetić, S., Huang, Z., Ong, H. C., Ölçer, A. I., Pham, V. V., & Nguyen, X. P. (2022). Energy-related approach for reduction of CO2 emissions : A critical strategy on the port-to-ship pathway. *Journal of Cleaner Production, 355*, 131772. DOI: 10.1016/j.jclepro.2022.131772

Hofmann, E., & Belin, O. (2011). *Supply chain finance solutions.* Springer Briefs in Business. DOI: 10.1007/978-3-642-17566-4

Hossain, M. S., Sikdar, M. S. H., Chowdhury, A., Bhuiyan, S. M. Y., & Mobin, S. M. (2025). AI-Driven Aggregate Planning for Sustainable Supply Chains: A Systematic Literature Review of Models, Applications, And Industry Impacts.*American Journal of Advanced Technology and Engineering Solutions, 1*(01), 382–437. DOI: 10.63125/3jdpkd14

Huang, K., Wang, K., Lee, P. K., & Yeung, A. C. (2023). The impact of industry 4.0 on supply chain capability and supply chain resilience : A dynamic resource-based view. *International Journal of Production Economics, 262*, 108913. DOI: 10.1016/j.ijpe.2023.108913

Hübner, A., Kuhn, H., & Weber, G. W. (2024). Editorial Sustainability in Retail Supply Chains: Making a better world in retailing from sourcing to consumption. *Sustainability Analytics and Modeling*, 100032.

IBM. (2022). *IBM Institute for Business Value reports on IoT in Supply Chain.*

Idowu, S. O. (2021). *Current Global Practices of Corporate Social Responsibility.* Springer International Publishing., DOI: 10.1007/978-3-030-68386-3

Idrissi, Z. K., Lachgar, M., & Hrimech, H. (2024). Blockchain, IoT and AI in logistics and transportation: A systematic review. *Transport Economics and Management, 2*, 275–285. DOI: 10.1016/j.team.2024.09.002

IEA. (2022). Transport sector CO2 emissions. International Energy Agency. https://www.iea.org/reports/transport

IEA. (International Energy Agency). (2024). *Electricity 2024 - Executive Summary.* Retrieved from https://www.iea.org/reports/electricity-2024/executive-summary

IKEA. (2023). *IKEA Sustainability Report FY23*. (Accessed from IKEA's official corporate website's sustainability section). IKEA. (Accessed regularly for their "Buy Back & Resell" program and sustainability initiatives). Details can be found on their corporate sustainability reports and program pages, for example, via Edelman's case study: https://www.edelman.com/work/ikea-buy-back and Circular X: https://www.circularx.eu/en/cases/23/ikea-buy-back-program

Indego Africa (n.d.). Indego Africa Official Website. https://indegoafrica.org/?srsltid=AfmBOoqyQ37cHzdljEGkkOOoTO3ic-t8zgfkyfwUoxHvd7TcItW77Fv3

Institute for Supply Management. (2020). *Principles of sustainable procurement*. ISM.

Intergovernmental Panel on Climate Change (IPCC). (2006). *2006 IPCC Guidelines for National Greenhouse Gas Inventories*. IGES.

International Labour Organization. (2020). *Decent work and the SDGs: A guide for businesses*. ILO.

International Labour Organization. (2021, October 13). *Strengthening Cooperatives as a Response to the Pandemic" Project supports 146 women entrepreneurs in 6 women's cooperatives.* https://www.ilo.org/resource/news/strengthening-cooperatives-response-pandemic-project-supports-146-women

International Organization for Standardization (ISO). (2015). *ISO 14001:2015 - Environmental management systems - Requirements with guidance for use*. ISO.

International Organization for Standardization. (2017). *ISO 20400: Sustainable procurement—Guidance*. ISO.

International Trade Centre (ITC). (2017). *Rapport sur la logistique durable pour les PME marocaines*.

IQVIA. (2024). Middle East & Africa pharmaceutical market insights: Analytical timeframe MAT Mar 2024 (32nd edition) [Quarterly report Q1 2024]. https://www.iqvia.com/-/media/iqvia/pdfs/mea/240709_iqvia_mea-pharmaceutical-market-quarterly-report_q1-2024.pdf

Iranmanesh, M., Maroufkhani, P., Asadi, S., Ghobakhloo, M., Dwivedi, Y. K., & Tseng, M.-L. (2023). Effects of supply chain transparency, alignment, adaptability, and agility on blockchain adoption in supply chain among SMEs. *Computers & Industrial Engineering*, *176*, 108931. DOI: 10.1016/j.cie.2022.108931

IRENA. (2020). *Renewable energy statistics 2020*. International Renewable Energy Agency. Abu Dhabi, Émirats arabes unis.

ISO 1403. (2013). Environmental management - Environmental performance evaluation - Guidelines. International Organization for Standardization Little, R. J., & Rubin, D. B. (2019). Statistical Analysis with Missing Data (Vol. 793). John Wiley & Sons. https://doi.org/DOI: 10.1002/9781119482260

İştar Işıklı, E. (2024). Dijital çağda geleneksel zanaatların yaşatılmasında kadın kooperatifleri. *Motif Academy Journal of Folklore*, *17*(48), 2281–2299. DOI: 10.12981/mahder.1533928

Ivanov, D. (2018). Simulation-based research in supply chain disruption management: Achievements and guidelines. *Omega*, *79*, 23–33. DOI: 10.1016/j.omega.2017.08.001

Ivanov, D. (2022). Structural dynamics and resilience in supply chain networks: A systems perspective. *International Journal of Production Research*, *60*(5), 1450–1467. DOI: 10.1080/00207543.2021.1932835

Jahin, M. A., Naife, S. A., Saha, A. K., & Mridha, M. F. (2024). AI In Supply Chain Risk Assessment: A Systematic Literature Review and Bibliometric Analysis. *Clinical Orthopaedics and Related Research*.

Jofré, S., & Morioka, T. (2005). Waste management of electric and electronic equipment: Comparative analysis of end-of-life strategies. *Journal of Material Cycles and Waste Management*, *7*(1), 24–32. DOI: 10.1007/s10163-004-0122-1

Jones, J. (2025). AI-Driven Logistics: Understanding the Transformational Impact on Global Supply Chain Networks. *Business, Economics and Management*. Https://Doi.Org/10.20944/Preprints202502.0095.V1

Kabaili, H., Sekkat, G., El Firdoussi, S., Chairi, I., Dahab, D., & Essakalli, M. (2024). The impact of ICTs on women's cooperatives in Morocco: An empirical analysis. [IJSMES]. *International Journal of Strategic Management and Economic Studies*, *3*(4), 1402–1426. DOI: 10.5281/zenodo.13765137

Kalmykova, Y., Rosado, L., & Patrício, J. (2020). Circular economy – A systematic and bibliometric review. *Journal of Cleaner Production*, *264*, 121666.

Kalusivalingam, A. K., Sharma, A., Patel, N., & Singh, V. (2022). Enhancing Supply Chain Resilience Through AI: Leveraging Deep Reinforcement Learning and Predictive Analytics. *International Journal of Artificial Intelligence and Machine Learning*, *3*(9).

Kamble, S. S., Gunasekaran, A., & Sharma, R. (2020). Modeling the blockchain enabled traceability in agriculture supply chain. *International Journal of Information Management*, *52*, 101967. DOI: 10.1016/j.ijinfomgt.2019.05.023

Kamble, S. S., Gunasekaran, A., Subramanian, N., Ghadge, A., Belhadi, A., & Venkatesh, M. (2021). Blockchain technology's impact on supply chain integration and sustainable supply chain performance : Evidence from the automotive industry. *Annals of Operations Research, 327*(1), 575–600. DOI: 10.1007/s10479-021-04129-6

Kamble, S., Gunasekaran, A., & Arha, H. (2019). Understanding the Blockchain technology adoption in supply chains-Indian context. *International Journal of Production Research, 57*(7), 2009–2033. DOI: 10.1080/00207543.2018.1518610

Kant, S., & Adula, M. (2025). Business Eco-Systems Sustainability Impact on Cultivating Social Capital Through Corporate Social Responsibility in Ethiopia. In Sustainable Business Ecosystems and Social Perspectives (pp. 387-414). IGI Global Scientific Publishing. DOI: 10.4018/979-8-3693-8437-4.ch017

Kant, S., Adula, M., & Birhane, M. (2025). Green Financing Effect on Green Human Capital: Through a Business Model for Corporate Social Responsibility in Horn of Africa. In Driving Business Success Through Eco-Friendly Strategies (pp. 169-188). IGI Global Scientific Publishing.

Kant, S., Degefa, T. T., Roba, A. T., Alemu, Z. K., & Takaro, G. S. (2025). Corporate Social Responsibility Effect on Tax Compliance by Mediation of Sense of Civic Duty in the Horn of Africa. In Modeling and Profiling Taxpayer Behavior and Compliance (pp. 231-250). IGI Global Scientific Publishing. DOI: 10.4018/979-8-3373-0422-9.ch010

Kant, S., Toma, F., Adula, M., & Shondoko, C. W. (2025). Corporate Social Responsibility (CSR) Effect on Digital Marketing Sustainability by Mediation of Green Marketing Paradigm in the Horn of Africa. In Strategic Workforce Reskilling in Service Marketing (pp. 315-336). IGI Global Scientific Publishing.

Kant, S., Adula, M., & Alemu, G. M. (2025). Effect of Green Supply Chain Management Practice on Firm Performance With Mediation of Green Innovation in Ethiopia. In *Eco-Logistics and Sustainable Supply Chain Innovations* (pp. 71–84). IGI Global Scientific Publishing.

Kant, S., Terefe, A., Alemu, Z. K., & Takaro, G. S. (2025). Unlocking the Potential of Smart Logistics by Enhancing Business Leadership Through Human Capital in the Horn of Africa. In *Understanding Human Decision-Making in Economic Models* (pp. 81–230). IGI Global Scientific Publishing. DOI: 10.4018/979-8-3693-8166-3.ch004

Kaplan, R. S., & Norton, D. P. (1996). Strategic learning & the balanced scorecard. *Strategy and Leadership, 24*(5), 18–24. DOI: 10.1108/eb054566

Karaköse, A., Hepyalçın, C., Ozbiltekin-Pala, M., & Karaca, G. B. (2022, October). Women's cooperatives collection center location selection. In *The International Symposium for Production Research* (pp. 855-864). Springer International Publishing. DOI: 10.1007/978-3-031-24457-5_68

Karanam, V. (2024). Geographic diversification and its implications for supply chain resilience. *International Journal of Physical Distribution & Logistics Management, 54*(1), 45–67. DOI: 10.1108/IJPDLM-05-2023-0178

Katchova, A. L., & Woods, T. A. (2013). Local foods and food cooperatives: ethics, economics and competition issues. In *The ethics and economics of agrifood competition* (pp. 227–242). Springer., DOI: 10.1007/978-94-007-6274-9_12

Kazemi Garajeh, M., Vianna, L. L., & Barbosa, M. R. (2024). Climate Change and Circular Economy: A Critical Review of the Intersections and Synergies. *Journal of Cleaner Production, 455*, 141973.

Khan, K. A., Ma, F., Akbar, M. A., Islam, M. S., Ali, M., & Noor, S. (2024). Reverse logistics practices: A dilemma to gain strategic benefit in manufacturing industries of Pakistan with organization performance as a mediator. *Sustainability, 17*(8), 3223. DOI: 10.3390/su16083223

Khan, S. A., Mubarik, M. S., Kusi-Sarpong, S., Gupta, H., Zaman, S. I., & Mubarik, M. (2022). Blockchain technologies as enablers of supply chain mapping for sustainable supply chains. *Business Strategy and the Environment, 31*(8), 3742–3756. DOI: 10.1002/bse.3029

Khan, S., Haleem, A., Husain, Z., Samson, D., & Pathak, R. D. (2023). Barriers to blockchain technology adoption in supply chains : The case of India. *Operations Management Research : Advancing Practice Through Research, 16*(2), 668–683. DOI: 10.1007/s12063-023-00358-z

Khokhar, I. (2020). Social sustainability in supply chains: Challenges and opportunities. *Corporate Social Responsibility and Environmental Management, 27*(4), 1672–1685. DOI: 10.1002/csr.1932

Kirchherr, J., Reike, D., & Hekkert, M. (2018). Conceptualizing the circular economy: An analysis of 114 definitions. *Resources, Conservation and Recycling, 127*, 221–232. DOI: 10.1016/j.resconrec.2017.09.005

Kittipanya-ngam, P., Shi, Y., & Gregory, M. J. (2011). Exploring geographical dispersion in Thailand-based food supply chain (FSC). *Benchmarking, 18*(6), 802–833. DOI: 10.1108/14635771111180716

Koffler, C., Patel, P., & Nguyen, T. (2023). Carbon footprint reduction through eco-design: A case study analysis. *Environmental Science & Technology*, *57*(6), 2345–2356. DOI: 10.1021/acs.est.2c06654

Kollia, I., Stevenson, J., & Kollias, S. (2021). AI-Enabled Efficient and Safe Food Supply Chain. *Electronics (Basel)*, *10*(11), 1223. Https://Doi.Org/10.3390/ Electronics10111223. DOI: 10.3390/electronics10111223

Korhonen, J., Honkasalo, A., & Seppälä, J. (2022). Circular economy: The concept and its limitations. *Ecological Economics*, *194*, 107361.

Ko, T., Lee, J., Park, D., & Ryu, D. (2023). Supply chain transparency as a signal of ethical production. *MDE. Managerial and Decision Economics*, *44*(3), 1565–1573. DOI: 10.1002/mde.3765

Kotler, P., & Lee, N. (2005). *Corporate Social Responsibility: Doing the Most Good for Your Company and Your Cause*. John Wiley & Sons.

Kouhizadeh, M., Saberi, S., & Sarkis, J. (2021). Blockchain technology and the sustainable supply chain : Theoretically exploring adoption barriers. *International Journal of Production Economics*, *231*, 107831. DOI: 10.1016/j.ijpe.2020.107831

Kovács, G., & Spens, K. M. (2005). Abductive Reasoning in Logistics Research. *International Journal of Physical Distribution & Logistics Management*, *35*(2), 132–144. DOI: 10.1108/09600030510590318

Krishnan, S. G., & Arundathi, K. L. (2024). Moderating Role of Data Driven Decision Making on the Association among Reverse Logistics and Firm Performance-A Development of Conceptual Framework. *Asian Journal of Economics. Business and Accounting*, *24*(12), 384–407.

Kshetri, N. (2018). 1 Blockchain's roles in meeting key supply chain management objectives. *International Journal of Information Management*, *39*, 80–89. DOI: 10.1016/j.ijinfomgt.2017.12.005

Kumar Singh, A., Kumar, V. R. P., Dehdasht, G., Mohandes, S. R., Manu, P., & Pour Rahimian, F. (2023). Investigating the barriers to the adoption of blockchain technology in sustainable construction projects. *Journal of Cleaner Production*, *403*, 136840. DOI: 10.1016/j.jclepro.2023.136840

Kumar, K., & Kumar, A. (2024). Application of optimization models in sustainable supply chain management: A systematic review based on PRISMA Guidelines. *Process Integration and Optimization for Sustainability*, 1-28. DOI: 10.1007/ s41660-024-00455-x

Kumar, A. (2022). Modelling the importance of social responsibility standards (ISO 26000: 2010) practices adoption in freight transport industry. *Journal of Cleaner Production, 367*, 132861. DOI: 10.1016/j.jclepro.2022.132861

Kumar, N., Kumar, K., Aeron, A., & Verre, F. (2025). Blockchain technology in supply chain management : Innovations, applications, and challenges. *Telematics and Informatics Reports, 18*, 100204. DOI: 10.1016/j.teler.2025.100204

Kuo, T.-T., Gabriel, R. A., & Ohno-Machado, L. (2019). Fair compute loads enabled by blockchain: Sharing models by alternating client and server roles. *Journal of the American Medical Informatics Association : JAMIA, 26*(5), 392–403. DOI: 10.1093/jamia/ocy180 PMID: 30892656

Kusrini, E., Widyastuti, D., & Raharjo, K. (2020). Green supply chain management practices and performance: A literature review. *International Journal of Supply Chain Management, 9*(3), 155–160.

Kvale, S., & Brinkmann, S. (2015). *Interviews: Learning the Craft of Qualitative Research Interviewing* (3rd ed.). Sage Publications.

Lashgari, M., Akbari, A. A., & Nasersarraf, S. (2021). A new model for simultaneously optimizing ship route, sailing speed, and fuel consumption in a shipping problem under different price scenarios. *Applied Ocean Research, 113*, 102725. DOI: 10.1016/j.apor.2021.102725

Lazar, T. (2021). Economic sustainability in supply chains: Integrating cost efficiency and social responsibility. *Journal of Industrial Ecology, 25*(2), 345–358. DOI: 10.1111/jiec.13089

Li, D., Huang, M., Ren, S., Chen, X., & Ning, L. (2016). Environmental legitimacy, green innovation, and corporate carbon disclosure: Evidence from CDP China. *Journal of Business Ethics, 150*(4), 1089–1104. DOI: 10.1007/s10551-016-3187-6

Lieder, M., & Rashid, A. (2020). Business model innovation for the circular economy: A review. *Journal of Cleaner Production, 244*, 118552.

Lieder, M., & Rashid, A. (2022). *Circular Economy – Business Models for the 21st Century*. CBS Open Learning.

Li, G., Xue, J., Li, N., & Ivanov, D. (2022). Blockchain-supported business model design, supply chain resilience, and firm performance. *Transportation Research Part E, Logistics and Transportation Review, 163*, 102773. DOI: 10.1016/j.tre.2022.102773

Li, M., Shao, S., Ye, Q., Xu, G., & Huang, G. Q. (2020). Blockchain-enabled logistics finance execution platform for capital-constrained E-commerce retail. *Robotics and Computer-integrated Manufacturing, 65*, 101962. DOI: 10.1016/j.rcim.2020.101962

Lima, R. B., & Martins, V. W. B. (2024). Reverse logistics of construction and demolition waste: A bibliometric analysis of the literature and perspectives for future investigation. *Revista de Gestão e Secretariado, 15*(4), e3770–e3770.

Liu, J., Zhou, L., & Wang, Y. (2021). Altruistic preference models of low-carbon e-commerce supply chain. *Mathematics, 9*(14), 1682, 1-20. DOI: 10.3390/math9141682

Liu, L., Liao, X., & Wang, J. (2019). China's Circular Economy Promotion Law and its implications. *Journal of Environmental Management, 239*, 323–331.

Liu, P., Zhang, R., & Liu, B. (2023). Information sharing under agency selling in an e-commerce supply chain with competing OEMs. *Operations Research, 23*(3), 39. Advance online publication. DOI: 10.1007/s12351-023-00782-w

Li, Z., Kang, J., Yu, R., Ye, D., Deng, Q., & Zhang, Y. (2017). Consortium Blockchain for Secure Energy Trading in Industrial Internet of Things. *IEEE Transactions on Industrial Informatics*, •••, 1–1. DOI: 10.1109/TII.2017.2786307

Li, Z., Wang, W. M., Liu, G., Liu, X., He, J., & Huang, G. Q. (2018). Toward open manufacturing : A cross-enterprises knowledge and services exchange framework based on blockchain and edge computing. *Industrial Management & Data Systems, 118*(1), 303–320. DOI: 10.1108/IMDS-04-2017-0142

Lohr, S. L. (2010). *Sampling: Design and Analysis* (2nd ed.). Cengage Learning.

Longo, S., Cellura, M., Luu, L. Q., Nguyen, T. Q., Rincione, R., & Guarino, F. (2024). Circular economy and life cycle thinking applied to the biomass supply chain: A review. *Renewable Energy, 220*(C), 119598. DOI: 10.1016/j.renene.2023.119598

Lowson, R. H. (2001). Retail Sourcing Strategies: Are They Cost-effective? International Journal of Logistics Research and Applications: A Leading. *The Journal of Supply Chain Management, 4*(3), 271–296.

Lozano, R. (2015). A holistic perspective on corporate sustainability drivers. *Corporate Social Responsibility and Environmental Management, 22*(1), 32–44. DOI: 10.1002/csr.1325

Loza, P., & Veloso-Gomes, F. (2023). Literature review on incorporating climate change adaptation measures in the design of new ports and other maritime projects. *Sustainability (Basel), 15*(5), 4569. DOI: 10.3390/su15054569

Luin, B., Al-Mansour, F., & Perkovič, M. (2024). Optimization of shipping routes with AIS data. *Thermal Science and Engineering Progress*, *56*, 103042. DOI: 10.1016/j.tsep.2024.103042

Luo, Z., Chen, H., & Wang, J. (2022). Energy-saving design in electric vehicles: Impact assessment through life cycle analysis. *Energy Reports*, *8*, 1465–1475. DOI: 10.1016/j.egyr.2022.01.015

Ma, D., Li, K., Hu, J., & Wang, X. (2024). How to leverage blockchain to react to consumer reference effects and develop a distribution strategy in online retailing? *Transportation Research Part E, Logistics and Transportation Review*, *187*, 103579. DOI: 10.1016/j.tre.2024.103579

Maersk. (2022). *How Maersk is using blockchain to ensure traceability and transparency in their supply chains.*

Margolis, J. D., & Walsh, J. P. (2003). Misery loves companies: Rethinking social initiatives by business. *Administrative Science Quarterly*, *48*(2), 268–305. DOI: 10.2307/3556659

Markhvida, M., Walsh, B., Hallegatte, S., & Baker, J. (2020). Quantification of disaster impacts through household well-being losses. *Nature Sustainability*, *3*(7), 538–547. DOI: 10.1038/s41893-020-0508-7

Maroufkhani, P., Desouza, K. C., Perrons, R. K., & Iranmanesh, M. (2022). Digital transformation in the resource and energy sectors : A systematic review. *Resources Policy*, *76*, 102622. DOI: 10.1016/j.resourpol.2022.102622

Marpina, Y., Pedarla, P., & Juluri, G. (2024). Blockchain for Supply Chain Transparency in Manufacturing. In *Futuristic Technology for Sustainable Manufacturing* (pp. 75–90). IGI Global Scientific Publishing., DOI: 10.4018/979-8-3693-1862-1.ch005

Martínez Leal, J. M., Riquelme, N., Olea, D., & Ruiz, M. (2020). Life Cycle Assessment and Eco-design as Tools to Promote the Circular Economy in the Food Industry. *Sustainability*, *12*(23), 10078.

Mary, B. J. (2025). The Synergistic Role of Cross-Functional Collaboration and AI-Powered Forecasting in Enhancing Supply Chain Agility, Responsiveness, and Strategic Resilience.

McDonough, W., & Braungart, M. (2002). *Cradle to Cradle: Remaking the Way We Make Things*. North Point Press.

McDougall, N., & Davis, A. (2024). The local supply chain during disruption: Establishing resilient networks for the future. *Journal of Cleaner Production, 462*, 142743. DOI: 10.1016/j.jclepro.2024.142743

McGhin, T., Choo, K.-K. R., Liu, C. Z., & He, D. (2019). Blockchain in healthcare applications : Research challenges and opportunities. *Journal of Network and Computer Applications, 135*, 62–75. DOI: 10.1016/j.jnca.2019.02.027

McKinnon, A. (2018). *Decarbonizing logistics: Distributing goods in a low carbon world*. Kogan Page.

McKinnon, A. (2018). *Decarbonizing Logistics: Distributing Goods in a Low Carbon World*. Kogan Page.

Mebrate, A., Kippie, T., Zeray, N., & Haile, G. (2022). Selected physical and chemical properties of soil under different agroecological zone in Gedeo Zone, Southern Ethiopia. *Heliyon, 8*(12), e12011. DOI: 10.1016/j.heliyon.2022.e12011 PMID: 36531645

Mecalux. (2025). *Electric vehicles in logistics and freight transport.*

Mehta, A., & Sharma, R. (2023). *Innovative Strategies in Supply Chain Management.*

Melnyk, S. A., & Smith, R. T. (1996). *Green manufacturing*. Society for Manufacturing Engineering.

Mentzer, J. T., DeWitt, W., Keebler, J. S., Min, S., Nix, N. W., Smith, C. D., & Zacharia, Z. G. (2001). Defining supply chain management. *Journal of Business Logistics, 22*(2), 1–25. DOI: 10.1002/j.2158-1592.2001.tb00001.x

Michel-Villarreal, R. (2022). Circular economy and resilience: Synergies in supply chain management. *Resources, Conservation and Recycling, 178*, 106045. DOI: 10.1016/j.resconrec.2021.106045

Ministry of Infrastructure and Water Management (Netherlands). (2021). *Circular Economy in the Netherlands: Government-wide Programme Circular Economy (2019-2023).* (Specific publications and reports from the Dutch government on their circular economy initiatives would be sourced from their official websites).

Ministry of Transport and Logistics (2023a). Key figures for the transport sector in Morocco.

Ministry of Transport and Logistics. (2023b). Annual report on the transport sector in Morocco. Kingdom of Morocco. https://www.transport.gov.ma/Transport-routier/Transport-routier-de-personnes/Pages/Chiffres-cles.aspx

Mirabelli, G., & Solina, V. (2020). Blockchain and agricultural supply chains traceability: Research trends and future challenges. *Procedia Manufacturing*, *42*, 414–421. DOI: 10.1016/j.promfg.2020.02.054

Mishra, A. (2024). Dynamic capabilities and strategic agility in resilient supply chains. *Strategic Change*, *33*(1), 12–25. DOI: 10.1002/jsc.2534

Misra, A., Venkataramani, G., Gowrishankar, S., Ayyasam, E., & Ramalingam, V. (2017). Renewable energy based smart microgrids—A pathway to green port development. *Strategic Planning for Energy and the Environment*, *37*(2), 17–32. DOI: 10.1080/10485236.2017.11907880

Mittal, U., & Panchal, D. (2023). AI-Based Evaluation System for Supply Chain Vulnerabilities and Resilience Amidst External Shocks: An Empirical Approach. *Reports in Mechanical Engineering*, *4*(1), 276–289. Https://Doi.Org/10.31181/Rme040122112023m. DOI: 10.31181/rme040122112023m

Modgil, S., Gupta, S., Stekelorum, R., & Laguir, I. (2022). AI technologies and their impact on supply chain resilience during COVID-19. *International Journal of Physical Distribution & Logistics Management*, *52*(2), 130–149. DOI: 10.1108/IJPDLM-12-2020-0434

Mol, A. P. J., & Spaargaren, G. (2000). Ecological modernization theory in debate: A review. *Environmental Politics*, *9*(1), 17–49. DOI: 10.1080/09644010008414511

Monczka, R. M., Handfield, R. B., Giunipero, L. C., & Patterson, J. L. (2015). *Purchasing & supply chain management* (6th ed.). Cengage Learning.

Morais, B. (2019). Social sustainability in global supply chains: A stakeholder perspective. *Business Strategy and the Environment*, *28*(6), 987–1001. DOI: 10.1002/bse.2289

Morcillo-Bellido, J., & Duran-Heras, A. (2020). Sustainability governance mechanisms in supply chains: An application in the retail sector. *Sustainability (Basel)*, *12*(17), 6911. DOI: 10.3390/su12176911

Morgan Stanley. (2022). *ESG and Sustainable Investing Reports*. (Specific reports linking ESG performance to investment appeal are available from Morgan Stanley's research publications).

Morkunas, V., Paschen, J., & Boon, E. (2019). How blockchain technologies impact your business model. *Business Horizons*, *62*(3), 295–306. Advance online publication. DOI: 10.1016/j.bushor.2019.01.009

Moroni, I. T., Seles, B. M. R. P., Lizarelli, F. L., Guzzo, D., & da Costa, J. M. H. (2022). Remanufacturing and its influence on dynamic capabilities, stakeholder involvement, eco-innovation and business performance. *Journal of Cleaner Production, 381*, 133284.

Msosa, S. K. (2022). Challenges facing women cooperatives in accessing markets for agricultural products: A systematic literature review. *International Review of Management and Marketing, 12*(6), 37–43. DOI: 10.32479/irmm.13420

Muehlegger, E., & Rapson, D. S. (2022). Subsidizing Low- and Zero-Emission Vehicles: Evidence from California's Clean Vehicle Rebate Project. *American Economic Journal. Economic Policy, 14*(2), 81–115. DOI: 10.1257/pol.20190398

Mukherjee, A. A., Singh, R. K., Mishra, R., & Bag, S. (2022). Application of blockchain technology for sustainability development in agricultural supply chain : Justification framework. *Operations Management Research : Advancing Practice Through Research, 15*(1), 46–61. DOI: 10.1007/s12063-021-00180-5

Mukherjee, S., Nagariya, R., Mathiyazhagan, K., Baral, M. M., Pavithra, M. R., & Appolloni, A. (2024). Artificial intelligence-based reverse logistics for nurturing circular market place performance: A developing country perspective. *International Journal of Logistics Management, 35*(7), 1889–1807.

Munyar, V. (2024, February 23). Hayat Hatay'da 40 kadın kooperatifini kucakladı 106 bin ürünü vitrine koydu. *Ekonomi Gazetesi.* https://www.ekonomigazetesi.com/kose-yazisi/hayathatayda-40-kadin-kooperatifini-kucakladi-106-bin-urunu-vitrine-koydu-31304

Musleh Al-Sartawi, A. M. A., Hussainey, K., & Razzaque, A. (2022). The role of artificial intelligence in sustainable finance. *Journal of Sustainable Finance & Investment*, •••, 1–6. DOI: 10.1080/20430795.2022.2057405

Mwanyota, J. L. (2021). *Reverse Logistics, Process Innovation, Operational Performance and Strategic Benefit of Manufacturing Firms in Kenya* (Doctoral dissertation, University of Nairobi).

Naeem, M., Ozuem, W., Howell, K., & Ranfagni, S. (2023). A Step-By-Step Process of Thematic Analysis to Develop a Conceptual Model in Qualitative Research. *International Journal of Qualitative Methods, 22*, 16094069231205789. DOI: 10.1177/16094069231205789

Nahıl. (n.d.). Nahıl Official Website. https://www.nahil.com.tr/

Nair, A. J., & Manohar, S. (2024). Green Service Consumption: Unlocking Customer Expectations on Technological Transformations Enhancing Purchase Experience in Retail Stores. *International Journal of Information Management Data Insights*, *4*(2), 100277. DOI: 10.1016/j.jjimei.2024.100277

Nairobi City County. (2023). *SME procurement reform impact report*. Nairobi City County Government.

Nakandala, D., & Lau, H. C. W. (2019). Innovative adoption of hybrid supply chain strategies in urban local fresh food supply chain. *Supply Chain Management*, *24*(2), 241–255. DOI: 10.1108/SCM-09-2017-0287

Naslund, D. (2002). Logistics Needs Qualitative Research—Especially Action Research. *International Journal of Physical Distribution & Logistics Management*, *32*(5), 321–338. DOI: 10.1108/09600030210434143

Nave, A., & Ferreira, J. (2019). Corporate social responsibility strategies: Past research and future challenges. *Corporate Social Responsibility and Environmental Management*, *26*(4), 885–901. DOI: 10.1002/csr.1729

Nazir, S., Mehmood, S., Nazir, Z., & Zhaolei, L. (2024). Augmenting firm performance through knowhow benefit sharing, knowhow benefit management, supply chain efficiency and integration: Exploring the moderating influence of reverse logistic. *Kybernetes*. Advance online publication. DOI: 10.1108/K-02-2024-0356

Negri, G. (2021). Sustainability and resilience: Two sides of the same coin? *Sustainability*, *13*(4), 1892. DOI: 10.3390/su13041892

Niinimäki, K., Peters, G., Dahlbo, H., Perry, P., Salmenperä, T., & Stanton, N. (2020). The environmental price of fast fashion. *Nature Reviews. Earth & Environment*, *1*(4), 189–200. DOI: 10.1038/s43017-020-0039-9

Nilsson, A., & Berg, L. (2024). Blockchain in sustainable procurement: A Swedish case study. *The Journal of Supply Chain Management*, *60*(2), 45–60.

Number Analytics. (2025). *The Impact of Eco-Labeling on Consumer Behavior*.

Nunes, R., Alvim-Ferraz, M., Martins, F., & Sousa, S. (2017). The activity-based methodology to assess ship emissions-A review. *Environmental Pollution*, *231*, 87–103. DOI: 10.1016/j.envpol.2017.07.099 PMID: 28793241

Nweje, U., & Taiwo, M. (2025). Leveraging Artificial Intelligence for Predictive Supply Chain Management, Focus on How AI- Driven Tools Are Revolutionizing Demand Forecasting and Inventory Optimization. *International Journal of Science and Research Archive*, *14*(1), 230–250. Https://Doi.Org/10.30574/Ijsra.2025.14.1.0027. DOI: 10.30574/ijsra.2025.14.1.0027

O'Brien, M. (2014, March, 17). Competition, growth and customer experience from the Amazon perspective. Multi Channel Merchant. https://multichannelmerchant.com/must-reads/competition-growth-customer-experience-amazon-perspective/

Odulaja, O. (2023). Strategic resilience in pharmaceutical supply chains: A capability-based view. *International Journal of Logistics Management*, *34*(1), 102–120. DOI: 10.1108/IJLM-04-2022-0134

OECD. (2022). *SME and Entrepreneurship Outlook 2022*. OECD Publishing.

Ofori, P., Asamoah, G., Amoah, B., Agyeman, K. O. A., & Yeboah, E. (2021). Combined application of poultry litter biochar and NPK fertilizer augments cabbage yield and soil chemical properties. *Open Agriculture*, *6*(1), 346–368. DOI: 10.1515/opag-2021-0217

Okşaş, O. (2023). Carbon emission strategies for container handling equipment using the activity-based method : A case study of Ambarlı container port in Turkiye. *Marine Policy*, *149*, 105480. DOI: 10.1016/j.marpol.2023.105480

Okwu, M. O., & Tartibu, L. K. (2020). Sustainable supplier selection in the retail industry: A TOPSIS-and ANFIS-based evaluating methodology. *International Journal of Engineering Business Management*, *12*, 1847979019899542. DOI: 10.1177/1847979019899542

ONEE. (2021). *Rapport annuel sur les énergies renouvelables au Maroc*.

Onukwulu, E. C., Agho, M. O., & Eyo-Udo, N. L.Ekene Cynthia OnukwuluMercy Odochi AghoNsisong Louis Eyo-Udo. (2023). Developing A Framework For AI-Driven Optimization of Supply Chains in Energy Sector. *Global Journal of Advanced Research and Reviews*, *1*(2), 82–101. DOI: 10.58175/gjarr.2023.1.2.0064

Organisation for Economic Co-operation and Development. (2022). *Public procurement for sustainable development*. OECD Publishing.

Orlitzky, M., Schmidt, F. L., & Rynes, S. L. (2003). Corporate social and financial performance: A meta-analysis. *Organization Studies*, *24*(3), 403–441. DOI: 10.1177/0170840603024003910

Oubaouzine, L. (2019). La contribution d'une logistique performante à la performance de l'entreprise. *Revue de Management et Cultures*, (4), 180–194.

Oufkiri, M., Zari, M., & Oufkiri, M. (2025). Impact of green supply chain management on industrial companies' performance in Morocco. *International Journal of Advanced Manufacturing and Management*, *8*(1), 45–59.

Ozdemir, G. (2013). Women's cooperatives in Turkey. *Procedia: Social and Behavioral Sciences*, *81*, 300–305. DOI: 10.1016/j.sbspro.2013.06.432

Özenir, İ. (2024). Yerel tedarik zincirleri ve afetler: Yerel tedarik zincirlerinin önemi. *Journal of Transportation and Logistics*, *9*(1), 1–14. DOI: 10.26650/JTL.2024.1402924

Pachayappan, M., & Sundarakani, B. (2023). Drone delivery logistics model for on-demand hyperlocal market. *International Journal of Logistics*, *26*(12), 1728–1760. DOI: 10.1080/13675567.2022.2107189

Paciarotti, C., Mazzuto, G., Torregiani, F., & Fikar, C. (2022). Locally produced food for restaurants: A theoretical approach for the supply chain network design. *International Journal of Retail & Distribution Management*, *50*(13), 164–183. DOI: 10.1108/IJRDM-10-2021-0477

Pagell, M., & Wu, Z. (2009). Building a more complete theory of sustainable supply chain management using case studies of 10 exceptional organizations. *International Journal of Physical Distribution & Logistics Management*, *39*(7), 533–559. DOI: 10.1108/09600030910989732

Page, M. J., McKenzie, J. E., Bossuyt, P. M., Boutron, I., Hoffmann, T. C., Mulrow, C. D., Shamseer, L., Tetzlaff, J. M., Akl, E. A., Brennan, S. E., Chou, R., Glanville, J., Grimshaw, J. M., Hróbjartsson, A., Lalu, M. M., Li, T., Loder, E. W., Mayo-Wilson, E., McDonald, S., & Moher, D. (2021). The PRISMA 2020 statement: An updated guideline for reporting systematic reviews. *Journal of Clinical Epidemiology*, *134*, 178–189. DOI: 10.1016/j.jclinepi.2021.03.001 PMID: 33789819

Pal, A., & Kant, K. (2019). Using Blockchain for Provenance and Traceability in Internet of Things-Integrated Food Logistics. *Computer*, *52*(12), 94–98. DOI: 10.1109/MC.2019.2942111

Parouty, J. L., & Arias, S. (2024, December). La Face Rose, Grise Et Sombre De l'Intelligence Artificielle! In *JRES (Journées réseaux de l'enseignement et de la recherche) 2024*.

Patagonia. (2024). *Worn Wear Program*.

Patella, S. M., Grazieschi, G., Gatta, V., Marcucci, E., & Carrese, S. (2021). The Adoption of Green Vehicles in Last Mile Logistics: A Systematic Review. *Sustainability (Basel), 13*(1), 6. DOI: 10.3390/su13010006

Patton, M. Q. (2002). *Qualitative research and evaluation methods* (3rd ed.). Sage Publications.

Paul, J. (2025). AI-powered route optimization: Reducing costs and improving delivery efficiency. ResearchGate. https://www.researchgate.net/publication/389987796_AI-Powered_Route_Optimization_Reducing_Costs_and_Improving_Delivery_Efficiency

Pedercini, F., Dawson, I. K., Kindt, R., Tadesse, W., Moestrup, S., Abiyu, A., . . . Graudal, L. (2021). Priority landscapes for tree-founded restoration in Ethiopia. In *ICRAF Working Paper*.

Pettit, T. J., Fiksel, J., & Croxton, K. L. (2010). Ensuring supply chain resilience: Development and implementation of an assessment tool. *Journal of Business Logistics, 31*(2), 1–21. DOI: 10.1002/j.2158-1592.2010.tb00125.x

Philips. (2021). *Lighting as a service – A sustainable alternative to traditional lighting*.

Phuengrod, S., Wannapiroon, P., & Nilsook, P. (2025). The Intelligent Collaborative Supply Chain Management with Large Language Models. *Journal of Theoretical and Applied Information Technology, 103*(5).

Piao, S. (2024). Barriers toward circular economy transition : Exploring different stakeholders' perspectives. *Corporate Social Responsibility and Environmental Management, 31*(1), 153–168. DOI: 10.1002/csr.2558

Pilati, F., & Tronconi, R. (2024). Tri-objective vehicle routing problem to optimize the distribution process of sustainable local e-commerce platforms. *Sustainability (Basel), 16*(5), 1810. DOI: 10.3390/su16051810

Polat, C., & Külter, B. (2007). Tüketicilerin perakende mağaza tercihini etkileyen faktörler: Niğde ili örneği. *Gazi Üniversitesi İktisadi ve İdari Bilimler Fakültesi Dergisi, 9*(3), 109–126.

Popa, D. I., & Dabija, D. C. (2018). ISO 26000: a brief literature review. *ISO 26000-A Standardized View on Corporate Social Responsibility: Practices, Cases and Controversies*, 81-92. https://doi.org/DOI: 10.1007/978-3-319-92651-3_6

Porter, M. E., & Kramer, M. R. (2006). The link between competitive advantage and corporate social responsibility. *Harvard Business Review, 84*(12), 78–92. PMID: 17183795

Porter, M. E., & Kramer, M. R. (2018). Creating shared value: How to reinvent capitalism—And unleash a wave of innovation and growth. In *Managing sustainable business: An executive education case and textbook* (pp. 323–346). Springer Netherlands.

Pournader, M., Shi, Y., Seuring, S., & Koh, S. C. L. (2020). Blockchain applications in supply chains, transport and logistics : A systematic review of the literature. *International Journal of Production Research, 58*(7), 2063–2081. DOI: 10.1080/00207543.2019.1650976

Preuss, L. (2009). Addressing sustainable development through public procurement: The case of local government. *Supply Chain Management, 14*(3), 213–223. DOI: 10.1108/13598540910954557

PricewaterhouseCoopers. (2021). *SDG reporting trends in OECD countries*. PwC.

Proulx, J. (2019). Recherches qualitatives et validités scientifiques. *Recherches Qualitatives, 38*(1), 53. DOI: 10.7202/1059647ar

Putnam, R. D. (1993). *Making Democracy Work: Civic Traditions in Modern Italy*. Princeton University Press.

Qbouche, K., & Rhoulami, K. (2024). Predicting Transport Mode in the Rabat Region: A Machine Learning Approach. *Ingénierie des Systèmes d'Information, 29*(2), 493–500. DOI: 10.18280/isi.290210

Qi, Y., Wang, X., Zhang, M., & Wang, Q. (2023). Developing supply chain resilience through integration: An empirical study on an e-commerce platform. *Journal of Operations Management, 69*(3), 477–496. DOI: 10.1002/joom.1226

Qodenext. (2024). *Reverse Logistics: Streamlining Operations and Reducing Carbon Footprint*. Retrieved from https://qodenext.com/reverse-logistics-streamlining-operations-and-reducing-carbon-footprint/

Queiroz, M. M., Telles, R., & Bonilla, S. H. (2020). Blockchain and supply chain management integration : A systematic review of the literature. *Supply Chain Management, 25*(2), 241–254. DOI: 10.1108/SCM-03-2018-0143

Queiroz, M. M., Telles, R., & Bonilla, S. H. (2020). Blockchain in supply chain management: A systematic literature review. *Journal of Cleaner Production, 264*, 121852.

Radebe, N., & Chipangamate, N. (2024). Mining industry risks, and future critical minerals and metals supply chain resilience in emerging markets. *Resources Policy, 91*, 104887. DOI: 10.1016/j.resourpol.2024.104887

Ramasamy, N., & Anand, M. (2024). A Review Investigation on Advancements in Reverse Supply Chain Management for Industrial Waste Management Process. *International Journal of Supply & Operations Management, 23*(4).

Redżeb, K. (2024). The Transformative Role of AI In Modern Supply Chains: A Study on Collaboration and Efficiency.

Rejeb, A., & Rejeb, K. (2020). Blockchain and supply chain sustainability. *Logforum, 16*(3). https://yadda.icm.edu.pl/baztech/element/bwmeta1.element.baztech-3e949e5f-8a67-4d5a-ad4f-342c28a6cb32

Renault Group. (2020). *Renault's commitment to circular economy practices and CO2 reduction.*

Renewcell. (2023). *Renewcell's production development in December 2023.*

Republic of Cyprus. (2022). *Green public procurement action plan.* Ministry of Agriculture.

Republic of Türkiye, Ministry of Industry and Technology, General Directorate of Development Agencies. (2023). *Hatay Saha Raporu* https://seeco.org.tr/upload_cms/files/rapor/Hatay_Saha_Durum_Raporu-Nisan-2023_R1.pdf

Republic of Türkiye, Presidency of the Legislation Information System. (2017). *Turkish Food Codex Regulation on Food Labelling and Consumer Information.* https://www.mevzuat.gov.tr/mevzuat?MevzuatNo=23282&MevzuatTur=7&MevzuatTertip=5

Resilinc. (2023). Annual report on global supply chain disruptions: 2023 trends and insights. Resilinc Corporation. https://www.resilinc.com/

Resmi, A. G., & Popovs, A. (2024). Big Data Analytics and Innovation in the Sustainable Performance of Electronic Waste Reverse Logistics: An Empirical Investigation in India. *Acta Prosperitatis, 15*(1), 120–132. DOI: 10.2478/acpro-2024-0011

Riad, M., Naimi, M., & Okar, C. (2024). Enhancing Supply Chain Resilience Through Artificial Intelligence: Developing A Comprehensive Conceptual Framework for AI Implementation and Supply Chain Optimization. *Logistics (Basel), 8*(4), 111. Https://Doi.Org/10.3390/Logistics8040111. DOI: 10.3390/logistics8040111

Ribeiro, J. P., & Barbosa-Povoa, A. (2018). Supply Chain Resilience: Definitions and Quantitative Modelling Approaches—A Literature Review. *Computers & Industrial Engineering, 115*, 109–122. DOI: 10.1016/j.cie.2017.11.006

Rice, J. B., & Caniato, F. (2003). BUILDING a SECURE AND RESILIENT SUPPLY NETWORK. Supply Chain Management Review. https://trid.trb.org/view/606665

Riggan, J., & Poole, A. (2024). *Hosting states and unsettled guests: Eritrean refugees in a time of migration deterrence*. Indiana University Press. DOI: 10.2307/jj.7249513

Rogers, D. S., & Tibben-Lembke, R. S. (1999). *Going backwards: Reverse logistics trends and practices*. Reverse Logistics Executive Council.

Saberi, S., Kouhizadeh, M., Sarkis, J., & Shen, L. (2019a). Blockchain technology and its relationships to sustainable supply chain management. *International Journal of Production Research*, *57*(7), 2117–2135. DOI: 10.1080/00207543.2018.1533261

Safaai, S., Sopian, K., & Zaidi, S. (2010). Life cycle assessment of biofuels and fossil fuels for road transport. *Renewable & Sustainable Energy Reviews*, *14*(6), 1839–1845.

Sah, B. P., & Shaikh, S. (2024). AI-Driven Iot and Blockchain Integration in Industry 5.0: A Systematic Review of Supply Chain Transformation. *Innovatech Engineering Journal*, *1*(01), 99–116. DOI: 10.70937/itej.v1i01.12

Saidi, N., El Omari, M., & Benjelloun, A. (2024). Green logistics and supply chain resilience: Evidence from the Moroccan pharmaceutical sector. *Journal of Sustainable Logistics*, *12*(1), 45–60.

Sakib, S. M. N. (2024). Blockchain Technology for Smart Contracts : Enhancing Trust, Transparency, and Efficiency in Supply Chain Management. In *Achieving Secure and Transparent Supply Chains With Blockchain Technology* (p. 246-266). IGI Global Scientific Publishing. DOI: 10.4018/979-8-3693-0482-2.ch014

Salburg, H. (2000). *Green logistics: Improving the environmental sustainability of logistics*. Kogan Page.

Samandağ Women's Cooperative. (n.d.) Samandağ Women's Cooperative Official Instagram Account. Instagram. https://www.instagram.com/samandagkadinkooperatifi/?hl=tr

São Paulo Municipality. (2023). *Blockchain for ethical food supply chains*. São Paulo Municipal Government.

SAP. (2024). *SAP Ariba sustainability analytics*. https://www.sap.com/products/ariba

Sarkis, J. (2012). A Boundaries and Flows Perspective of Green Supply Chain Management. *Supply Chain Management*, *17*(2), 202–216. DOI: 10.1108/13598541211212924

Sarkis, J., Cohen, M. J., Dewick, P., & Schröder, P. (2020). A brave new world: Lessons from the COVID-19 pandemic for transitioning to sustainable supply and production. *Resources, Conservation and Recycling*, *159*, 104894. DOI: 10.1016/j.resconrec.2020.104894 PMID: 32313383

Sartori, S., Witjes, S., & Campos, L. M. (2017). Sustainability performance for Brazilian electricity power industry: An assessment integrating social, economic and environmental issues. *Energy Policy, 111*, 41–51. DOI: 10.1016/j.enpol.2017.08.054

Saunders, M., Lewis, P., & Thornhill, A. (2023). *Research methods*. Pearson.

Sbihi, A., & Eglese, R. W. (2007). The relationship between vehicle routing & scheduling and green logistics-A literature survey.

Schneider, L., & Wallenburg, C. M. (2012). Implementing sustainable sourcing, Does purchasing need to change? *Journal of Purchasing and Supply Management, 18*(4), 243–257. DOI: 10.1016/j.pursup.2012.03.002

Sebastian, B., & Mathew, T. (2018). Corporate social responsibility practices of public sector and private sector companies in India. *SMART Journal of Business Management Studies, 14*(1), 32–41. DOI: 10.5958/2321-2012.2018.00004.0

Seghyar, N., Gouiouez, M., Nafzaoui, M. A., & Amane, M. (2025). Influence Of Social Responsibility Management On The Configuration Of Management Control In Moroccan Companies. http://dx.doi.org/DOI: 10.21511/ppm.23(1).2025.03

Seuring, S., & Müller, M. (2008). From a Literature Review to a Con ceptual Framework for Sustainable Supply Chain Management. *Journal of Cleaner Production, 16*(15), 1699–1710. DOI: 10.1016/j.jclepro.2008.04.020

Shaikh, E. (Ed.). (2024). *Strategies and Approaches of Corporate Social Responsibility Toward Multinational Enterprises*. IGI Global. DOI: 10.4018/979-8-3693-0363-4

Shamsuddoha, M., Khan, E. A., Chowdhury, M. M. H., & Nasir, T. (2025). Revolutionizing Supply Chains: Unleashing The Power Of AI-Driven Intelligent Automation and Real-Time Information Flow. *Information (Basel), 16*(1), 26. Https://Doi.Org/10.3390/Info16010026. DOI: 10.3390/info16010026

Sharma, S. (2000). Managerial interpretations and organizational context as predictors of corporate choice of environmental strategy. *Academy of Management Journal, 43*(4), 681–697. DOI: 10.2307/1556361

Sharma, V. K., Chandna, P., & Bhardwaj, A. (2017). GSCM related performance indicators in agro industry: A review. *Journal of Cleaner Production, 141*, 1194–1208. DOI: 10.1016/j.jclepro.2016.09.103

Shashi, K. (2022). Sustainability across the value chain: A holistic approach. *Business Horizons, 65*(3), 345–357. DOI: 10.1016/j.bushor.2021.11.003

Shitu, S., & Popoola, O. M. J. (2017). Investigating socially sustainable behaviour of local actors in the supply chain of shea butter in Rural Borgu, Nigeria: From a role theory perspective. *Indian-Pacific Journal of Accounting and Finance, 1*(1), 3–16. DOI: 10.52962/ipjaf.2017.1.1.2

Shrivastav, M. (2022). Barriers Related to AI Implementation in Supply Chain Management. *Journal of Global Information Management, 30*(8), 1–19. Https://Doi.Org/10.4018/JGIM.296725. DOI: 10.4018/JGIM.296725

Shujaat Mubarik, M., Ahmed Khan, S., Kusi-Sarpong, S., & Mubarik, M. (2023). Supply chain sustainability in VUCA : Role of BCT-driven SC mapping and 'Visiceability'. *International Journal of Logistics, 0*(0), 1–19. DOI: 10.1080/13675567.2023.2222660

Silva, C., Martins, J., & Pinto, R. (2023). Absorptive and adaptive capacities in supply chain resilience: A systematic review. *International Journal of Production Economics, 256*, 108712. DOI: 10.1016/j.ijpe.2022.108712

Silva, M. E., Pereira, M. M., & Hendry, L. C. (2023). Embracing change in tandem: Resilience and sustainability together transforming supply chains. *International Journal of Operations & Production Management, 43*(1), 166–196. DOI: 10.1108/IJOPM-09-2022-0625

Silverman, D. (2013). *Doing Qualitative Research: A Practical Handbook*. Sage Publications.

Sindakis, S., Showkat, S., & Su, J. (2023). Unveiling the influence: Exploring the impact of interrelationships among e-commerce supply chain members on supply chain sustainability. *Sustainability, 15*(24), 16642, 1-27. DOI: 10.3390/su152416642

Singapore Government. (2023). *Smart procurement strategy report*. Ministry of Finance.

Smith, J., Brown, K., & Lee, M. (2025). AI in public procurement: Opportunities and challenges. *Public Administration Review, 85*(1), 22–35.

Sodhi, M. S., Seyedghorban, Z., Tahernejad, H., & Samson, D. (2022). Why Emerging Supply Chain Technologies Initially Disappoint: Blockchain, Iot, And AI. *Production and Operations Management, 31*(6), 2517–2537. Https://Doi.Org/10.1111/Poms.13694. DOI: 10.1111/poms.13694

Sodhi, M. S., & Tang, C. S. (2019). Research Opportunities in Supply Chain Transparency. *Production and Operations Management, 28*(12), 2946–2959. DOI: 10.1111/poms.13115

Solomon, D. B. (2021). *Identity Politics and Its Influence on the Spread of Digital Market placeing:(A Framework to Manage Country Level Political Risk in Ethiopia)*. Balboa Press.

Souhli, K., & Ennadi, A. (2023). Adoption of GSCM practices and sensitivity/influencing factors: An empirical study at the Moroccan firm level. *International Journal of Industrial Engineering and Management*, *14*(3), 214–231. DOI: 10.24867/IJIEM-2023-3-334

South African Government. (2023). *Black Economic Empowerment procurement report*. Department of Trade and Industry.

Srivastava, S. K. (2007). Green supply-chain management: A state of the art literature review. *International Journal of Management Reviews*, *9*(1), 53–80. DOI: 10.1111/j.1468-2370.2007.00202.x

State of California. (2022). *Senate Bill No. 54 (SB 54) - Plastic Pollution Prevention and Packaging Producer Responsibility Act.*

Styhre, L., Winnes, H., Black, J., Lee, J., & Le-Griffin, H. (2017). Greenhouse gas emissions from ships in ports–Case studies in four continents. *Transportation Research Part D, Transport and Environment*, *54*, 212–224. DOI: 10.1016/j.trd.2017.04.033

Suali, A. S., Srai, J. S., & Tsolakis, N. (2024). The role of digital platforms in e-commerce food supply chain resilience under exogenous disruptions. *Supply Chain Management*, *29*(3), 573–601. DOI: 10.1108/SCM-02-2023-0064

Suhara, A., & Muttaqien, Z. (2025). Adaptation strategies in supply chains based on local wisdom in the manufacturing industry. *International Journal for Science Review*, *2*(4), 1–13. DOI: 10.71364/ijfsr.v2i4.29

Sun, Z., Xu, Q., & Liu, J. (2021). Pricing and replenishment decisions for seasonal and nonseasonal products in a shared supply chain. *International Journal of Production Economics*, *233*(108011), 1–17. DOI: 10.1016/j.ijpe.2020.108011

Suseno, G. P., & Nataliningsih, N. (2023). The role of women's cooperatives as an institution to empower women in supporting the achievement of the SDGs. *International Journal of Ethno-Sciences and Education Research*, *3*(2), 45–53. DOI: 10.46336/ijeer.v3i2.418

Sustainalytics and MSCI. (Ongoing). *ESG Ratings Methodologies and Company Reports*.

Tabachnick, B. G., & Fidell, L. S. (2013). *Using Multivariate Statistics* (6th ed.). Pearson Education.

Taleizadeh, A. A., Darabi, E. S., & Thaichon, P. (2024). Optimizing online selling through an online-to-offline platform: Strategic ramifications for local n stores. *Journal of Revenue and Pricing Management, 23*(6), 554–574. DOI: 10.1057/s41272-023-00448-4

Tang, D., & Lei, J. (2016). B2C e-commerce platform competition strategy research based on the network externality. *Management Science and Engineering, 10*(2), 70–76. DOI: 10.3968/8420 PMID: 35886272

Tao, X., & Wu, Q. (2021). Energy consumption and CO2 emissions in hinterland container transport. *Journal of Cleaner Production, 279*, 123394. DOI: 10.1016/j.jclepro.2020.123394

Technology Review Insights, M. I. T. (2021). *Climate performance index 2021.* Massachusetts Institute of Technology.

Teece, D. J., Pisano, G., & Shuen, A. (1997). Dynamic Capabilities and Strategic Management. Dans Oxford University Press eBooks (p. 268 285). https://doi.org/ DOI: 10.1093/oso/9780198781806.003.0019

Teece, D. J. (2018). Business models and dynamic capabilities. *Long Range Planning, 51*(1), 40–49. DOI: 10.1016/j.lrp.2017.06.007

Teixeira, A. R., Ferreira, J. V., & Ramos, A. L. (2025). Intelligent Supply Chain Management: A Systematic Literature Review on Artificial Intelligence Contributions. *Information (Basel), 16*(5), 399. Https://Doi.Org/10.3390/Info16050399. DOI: 10.3390/info16050399

Teng, J. T. C., & Jaramillo, P. (2006). Analyzing sustainable supply chain management using system dynamics modeling. *Systems Research and Behavioral Science, 23*(5), 647–661. DOI: 10.1002/sres.793

Testa, F., Iovino, R., & Iraldo, F. (2020). The circular economy and consumer behaviour: The mediating role of information seeking in buying circular packaging. *Business Strategy and the Environment, 29*(8), 3435–3448.1. DOI: 10.1002/bse.2587

Thilmany, D., Canales, E., Low, S. A., & Boys, K. (2021). Local food supply chain dynamics and resilience during COVID-19. *Applied Economic Perspectives and Policy, 43*(1), 86–104. DOI: 10.1002/aepp.13121

Thornton, P.. (2012). *The Institutional Logics Perspective.* Oxford UP. DOI: 10.1093/acprof:oso/9780199601936.001.0001

Tian, X., Wang, M., & Xu, Y. (2024). Investment strategies of information-provision technology in the platform-based supply chain. [NRL]. *Naval Research Logistics, 71*(3), 351–364. DOI: 10.1002/nav.22155

Tijan, E., Aksentijević, S., Ivanić, K., & Jardas, M. (2019). Blockchain Technology Implementation in Logistics. *Sustainability (Basel), 11*(4), 1185. DOI: 10.3390/su11041185

Tönnissen, S., & Teuteberg, F. (2020). Analysing the impact of blockchain-technology for operations and supply chain management : An explanatory model drawn from multiple case studies. *International Journal of Information Management, 52*, 101953. DOI: 10.1016/j.ijinfomgt.2019.05.009

Touboulic, A., & Walker, H. (2015). Theoretical frameworks for implementing sustainability in organizations. *Journal of Cleaner Production, 105*, 215–225. DOI: 10.1016/j.jclepro.2014.05.063

Toz, A. C., Buber, M., Koseoglu, B., & Sakar, C. (2021). An estimation of shipping emissions to analysing air pollution density in the Izmir Bay. *Air Quality, Atmosphere & Health, 14*(1), 69–81. DOI: 10.1007/s11869-020-00914-7

Trabucchi, D., & Buganza, T. (2020). Fostering digital platform innovation: From two to multi-sided platforms. *Creativity and Innovation Management, 29*(2), 345–358. DOI: 10.1111/caim.12320

Transparency International. (2023). *ProZorro impact assessment.* Transparency International Ukraine.

Tripathi, G., Ahad, M. A., & Casalino, G. (2023). A comprehensive review of blockchain technology : Underlying principles and historical background with future challenges. *Decision Analytics Journal, 9*, 100344. DOI: 10.1016/j.dajour.2023.100344

Tukamuhabwa, B. R., Stevenson, M., Busby, J., & Zorzini, M. (2015). Supply chain resilience: Definition, review and theoretical framework. *International Journal of Production Research, 53*(18), 5592–5623. DOI: 10.1080/00207543.2015.1037934

Tümer, Ö. (2024). Niğde kadın emek pazarı projesi ve Niğde'de yerel yönetimlerin kadın girişimcilere verdiği desteklerin sürdürülebilir kalkınma amaçları kapsamında değerlendirilmesi. *Ömer Halisdemir Üniversitesi İktisadi Ve İdari Bilimler Fakültesi Dergisi, 17*(4), 1020-1035. DOI: 10.25287/ohuiibf.1539913

Türkiye, U. N. D. P. (2022). *Kadın Dostu Dijital Yol: Türkiye'de Turizm ve İmalat Sektörlerinde Dijitalleşme ve Kadın İş Gücü Araştırması Raporu.* https://www.undp.org/tr/turkiye/publications/kadin-dostu-dijital-yol

Ugur-Cinar, M., Cinar, K., Onculer-Yayalar, E., & Akyuz, S. (2024). The political economy of women's cooperatives in Turkey: A social reproduction perspective. *Gender, Work and Organization, 31*(4), 1268–1289. DOI: 10.1111/gwao.12925

UK Government. (2023). *Contracts Finder transparency report*. Cabinet Office.

UNCTAD. (2021). *Emerging technologies in sustainable logistics: Opportunities for developing countries. United Nations Conference on Trade and Development.*

Ungerman, O., & Dědková, J. (2024). Consumer behavior in the model of the circular economy in the field of handling discarded items. *PLoS One, 19*(3), e0300707. DOI: 10.1371/journal.pone.0300707 PMID: 38512927

UN-Habitat. (2022). *Sustainable cities and communities: The role of procurement*. United Nations Human Settlements Programme.

Unilever. (2023). *Sustainable sourcing report*. Unilever.

Unilever. (Accessed regularly for their sustainable packaging commitments and initiatives). Information on their programs can be found on their official sustainability pages

United Nations Conference on Trade and Development (UNCTAD). (2020). Trade and Development Report 2020: From Global Pandemic to Prosperity for All - Avoiding Another Lost Decade. United Nations.

United Nations Environment Programme. (2019). *Sustainable public procurement: A global review*. UNEP.

United Nations. (2015). *Transforming our world: The 2030 Agenda for Sustainable Development*. United Nations General Assembly.

Ünsalan, E. (2024). Kadın girişimci kooperatiflerinin pazarlama stratejilerinin belirlenmesi: Amasya ili örneği. *Third Sector Social Economic Review, 59*(4), 2104–2129. DOI: 10.15659/3.sektor-sosyal-ekonomi.24.10.2476

Upreti, B. R., Ghale, Y., Shivakoti, S., & Acharya, S. (2018). Feminization of agriculture in the Eastern Hills of Nepal: A study of women in cardamom and ginger farming. *SAGE Open, 8*(4), 2158244018817124. DOI: 10.1177/2158244018817124

Urbinati, A., Chiaroni, D., & Chiesa, V. (2020). Remanufacturing in the automotive industry: A review of current practices and future challenges. *Journal of Cleaner Production, 270*, 122502.

Urbinati, A., Chiaroni, D., & Chiesa, V. (2020). The role of digital technologies in circular economy transition: A systematic literature review. *Journal of Cleaner Production, 273*, 122822.

US Government Accountability Office. (2023). *Cybersecurity in procurement systems*. GAO Report No. 23-123.

Vadakkepatt, G. G., Winterich, K. P., Mittal, V., Zinn, W., Beitelspacher, L., Aloysius, J., Ginger, J., & Reilman, J. (2021). Sustainable retailing. *Journal of Retailing, 97*(1), 62–80. DOI: 10.1016/j.jretai.2020.10.008

Vakoufaris, H., Kizos, T., Spilanis, I., Koulouri, M., & Zacharaki, A. (2007). Women's cooperatives and their contribution to the local development of the North Aegean Region, Greece. *Journal of Rural Cooperation, 35*(1), 19–41. DOI: 10.22004/ag.econ.58681

Valmohammadi, C. (2014). Impact of corporate social responsibility practices on organizational performance: An ISO 26000 perspective. *Social Responsibility Journal, 10*(3), 455–479. DOI: 10.1108/SRJ-02-2013-0021

Van Twiller, J.. (2025). Navigating Demand Uncertainty in Container Shipping: Deep Reinforcement Learning for Enabling Adaptive and Feasible Master Stowage Planning. arXiv. https://arxiv.org/abs/2502.12756

Vandana, M., Naveena, M., Ellaturu, N., Kumari, T. L., Bambuwala, S., & Rajalakshmi, M. (2024). Ai-Driven Solutions for Supply Chain Management. *Journal Of Informatics Education and Research, 4*(2), 1526–4726.

Vann Yaroson, E., Abadie, A., & Roux, M. (2025). Human-Artificial Intelligence Collaboration in Supply Chain Outcomes: The Mediating Role of Responsible Artificial Intelligence. *Annals Of Operations Research*. Https://Doi.Org/10.1007/S10479-025-06534-7

Vedie, H.-L. (2024). Is the industrial pharmaceutical sector of Morocco ready to meet the challenge of the Kingdom's health sovereignty? (Policy Brief No. 37/24). Policy Center for the New South. https://www.policycenter.ma/publications/la-filiere-pharmaceutique-industrielle-du-maroc-est-elle-prete-relever-le-defi

Venkat, K., & Wakeland, W. (2006). Is lean necessarily green? In *Proceedings of the 50th Annual Meeting of the International Society for the Systems Sciences (ISSS)*. https://journals.isss.org/index.php/proceedings50th/article/view/401

Villalba, G., & Gemechu, E. D. (2011). Estimating GHG emissions of marine ports—The case of Barcelona. *Energy Policy, 39*(3), 1363–1368. DOI: 10.1016/j.enpol.2010.12.008

Villena Verónica, H.. (2018). The Missing Link? The Strategic Role of Procurement in Building Sustainable Supply Networks. *Production and Operations Management.*

Voss, C., Tsikriktsis, N., & Frohlich, M. (2002). Case Research in Operations Management. *International Journal of Operations & Production Management, 22*(2), 195–219. DOI: 10.1108/01443570210414329

Walker, H., Miemczyk, J., Johnsen, T., & Spencer, R. Sustainable procurement: Past, present and future, Journal of Purchasing and Supply Management, Volume 18, Issue 4, 2012, Pages 201-206, ISSN 1478-4092.

Walker, H., & Brammer, S. (2012). The relationship between sustainable procurement and e-procurement in the public sector. *International Journal of Production Economics, 140*(1), 256–268. DOI: 10.1016/j.ijpe.2012.01.008

Wamba, J. H. T., Bekono, S. S., & Fouda, J. B. I. (2020). Les déterminants de la performance financière des petites entreprises camerounaises. *Finance & Finance Internationale, 1*(20). https://doi.org/DOI: 10.34874/IMIST.PRSM/ffi-v1i20.21818

Wamba, S. F., & Queiroz, M. M. (2022). Industry 4.0 and the supply chain digitalisation : A blockchain diffusion perspective. *Production Planning and Control, 33*(2-3), 193–210. DOI: 10.1080/09537287.2020.1810756

Wang, H. F., & Gupta, S. M. (2011). *Green supply chain management A product life cycle approach.* McGraw-Hill Education.

Wang'ombe, S. N. (2021). *Corporate Change and Sustainability of Coffee Commercial Market placeing Agents in Nairobi City County, Kenya* (Doctoral dissertation, Kenyatta University).

Wang, B., Liu, Q., Wang, L., Chen, Y., & Wang, J. (2023). A review of the port carbon emission sources and related emission reduction technical measures. *Environmental Pollution, 320*, 121000. DOI: 10.1016/j.envpol.2023.121000 PMID: 36621720

Wang, J., Wang, S., & Li, R. (2020). Servitization in manufacturing: The transition from linear to circular economy. *Journal of Cleaner Production, 247*, 119077.

Wang, L., & Li, Y. (2023). Estimation methods and reduction strategies of port carbon emissions—What literatures say? *Marine Pollution Bulletin, 195*, 115451. DOI: 10.1016/j.marpolbul.2023.115451 PMID: 37659387

Wang, L., Rabinovich, E., & Richards, T. J. (2022). Scalability in platforms for local groceries: An examination of indirect network economies. *Production and Operations Management, 31*(1), 318–340. DOI: 10.1111/poms.13536

Wang, M., Wu, Y., Chen, B., & Evans, M. (2020). Blockchain and supply chain management : A new paradigm for supply chain integration and collaboration. *Operations and Supply Chain Management: An International Journal, 14*(1), 111–122. DOI: 10.31387/oscm0440290

Wang, P., Du, S., Hu, L., & Tang, W. (2023). Logistics choices in a platform supply chain: A co-opetitive perspective. *The Journal of the Operational Research Society, 74*(1), 93–111. DOI: 10.1080/01605682.2021.2023675

Wang, Y., Han, J. H., & Beynon-Davies, P. (2018). Understanding blockchain technology for future supply chains : A systematic literature review and research agenda. *Supply Chain Management, 24*(1), 62–84. Advance online publication. DOI: 10.1108/SCM-03-2018-0148

Wang, Y., Wu, B., Li, J., & Guo, Q. (2020). Enhancing Eco-design and Competitive Sustainability in Circular Supply Chains: The Role of AI, IoT and Blockchain. *Resources, Conservation and Recycling, 163*, 105072.

Warrior, D., Chandhok, V., Khajuria, A., Gauba, S., & Jain, R. (2024). *Strengthening India's Clean Energy Supply Chains: Building Manufacturing Competitiveness in a Globally Fragmented Market*. Council on Energy, Environment and Water (CEEW).

Wei, R., & Pardo, C. (2024). Looking for resource density in the platform supply network. *Journal of Purchasing and Supply Management, 30*(3), 100938. DOI: 10.1016/j.pursup.2024.100938

Weisz, E., Herold, D. M., Ostern, N. K., Payne, R., & Kummer, S. (2025). Artificial Intelligence (AI) For Supply Chain Collaboration: Implications on Information Sharing and Trust. *Online Information Review, 49*(1), 164–181. Https://Doi.Org/10.1108/OIR-02-2024-0083. DOI: 10.1108/OIR-02-2024-0083

Wei, Y. D., Lin, J., & Zhang, L. (2020). E-commerce, Taobao villages and regional development in China. *Geographical Review, 110*(3), 380–405. DOI: 10.1111/gere.12367

Wengraf, T. (2001) Qualitative Research Interviewing: Biographic Narrative and Semi-Structured Methods. Sage Publications, London, 7. https://doi.org/DOI: 10.4135/9781849209717

Winnes, H., Styhre, L., & Fridell, E. (2015). Reducing GHG emissions from ships in port areas. *Research in Transportation Business & Management, 17*, 73–82. DOI: 10.1016/j.rtbm.2015.10.008

Wirth, N. (2018). Hello marketing, what can artificial intelligence help you with? *International Journal of Market Research, 60*(5), 435–438. DOI: 10.1177/1470785318776841

Woldekiros, H. S. (2023). *The boundaries of ancient trade: kings, commoners, and the Aksumite salt trade of Ethiopia*. University Press of Colorado. DOI: 10.1353/book.131401

Wold, H., Sjöström, M., & Eriksson, L. (2001). PLS-regression: A basic tool of chemometrics. *Chemometrics and Intelligent Laboratory Systems*, *58*(2), 109–130. DOI: 10.1016/S0169-7439(01)00155-1

Women Watch. (2012). *Empower Rural Women-End Hunger and Poverty*. https://www.un.org/womenwatch/feature/ruralwomen/index.html

Wong, L.-W., Leong, L.-Y., Hew, J.-J., Tan, G. W.-H., & Ooi, K.-B. (2020). Time to seize the digital evolution : Adoption of blockchain in operations and supply chain management among Malaysian SMEs. *International Journal of Information Management*, *52*, 101997. DOI: 10.1016/j.ijinfomgt.2019.08.005

World Bank Group. (2022). *Morocco Country Climate and Development Report*. World Bank Group.

World Bank. (2015). Green growth: Putting Morocco in the lead against climate change. World Bank.https://documents.worldbank.org/en/publication/documents-reports/documentdetail/368381638794966349/

World Commission on Environment and Development (WCED). (1987). *Our common future*. World Commission on Environment and Development.

World Economic Forum. (2022). *Reports on Circular Economy*.

World Food Programme. (2020). *School feeding and the SDGs*. WFP.

World Health Organization (WHO). (2018). *Environmental Noise Guidelines for the European Region*. WHO Regional Office for Europe.

Wright, P. (2013). Impacts of climate change on ports and shipping. *MCCIP Science Review*, *2013*, 263–270.

Xia, J., Zhang, W., Zhang, W., Yang, Y., Hu, G., Ge, D., Liu, H., & Cao, H. (2021). A cloud computing-based approach using the visible near-infrared spectrum to classify greenhouse tomato plants under water stress. *Computers and Electronics in Agriculture*, *181*, 105966. DOI: 10.1016/j.compag.2020.105966

Xie, N., He, H., & Tong, Y. (2022). Game model of green financial supply chain based on government subsidy analysis. *IEEE Access : Practical Innovations, Open Solutions, 10*, 60929–60945. DOI: 10.1109/ACCESS.2022.3181044

Xue, W., & Te Chuan, L. (2024). The Application of Artificial Intelligence in China's Cross Border E-Commerce Field.

Xu, H., Zhang, Y., & Liu, Y. (2024). Digital platform-based domestic supply chain: International comptitiveness and government regulation. *Economic Theory and Business Management, 44*(2), 133.

Xu, P., Lee, J., Barth, J. R., & Richey, R. G. (2021). Blockchain as supply chain technology : Considering transparency and security. *International Journal of Physical Distribution & Logistics Management, 51*(3), 305–324. DOI: 10.1108/IJPDLM-08-2019-0234

Xu, S., Govindan, K., Wang, W., & Yang, W. (2024). Supply chain management under cap-and-trade regulation: A literature review and research opportunities. *International Journal of Production Economics, 109199*, 109199. Advance online publication. DOI: 10.1016/j.ijpe.2024.109199

Xu, X., He, P., Zhou, L., & Cheng, T. C. E. (2023). Coordination of a platform-based supply chain in the marketplace or reselling mode considering cross-channel effect and blockchain technology. *European Journal of Operational Research, 309*(1), 170–187. DOI: 10.1016/j.ejor.2023.01.057

Yalçın, S. E. (2025). Estimation of CO_2 Emissions in Transportation Systems Using Artificial Neural Networks, Machine Learning, and Deep Learning: A Comprehensive Approach. *Systems, 13*(3), 194. https://www.mdpi.com/2079-8954/13/3/194. DOI: 10.3390/systems13030194

Yandrapalli, V. (2023). Revolutionizing Supply Chains Using Power of Generative AI. *International Journal of Research Publication and Reviews, 4*(12), 1556–1562. DOI: 10.55248/gengpi.4.1223.123417

Yang, C. S. (2018). The effect of CSR practices on organisational performance in the container shipping in Taiwan. *International Journal of Shipping and Transport Logistics, 10*(2), 181–201. DOI: 10.1504/IJSTL.2018.090098

Yang, R., Wakefield, R., Lyu, S., Jayasuriya, S., Han, F., Yi, X., Yang, X., Amarasinghe, G., & Chen, S. (2020). Public and private blockchain in construction business process and information integration. *Automation in Construction, 118*, 103276. DOI: 10.1016/j.autcon.2020.103276

Yang, X., Jiang, Y., Chen, X., & Jia, F. (2024). ICT-empowered rural e-commerce development in China: An adaptive structuration perspective. *International Journal of Technology Management*, *95*(1-2), 90–119. DOI: 10.1504/IJTM.2024.137016

Yang, Y.-C. (2017). Operating strategies of CO2 reduction for a container terminal based on carbon footprint perspective. *Journal of Cleaner Production*, *141*, 472–480. DOI: 10.1016/j.jclepro.2016.09.132

Yarın, E. (2024, September 11). Bölgesel kalkınmanın güç kaynağı, kadın kooperatifleri. https://egiadyarin.com/bolgesel-kalkinmanin-guc-kaynagi-kadin-kooperatifleri/

Yavaprabhas, K., Pournader, Mehrdokht, & and Seuring, S. (2024). Blockchain and trust in supply chains : A bibliometric analysis and trust transfer perspective. *International Journal of Production Research*, *0*(0), 1–28. DOI: 10.1080/00207543.2024.2389544

Yenilmez Gürsoy, Ö. B. (2022). Türkiye'de e-ticaret platformlarında satış yapan kırsal kadın kooperatiflerinin incelemesi. *Beykoz Akademi Dergisi*, *10*(1), 265–279. DOI: 10.14514/BYK.m.26515393.2022.10/1.265-279

Yeshiwas, Y., Alemayehu, M., & Adgo, E. (2024). Strategic mapping of onion supply chains: A comprehensive analysis of production and post-harvest processes in Northwest Ethiopia. *Frontiers in Sustainability*, *5*(1387907), 1–13. DOI: 10.3389/frsus.2024.1387907

Yıldırım, A., & Şimşek, H. (2021). *Sosyal bilimlerde nitel araştırma yöntemleri.* Seçkin Yayıncılık.

Yıldızcan, C. (2023). Deprem sonrasında afetin ve tarımın politik ekolojisi: Hatay örneği. *PRAKSIS*, *63*, 11–28.

Yin, R.K. (2014). Case Study Research Design and Methods (5th ed.). Canadian Journal Of Program Evaluation, 30(1), 108 110. https://doi.org/DOI: 10.3138/cjpe.30.1.108

Yoon, B., & Chung, Y. (2018). The effects of corporate social responsibility on firm performance: A stakeholder approach. *Journal of Hospitality and Tourism Management*, *37*, 89–96. DOI: 10.1016/j.jhtm.2018.10.005

Yu, H., Ge, Y.-E., Chen, J., Luo, L., Tan, C., & Liu, D. (2017). CO2 emission evaluation of yard tractors during loading at container terminals. *Transportation Research Part D, Transport and Environment*, *53*, 17–36. DOI: 10.1016/j.trd.2017.03.014

Yu-Wei, L., Gui-Hua, L., & Chen, P. (2024). Settlement selection strategic analysis for self-operated e-commerce platforms under market competition. *Systems, 12*(8), 293, 1-22. DOI: 10.3390/systems12080293

Yu, Y., Huang, G., & Guo, X. (2021). Financing strategy analysis for a multi-sided platform with blockchain technology. *International Journal of Production Research, 59*(15), 4513–4532. DOI: 10.1080/00207543.2020.1766718

Zamri, N. M., Andrew, L., Borhani, N., Latiff, S. F. A., & Mathew, D. H. (2024). Women's cooperatives in Turkey: Challenges, success stories, and policy recommendations. *International Journal of Academic Reserach in Economics and Management Sciences, 13*(3). Advance online publication. DOI: 10.6007/IJAREMS/v13-i3/22853

Zeba, S., Suman, P., & Tyagi, K. (2023). Types of blockchain. In *Distributed Computing to Blockchain* (pp. 55–68). Elsevier., DOI: 10.1016/B978-0-323-96146-2.00003-6

Zejjari, I., & Benhayoun, I. (2024). The Use of Artificial Intelligence to Advance Sustainable Supply Chain: Retrospective and Future Avenues Explored Through Bibliometric Analysis. *Discover Sustainability, 5*(1), 174. Https://Doi.Org/10.1007/S43621-024-00364-6. DOI: 10.1007/s43621-024-00364-6

Zelbst, P. J., Green, K. W., Sower, V. E., & Bond, P. L. (2020). The impact of RFID, IIoT, and Blockchain technologies on supply chain transparency. In *Journal of Manufacturing Technology Management* (Vol. 31, Numéro 3, p. 441-457). Emerald Group Holdings Ltd. DOI: 10.1108/JMTM-03-2019-0118

Zhang, Q., Wong, C. W. Y., & Klassen, R. (2024). Carbon neutrality: Operations management research opportunities. *International Journal of Operations & Production Management, 70*(3), 344–354.

Zhang, T., Li, G., & Tayi, G. K. (2023). A strategic analysis of virtual showrooms deployment in online retail platforms. *Omega, 117*(102824), 1–12. DOI: 10.1016/j.omega.2022.102824

Zhang, Z., Garimella, A., & Fan, M. (2024). Leveraging the social fabric to ımprove rural e-commerce access. *Production and Operations Management*, 1–16. DOI: 10.1177/10591478231224974

Zhao, Y. (2025). Platform economy and global supply chain integration: A technology-driven approach to business model innovation. H. Luo, T. Yao, W. Cui & H. Li (Eds.), *Advances in Economics, Business and Management Research* (ss. 612–620). Atlantis Press. DOI: 10.2991/978-94-6463-642-0_65

Zheng, Z., Xie, S., Dai, H. N., Chen, X., & Wang, H. (2018). Blockchain challenges and opportunities: A survey. *International Journal of Web and Grid Services, 14*(4), 352. DOI: 10.1504/IJWGS.2018.095647

Zhong, Y., Lai, I. K. W., Guo, F., & Tang, H. (2021). Research on government subsidy strategies for the development of agricultural products E-commerce. *Agriculture, 11*(11), 1152. DOI: 10.3390/agriculture11111152

Zhou, X., Liu, H., Li, J., Zhang, K., & Lev, B. (2023). Channel strategies when digital platforms emerge: A systematic literature review. *Omega, 120*(102919), 1–13. DOI: 10.1016/j.omega.2023.102919

Zhu, A., & Wen, Y. (2021). Green Logistics Location-Routing Optimization Solution Based on Improved Genetic Algorithm Considering Low-Carbon and Environmental Protection. *Journal of Mathematics, 2021*, 1–16. DOI: 10.1155/2021/6101194

Zhu, Q. (2024). Life-cycle assessment in sustainable supply chains: A strategic imperative. *Sustainable Production and Consumption, 55*, 123–135. DOI: 10.1016/j.spc.2023.11.007

Zhu, Q., & Geng, Y. (2001). Integrating environmental issues into supplier selection and management. *Greener Management International, 35*(35), 27–40. DOI: 10.9774/GLEAF.3062.2001.au.00005

Zhu, Q., Sarkis, J., & Lai, K. H. (2008). GSCM implications for closing the loop. *Transportation Research Part E, Logistics and Transportation Review, 44*(1), 1–18. DOI: 10.1016/j.tre.2006.06.003

Zhu, X., Zhao, L., & Liu, J. (2021). Sustainable packaging design: A review on the development of bio-based and biodegradable materials. *Journal of Cleaner Production, 280*, 124420. DOI: 10.1016/j.jclepro.2020.124420

Zou, H. (2024). Operationalizing supply chain resilience: Metrics, challenges, and opportunities. *International Journal of Logistics Management, 35*(1), 112–130. DOI: 10.1108/IJLM-05-2023-0198

About the Contributors

Badr Bentalha teaches Supply Chain and Operations Management at National School Of Business and Management – Fez, Sidi Mohamed Ben Abdellah University – Morocco. Dr. Bentalha investigates the structural dynamics and control of complex networks, applying his findings to supply chain management, Industry 4.0, risk analysis, and digital supply chains. His work emphasizes the intersection of supply chain management, operations research, industrial engineering, and digital technology. As a professor, he teaches undergraduate, graduate, and, doctoral courses in operations management, supply chain management, logistics, management information systems, and strategic management. Through guest lectures, webinars, and scholarly presentations, he engages students and fosters an active learning environment. Professor Bentalha specializes in supply chain and operations management, operations research, and service management. He is passionate about integrating knowledge across disciplines to solve real-world problems. He has delivered numerous invited plenary talks, keynotes, and panel discussions at conferences and global webinars.

Lhoussaine Alla is a professor in management sciences at the National School of Applied Sciences, and a researcher at the LAREMEF laboratory, Sidi Mohamed Ben Abdellah University, Fez, Morocco. He is a permanent professor of various marketing management modules at Sidi Mohamed Ben Abdellah University in Fez (Morocco) and in several public and private business graduate schools. He is an accredited professional expert in Training Engineering, Professional Coaching and Mentoring young entrepreneurs. After a PhD thesis on creating value for the customer and its impact on the overall performance of companies, Prof. Lhoussaine Alla invested more in scientific research in marketing, through various scientific contributions in the form of participation in international conferences and symposia and scientific publications, in various themes inherent to Marketing (customer value creation, customer behaviour, e-marketing, customer experience, sales performance,

Marketing Data Analytics, territorial marketing, territorial attractiveness, territorial economic intelligence, ...), finance (stock market performance, financial analysts, financing package,...), entrepreneurship (entrepreneur profile, startups, entrepreneurial resilience,...), tourism (tourist attractiveness, visitor behaviour, tourism, performance of tourist destinations,...), logistics (SC, SCM, SSCM, Green SCM, logistics performance,...). Prof. Lhoussaine ALLA is also (co)editor of the e-book "Integrating Intelligence and Sustainability into supply chains. IGI Global. DOI: 10.4018/979-8-3693-0225-5" and Editor-in-chief of the scientific journal Alternatives Managériales et Economiques (AME), indexed by IMIST-CNRST and Cairn. Prof. Lhoussaine ALLA is also (co)coordinator of many scientific events devoted to marketing research. As a scientific and educational accumulation, in his capacity as editor of numerous books, especially in collaboration with IGI Global, Prof ALLA has been able to mobilize a large community of imminent researchers and experts on a global scale, to share the fruits of their relevant and innovative research in the implementation, supervision, evaluation, control and reinvention of innovative solutions developed through data engineering and AI in marketing. He is co-editor of several collective works including: "Integrating Intelligence and Sustainability in Supply Chains", "AI and Data Engineering Solutions for Effective Marketing", "Data Collection and Analysis in Scientific Qualitative Research", Applying Qualitative Research Methods to Management Science, Utilizing Technology to Manage Territories, Sustainable and Intelligent Territorial Marketing and Entrepreneurship. Prof ALLA is the coordinator of a team of young researchers very dynamic and talented, with complementary profiles, creating a friendly space for high-level scientific research.

***

Joji Abey acting chairperson of Department of Business Management at the College of Business Administration, Kingdom University, Bahrain. She holds the prestigious Chartered Manager designation from the Chartered Management Institute (CMI), UK, and a Fellow of the Higher Education Academy (HEA), recognized for her attainment against the UK Professional Standards Framework for teaching and learning support in higher education. To her credit she published more than 25 articles in International Peer Reviewed Journals.

K. Balaji is an Assistant Professor in Business and Management, Christ University, Bangalore. He did Ph.D. from KLU Business School, KL University,Vijayawada. He is having 12 years of teaching experience. He has to his credit 23 Research Articles published in reputed National and International Journals. He has participated and presented more than 12 papers in both National and International Seminars/Conferences and also attended more than 50 workshops/ FDPS organized by various reputed organizations. He has also participated in more than 50 seminars / webinars organized by various reputed organizations. He has also attended in 6 Short Term Training programs.He has successfully conducted 3 webinars, one 5 day FDP and one International Conference on Reimaging marketing in New Normal. He is a ratified as Assistant Professor by both JNTUH, JNTU Anantapur and Sri Venkateswara University,Tirupati. His specialized areas include, Accounting and Financial management, Marketing Management, Retailing Management, Consumer Behavior studies, Entrepreneurship and Business Laws.

Badre Meziane is an Associate Professor specializing in the fields of economic intelligence and territorial development. Holding a master's degree in information systems management, he dedicates his research to crucial themes for strategic development and knowledge management at the territorial level. His research work primarily focuses on economic intelligence, a key domain for the competitiveness of organizations and territories. Prof. Meziane is particularly interested in applying these concepts in the context of territorial development, exploring synergies between economic, institutional, and academic actors to promote growth and innovation at local and regional scales. Territorial intelligence constitutes another major axis of his research, where he studies the processes of collecting, analyzing, and disseminating strategic information to support decision-making and territorial planning. In this framework, Prof. Meziane relies on his expertise in geographic information systems (GIS) to integrate the spatial dimension into the analysis and visualization of territorial data. Knowledge management also occupies an important place in his work. Prof. Meziane explores methods and tools for capitalizing, sharing, and valorizing knowledge within organizations and networks of territorial actors, thus contributing to strengthening their capacity for innovation and adaptation. Through his teaching and research, Prof. Badre Meziane strives to create bridges between theory and practice, developing innovative approaches to address the complex challenges of territorial development in the era of the knowledge and information economy

R. Bhuvaneswari working as an Assistant Professor of English at Dr.Mahalingam College of Engineering and Technology, Pollachi. To her credit she completed 10 NPTEL / SWAYAM and published few articles in Scopus indexed journals.

R. Velmurugan presently working as an Associate Professor in Commerce, Karpagam Academy of Higher Education, Coimbatore. He started his career as a Project Fellow under UGC Sponsored Major Research Project at NGM College, Pollachi. He has 18 years of teaching experience has 15 years of research experience.

To his credit he produced 22 Ph.D. Scholars and 5 M.Phil. Scholars. At present 1 scholar submitter his thesis, 6 Ph.D. Scholars are under his guidance. Moreover, to his credit he published 168 articles in various International Journals, of which 90 articles are published in Scopus Indexed Journals and 15 articles in Web of Science and 76 articles in edited volumes. He also participates and presented 27 papers in various International Seminar / Conference and 54 National Seminar / Conference. He acted as an external examiner for various M.Phil. and Ph.D. Viva voce examinations. He acted as a Vice Chancellor nominee for Doctoral Committee Expert at Bharathiar University, Coimbatore and Periyar University, Salem. He acted as a Chairperson at various National Level Conferences. He also acted as an Expert for Syllabus Scrutiny Committee, Question Paper Scrutiny Committee and Expert members for Board of Studies at various reputed Colleges. Considering his meritorious contribution, he received awards like Best Paper Presenter, Best Research Supervisor and Best Faculty Awards from various research foundations and councils. He too acted as a resource person for 27 Guest Lectures, 30 Workshops and 10 faculty development programmes.

Imad Terrah is an Associate Professor of Economics and Management since 2008 and a PhD candidate at the Laboratory of Applied Studies and Research in Economic Sciences, Faculty of Legal, Economic, and Social Sciences, Ibn Zohr University, Agadir, Morocco. His research focuses on supply chain management, contributing valuable insights to both academic and practical advancements in the field.

Index

I

L

M

P

Q

R

S

T

W

www.ingramcontent.com/pod-product-compliance
Lightning Source LLC
LaVergne TN
LVHW061525290926
849162LV00004B/666